Under the Shade of Thipaak

UNIVERSITY PRESS OF FLORIDA

Florida A&M University, Tallahassee
Florida Atlantic University, Boca Raton
Florida Gulf Coast University, Ft. Myers
Florida International University, Miami
Florida State University, Tallahassee
New College of Florida, Sarasota
University of Central Florida, Orlando
University of Florida, Gainesville
University of North Florida, Jacksonville
University of South Florida, Tampa
University of West Florida, Pensacola

UNDER THE SHADE OF THIPAAK

The Ethnoecology of Cycads in Mesoamerica and the Caribbean

Edited by
Michael D. Carrasco, Angélica Cibrián-Jaramillo,
Mark A. Bonta, and Joshua D. Englehardt

Foreword by Thomas C. Hart

University Press of Florida
Gainesville · Tallahassee · Tampa · Boca Raton
Pensacola · Orlando · Miami · Jacksonville · Ft. Myers · Sarasota

Copyright 2022 by Michael D. Carrasco, Angélica Cibrián-Jaramillo, Mark A. Bonta, and Joshua D. Englehardt
All rights reserved
Published in the United States of America

27 26 25 24 23 22   6 5 4 3 2 1

Library of Congress Cataloging-in-Publication Data
Names: Carrasco, Michael, editor. | Cibrián-Jaramillo, Angélica, editor.
 | Bonta, Mark, editor. | Englehardt, Joshua, editor.
Title: Under the shade of Thipaak : the ethnoecology of cycads in
 Mesoamerica and the Caribbean / edited by Michael D. Carrasco, Angélica
 Cibrián-Jaramillo, Mark A. Bonta, Joshua D. Englehardt.
Description: 1. | Gainesville : University Press of Florida, 2022. |
 Includes bibliographical references and index.
Identifiers: LCCN 2021055209 (print) | LCCN 2021055210 (ebook) | ISBN
 9780813069364 (hardback) | ISBN 9780813070179 (pdf)
Subjects: LCSH: Cycads—Central America—History. | Cycads—Caribbean
 Area—History. | Ethnoecology—Central America—History. |
 Ethnoecology—Caribbean Area—History. | Ethnoecology—Mexico—History.
 | Indians of North America—Agriculture—Caribbean Area—History. |
 Indians of Mexico—Agriculture—History. | BISAC: SOCIAL SCIENCE /
 Anthropology / Cultural & Social | SOCIAL SCIENCE / Archaeology
Classification: LCC SB413.C88 U93 2022 (print) | LCC SB413.C88 (ebook) |
 DDC 635.9/359—dc23/eng/20211221
LC record available at https://lccn.loc.gov/2021055209
LC ebook record available at https://lccn.loc.gov/2021055210

The University Press of Florida is the scholarly publishing agency for the State University System of Florida, comprising Florida A&M University, Florida Atlantic University, Florida Gulf Coast University, Florida International University, Florida State University, New College of Florida, University of Central Florida, University of Florida, University of North Florida, University of South Florida, and University of West Florida.

University Press of Florida
2046 NE Waldo Road
Suite 2100
Gainesville, FL 32609
http://upress.ufl.edu

Contents

List of Figures vii
List of Tables xi
Foreword xiii

Introduction: Elucidating Human-Cycad Relationships in the Indigenous Ethnoecological and Agroecological Systems of the Americas and Beyond 1
Michael D. Carrasco, Angélica Cibrián-Jaramillo, Mark A. Bonta, and Joshua D. Englehardt

1. An Overview of Cycadales in Mesoamerica and the Caribbean: Biology, Distribution, and Conservation 27
   Michael Calonje, Andrew P. Vovides, and José Said Gutiérrez-Ortega

2. Cycad Population Genetics in Mexico and the Caribbean 62
   Angélica Cibrián-Jaramillo, Francisco Pérez-Zavala, Naishla M. Gutiérrez-Arroyo, Dánae Cabrera-Toledo, Jorge González-Astorga, and Joshua D. Englehardt

3. *Zamia* in the Insular Caribbean: New Insights into the Historical Ecology of an Ancient Wild Food Plant 97
   Jaime R. Pagán-Jiménez

4. Ancient Mesoamerican Agricultural Strategies: The Role of Cycads and *Phaseolus* Beans in the Development of Intensive Field Cultivation Systems 125
   Amber M. VanDerwarker

5. The Archaeology of Cycad Use in Ancient Mesoamerica: Old Data, New Methods 141
   Joshua D. Englehardt, Edder D. Bustos-Díaz, Emanuel Bojórquez Quintal, Luis Rojas Abarca, Luis R. Velázquez Maldonado, Esteban Sánchez Rodríguez, and Francisco Barona-Gómez

6. The Maize God Revisited: Iconology of Cycads, Maize, and Fertility Deities in Mesoamerican Art   166
   Michael D. Carrasco

7. Tracking *Teosinte* across Mexico and Northern Central America   212
   Mark A. Bonta

8. The Sacred-Maize-Ancestor Concept: An Ethnographic Perspective   223
   Mark A. Bonta

Epilogue: What Have We Learned and Where Are We Going?   245
   Dennis William Stevenson

References Cited   259
List of Contributors   319
Index   327

Figures

0.1. Strobili of three Mexican cycad species   5
0.2. Pachycaul stem, rosette crown, and pinnate leaves of cycads   6
0.3. Approximate worldwide distribution of Cycadales in tropical and subtropical regions   7
0.4. Coralloid root structures in cycads   7
0.5. Australian Aboriginal rock art representations of cycads at the seed-producing stage   10
0.6. Vendor selling young microsporophylls and pollen-filled microsporangia of *Cycas pectinata* for consumption and use in medicinal preparations in Shillong, Meghalaya, India   11
0.7. Use of cycads in regional agroecology   12
0.8. Historical cycad use in the Amami Islands, Japan   13
0.9. Cycad miso paste available in Japan   13
0.10. Chief Seni Mau Tiruspe of the Malvatumauri Council of Chiefs, with Prince Charles in Vanuatu, dressed in chiefly regalia, including cycad leaves   14
0.11. Images of the Myōkoku-ji cycad   16
0.12. A ukiyo-e woodblock print by Yoshitsuya, 1864, detailing the legend of Oda Nobunaga and the sacred cycad of Myōkoku-ji   17
0.13. A formal portrait of Mukoma Mudjadji IV, the fourth Rain Queen, who reigned from 1960 to 1980, in royal attire and surrounded by Mudjadji cycads   18
0.14. The incorporation of cycads in Asian temples and shrines   19
1.1. Basic cycad morphology   28
1.2. Encircling leaf traces in cortex of *Zamia* stem   29
1.3. Basic cycad reproductive anatomy   31

1.4. Apex of the Permian fossil *Crossozamia* female scale compared with the living *Cycas revoluta* scale apex   34
1.5. Distribution of New World cycads   36
1.6. Species richness maps for three Mesoamerican genera, combined and for each genus individually   42
1.7. *Zamia furfuracea*   43
1.8. *Dioon edule* in tropical seasonally dry forest, Veracruz   44
1.9. *Ceratozamia brevifrons*   45
1.10. *Zamia inermis*   46
1.11. *C. miqueliana*   47
1.12. National Cycad Collection, Clavijero Botanic Garden, Xalapa, Veracruz   50
1.13. The Monte Oscuro nursery with a *Dioon edule* crop   56
1.14. Nursery-produced *Dioon edule, Beaucarnea recurvata,* and *Zamia furfuracea* plants delivered from the Monte Oscuro nursery for sale at the Clavijero Botanic Garden shop   56
1.15. Nursery-produced *Dioon edule* plant adorning urban gardens in the city of Xalapa, Veracruz   59
2.1. *Ceratozamia* species analyzed using SSRs   80
2.2. *Dioon* cycads in habitat   82
2.3. SSR analyses of *Dioon* cycads from throughout their range in Mexico and Honduras   86
2.4. Pairwise Fst between *Dioon* populations   89
2.5. Preparation of cycad seeds for human consumption   91
3.1. *Zamia* morphology   99
3.2. Territories of the Greater Caribbean where ancient *Zamia* spp. starches have been recovered from archaeological materials   100
3.3. Modern oval starch grains from the underground stem of a marunguey (*Zamia pumila* L.) specimen from Puerto Rico   102
3.4. Selected modern starch grains of *Zamia pumila, Z. erosa,* and *Z. portoricensis* showing some of its main morphometric features   105
3.5. Modern starch grains from manioc, *Zamia pumila,* and sweet potato   106
3.6. Presumed processing sequence of zamia underground stems and associated tools on which tentatively and securely identified zamia starches have been identified   109

3.7. Ancient zamia starches from Caribbean and northeastern South American sites   113
3.8. Modern pollen assemblage of *Zamia integrifolia*   119
3.9. Reconstruction of early to late colonial processing stages of zamia in the Greater Antilles   121
5.1. Dry cave sites in Mesoamerica that have yielded evidence of cycad remains in archaeological contexts   144
5.2. *Dioon edule* seed fragments recovered from MacNeish's excavations at the Ocampo cave sites   146
5.3. Approximate locations of cave sites excavated by MacNeish in the Tehuacán Valley, Puebla   148
5.4. Comparison of cycad cones and maize cobs   151
5.5. Multi-elemental analytic methods to visualize and quantify elements in soils and cycad tissues   157
5.6. Integrated methodological proposal for archaeobotanical, biological, paleogenomic, ethnoarchaeological, and experimental archaeology research on cycad use   160
5.7. Direct analysis real-time mass spectrometry (DART-MS) in cycad research   162
6.1. Cycads in agroecological mythological symbolic regimes over time   170
6.2. Codical, glyphic, and sculptural depictions of Cipactli   183
6.3. Tonsured Maize God   189
6.4. Maya Maize God as cacao tree   191
6.5. Maya Maize God emerging from bottle gourd   192
6.6. Olmec Maize God   193
6.7. Flora emerging from crocodilian   197
6.8. Equivalent Central Mexican and Maya day signs   198
6.9. Piscine-monster tails substituting for sprouting foliage   199
6.10. Scenes depicting ancestors sprouting/growing as trees   201
6.11. Late Formative period stone sculpture "Tenaspi Egg"   202
6.12. *Dioon spinulosum* seed germination   203
6.13. *Ceratozamia mexicana*   204
6.14. Similarities between cycad strobili and Olmec imagery   205
6.15. The relationship of agroecology, symbolism, and mythology   207
7.1. Teosinte regions in Mexico and northern Central America   215

8.1. Indigenous language areas in northeastern Mexico within the cycad/maize culture region   224

8.2. Principal locations for the SMA (sacred-maize-ancestor) concept related to cycads in Indigenous and mestizo regions of Mexico and northern Central America   225

E.1. The author with colleague and noted cycad expert Hiep T. Nguyen examining a handsome specimen of *Cycas pectinata* during fieldwork in Vietnam   246

E.2. Indigenous Colombian poem on cycads with photo of *Zamia obliqua* in front of a family's house in the Chocó region of Colombia   247

E.3. *Zamia stevensonii*, the cycad species named after the author   251

E.4. A cycad representation on the northern facade of the Basilica di San Marco, Venice, compared with a young megasporophyll of *Cycas revoluta*   256

Tables

2.1. Summary of genetic diversity and structure of cycad species   71
3.1. Main morphometric features of storage starch grains from the underground stem of three Caribbean *Zamia* species   104
3.2. Sites where tentatively and securely identified zamia starches have been identified   110
4.1. General ranges for major periods in Mesoamerican prehistory   134
5.1. Chronology of Tamaulipas cave sites and associated domesticate remains   145
5.2. Key elements in the diet of the hunter-gatherers in the Sierra de Tamaulipas cave sites, showing significance of cycads in early subsistence strategies   147
6.1. Tenochtitlan sequence of the five ages   174
6.2. Teenek world ages   177
6.3. Characteristics shared between Tonsured Maize God and Gulf Coast Maize Hero   190
8.1. Teenek cycad-maize terminology   227
8.2. Xi'iuy *Dioon edule* cycad terms   233
8.3. Mexican maize-cycad terminology derived from Nahuatl   236
8.4. Cycad terms relating to maize in Totonac, Zoque-Popoluca, Chontal, and Mazatec   239

Foreword

Maize is rightfully recognized as one of the most important crops in ancient Mesoamerica and was both a major source of food and a fundamental component of Mesoamerican cosmology. The dominance of maize in academic discourse, however, has potentially blinded scholars to other taxa that may have been equally as important. It is precisely this problem that Carrasco and colleagues seek to address in this volume through an interdisciplinary examination of a plant known as a cycad. Cycads belong to the order Cycadales and are an ancient taxon having first evolved 250 million years ago after the Permian-Triassic mass extinction. Like palms, with which they are often confused, cycads thrive at low altitudes in tropical and subtropical climates around the world. Cycads have starch-rich stems and seeds, a nitrogen-fixing root system, and broad, fan-shaped leaves, making them an appealing multiuse plant. The detailed analysis of this taxon is the result of a five-year interdisciplinary, multinational project titled *Cícadas y la Domesticación de Maíz en el Paisaje Mesoamericano: Elucidando una Relación Milenaria mediante la Genómica, la Arqueología y la Geografía Cultural*, funded by the Mexican National Council on Science and Technology (CONACYT) Fronteras de la Ciencia Grant program. The project succeeds in that it recognizes the importance of maize in ancient and modern Mesoamerica and the Caribbean, while also challenging common perceptions of agroecology, Indigenous foodways, and cultural practices through the introduction of new lines of biological, genetic, ethnographic, linguistic, and archaeological evidence and the reevaluation of older iconographic evidence using new perspectives, interpretations, and insights.

The biological, genetic, and conservation research of cycads in this volume provides the foundation for exploring human/cycad relationships. The earliest investigation of cycads in the Americas coincided with European colonization in the 1500s. Modern research indicates that there are four main genera in the Americas, *Ceratozamia*, *Dioon*, *Microcycas*, and *Zamia*, that contain 127 species. These taxa can be found in three distinct biogeographic groups: Mesoamerica, the Caribbean and Florida, and the Isthmus of Panama and northern South America. While researchers caution that no definitive genetic pattern has emerged for cycad population genetics, preliminary analysis suggests a human role in cycad genetic diversification and population structure when paired with other lines of evidence such as those presented in this volume. Efforts to keep these taxa from going extinct rely upon developing sustainable-management cycad nurseries. In these nurseries, cycads are grown for commercial sale, local use, and reintroduction into the wild so as to demonstrate the value of cultivated cycads while also helping to preserve forest ecosystems.

Archaeological evidence for cycads in Mesoamerica and the Caribbean is a rapidly growing area of research. The emerging data in this volume illustrate the importance of cycads in the transition from hunting and gathering to early agricultural societies. The earliest evidence for cycads is macrobotanical remains dating to ca. 4500 BCE from the Tamaulipas and Tehuacán caves in northeastern and central Mexico. Starch grain evidence from artifacts and dental calculus indicates widespread use of the *Zamia* genera throughout the Caribbean by at least 5800 BCE. Future archaeological projects should take cycads into consideration when planning a project and look for evidence of their use through macrobotanical, starch grain, phytolith, aDNA, chemical, and pollen analysis.

The archaeological research presented in this volume challenges many current ideas about the story of ancient Mesoamerica and the Caribbean. The research suggests that in Mesoamerica, cycads were widely used before the domestication of maize and that the techniques used in nixtamalization were originally developed for detoxifying cycads for consumption. Once maize was domesticated, cycads were intercropped because of their nitrogen fixing capabilities. Cycads were only later replaced by beans as the main field companion of maize. Cycads, however, would not disappear in Mesoamerica and may have been planted in agricultural terraces where they would thrive in the drier soil conditions. In the Caribbean, cycads emerged as an important food source and were frequently found alongside maize and sweet potatoes throughout the region.

The widespread linguistic, iconographic, and ethnographic evidence in Mesoamerican and northern Central American societies shows a deep history of cycad use that crosses cultural and linguistic boundaries. Reanalysis of sacred maize ancestor iconography points toward the possibility that, instead of a singular "Maize god" in Mesoamerican cosmology, the sacred maize ancestor should be viewed in broader terms because of the similarities between a maize cob and a cycad cone. The maize god is also more likely a fertility ancestor given the close relationship between maize and cycads in early agriculture.

While the iconographic evidence points toward an early Olmec origin for the sacred maize concept, modern ethnographic distribution of the sacred maize ancestor concept in communities throughout Mesoamerica links it with the later migration patterns of Nahuatl speaking peoples. Ethnographic accounts stretching from northeast Mexico to Honduras link the sacred maize ancestor concept with dual cycad and maize use. Linguistic and ethnographic accounts of cycads within Indigenous communities reimagine cycads as a bridge between pre- and post-maize foodways. These accounts suggest that maize is viewed as the mystical offspring of a cycad, because cycads are viewed as the guardians of the milpa and because they are dioecious; that is, they contain both male and female reproductive parts. Linguistic analysis of the Nahuatl term *teosinte,* a term that has become synonymous with the wild ancestor of maize, reveals two separate origins for the term: northeastern Mexico along the Sierra Madre Oriental and a region from southwestern Guatemala along the Pacific to Costa Rica. In the region starting in southwestern Guatemala, the term refers to wild grasses; however, in northeastern Mexico, the term refers to wild grasses *and* cycads. Indeed, research by Bonta in this volume reveals that maize and other wild grasses are viewed by native communities as a *type* of cycad, thereby once again reaffirming the closeness of the cycad/maize relationship.

The detailed chapters in this volume will be of great interest to a wide array of biologists, ecologists, geneticists, conservationist, archaeologists, linguists, cultural anthropologists, culinary scientists, and agricultural scientists, to name a few. This volume provides readers with a holistic picture of cycads and how they played an important, yet little understood role in ancient Mesoamerica and the Caribbean. It provides important information on the biology, genetics, and distribution of the taxa while also showing the deep connections between humans, cycads, and maize. Incorporating multiple lines of evidence into a volume permits the reader to go from chapter to chapter and see how evidence can complement or contradict one another. This volume not only opens new lines of inquiry across several disciplines working in the region, but also provides meaningful justification for the conservation of a whole order of plants whose emergence out of the greatest mass extinction event 250 million years ago, and whose continuing role in the story of humanity, should not be forgotten.

Thomas C. Hart
Director of Paleoethnobotanical Research for the Programme
for Belize Archaeological Project

Introduction

Elucidating Human-Cycad Relationships in the Indigenous Ethnoecological and Agroecological Systems of the Americas and Beyond

Michael D. Carrasco, Angélica Cibrián-Jaramillo, Mark A. Bonta, and Joshua D. Englehardt

Cycads: A Brief Introduction and History

Maize, beans, squash, and manioc initially come to mind as fundamental constituents of Mesoamerican and Caribbean foodways and agricultural systems—often viewed as the base on which regional cultures developed and flourished. Chilies, potatoes, and tomatoes emerge as other major American contributions to global foodways and agroecologies; however, cycads are essentially absent from this view. If recognized at all, these ancient plants might instead be regarded as ornamentals or associated with the intoxication of cattle and neurodegenerative diseases in humans (Cox and Sacks 2002; Patiño 1989; Thieret 1958; Whiting 1963, 1989). Yet, throughout Mesoamerica, the Caribbean, and the Americas more broadly, cycads figure prominently in regional culinary traditions, symbolism, and mythology from at least the Late Archaic period (ca. 3500–2000 BCE) to the present. These regional data parallel evidentiary patterns noted in other areas of the world, from Oceania to Asia to Africa, in which cycads held considerable alimentary and symbolic significance as a managed wild food since the Pleistocene–Holocene transition ca. 11 kya (see, e.g., Beck 1992; Bradley 2005; Hayward and Kuwahara 2012; Khuraijam and Singh 2012; Kira and Miyoshi 2000; Osborne et al. 1994, 2007; Patiño 1989; Smith 1951; Smith 1982; Thieret 1958; Tōyama and Ankei 2015, among numerous others). Their presence in the early archaeological records of Mesoamerica and the Caribbean, their prominent place in regional mythologies and symbolism, and their continued use and cultural significance point to the conceptual saliency and alimentary role of cycads.

In *Under the Shade of Thipaak: The Ethnoecology of Cycads in Mesoamerica and the Caribbean,* we begin a conversation on human-cycad relationships in Mesoamerica and the circum-Caribbean world that extends our understanding of the role of cycads and biocultural heritage in this and other regions. The title of this book refers to the child fertility deity who brought maize to the Teenek, a Mayan language-speaking cultural group centered along the western coast of the Gulf of Mexico in what are now the Mexican states of San Luis Potosí, Veracruz, and Tamaulipas. Crucially, "Thipaak" is also used in a number of Teenek terms that refer to cycads (Alcorn 1984; Bonta et al. 2019; Bonta and Osborne 2007; Englehardt et al. 2020). Thus, Thipaak as both deity and concept encapsulates the multiple relationships between cycads and people, as well as illustrating conceptual convergence between cycads and maize in Teenek epistemology (Englehardt and Carrasco 2020)—a pattern noted in other regional Indigenous groups (Bonta et al. 2019; Englehardt et al. 2020).

This volume brings together contributions from specialists in a broad range of fields—from genomics to art history—to document and describe the multifaceted roles that cycads have played in Mesoamerican and Caribbean agroecologies, foodways, and symbolism. Through this, we hope to bring the American data on cycads, their cultural significance, and conservation status into conversation with global patterns of cycad use and ideation. Our collaboration developed from an initial focus on the compelling association between certain cycads and maize, initially described by Janice Alcorn (1984, Alcorn et al. 2006) and developed more fully in the work of Mark Bonta and colleagues (Bonta, 2007, 2009, this volume; see also Bonta and Osborne, 2007; Bonta et al. 2006, 2019; Vite-Reyes 2012; Vite-Reyes et al. 2010, among others). Our consideration of this intriguing connection resulted in a range of exciting research avenues, eventually evolving into the wider set of concerns examined in this volume, which include cycad biology and genomics, classification, conservation, and distribution in Mesoamerica, as well as the use of these plants in Indigenous agroecological systems, foodways, and their roles in mythology and art.

From this holistic perspective we see evidence for a millennial human-cycad relationship that dates to at least the Late Archaic period in Mesoamerica (ca. 3500–2000 BCE), and likely earlier. Genomic evidence suggesting that American cycads were in part dispersed by humans, in tandem with their presence in the early archaeological record of the region, opens a vista onto a critical moment in Mesoamerican and wider Caribbean subsistence practices, at a time immediately prior to or concurrent with the domestication of maize and its rise to become the major staple crop of Mesoamerica. Considering cycads and similar managed wild plant resources encourages reflection on categories such as agriculture, horticulture, forest gardens, and managed resources, among others (see Ford 1985) that are fundamental to understanding pre-agricultural subsistence practices and attendant ideological systems. We find components of these Indigenous epistemologies in narratives of conflict, such as the opposition seen between cycads and the breadnut tree (*Brosimum alicastrum*) (Alcorn 1984; Alcorn et al. 2006; see also Lévi-Strauss 1969–1981).

The deep history of cycads in Mesoamerican foodways possibly explains their disproportionate role in regional mythologies and symbolic systems, the presence of shared ethnonyms with teosinte (*Zea mays parviglumis*) and maize, and the sustained use of cycads in the region, even when they are often considered now mainly as a "famine food"—a category to which many managed wild foods and resources have been relegated, often uncritically, in the academic literature. Nevertheless, human-cycad relationships appear to have shaped a rich symbolic world where plants parallel, manifest, and enact mythological structures—which themselves reflect Indigenous ecological perspectives. Unravelling these epistemologies offers another window on the significance of plants and their ecological contexts that is distinctly more complex than is typically recognized, and often posits somewhat uncomfortable challenges to traditional "Western" conceptions of human relationships with the natural world and its web of life, and the place of our own species in these wider ecologies.

At this point, it is perhaps necessary to clarify the status of cycads as "wild" plants. As many sources, and indeed, many chapters in the volume, attest, cycads can be discussed in reference and relation to a number of other plants that are not wild at all, such as maize, beans, and sweet potatoes. Further, it is possible that humans played a role in the genetic diversification and distribution of cycads across the landscape, perhaps by transporting seeds, transplanting live specimens, or even tending cycads in forest gardens or other horticultural, non-agricultural production systems (see Ford 1985). However, the crucial distinction is that cycads were never domesticated; at no point have these plants depended on humans—or artificial selection—for their biological reproduction. Moreover, as we have argued elsewhere (Bonta et al. 2019; Englehardt et al. 2020), the cultural links between cycads and other plants, particularly maize, are conceptual in nature—such as the maize-cycad convergence in Teenek culture discussed above. We therefore do not suggest that such associations reflect actual genetic or biological relationships, although it is certainly possible that they were conceived of this way within Indigenous classificatory systems (see Carrasco 2020, and chapter 6, this volume). In this sense, and although humans may have moved cycads, or cultivated them at small scales, such "management" does not change the biological status of the plant as wild, despite the fact that cycads were, and continue to be, utilized as a foodstuff and in a variety of cultural practices throughout Mesoamerica and the Caribbean.

In addition to speaking to the cultural and botanical place of cycads within Mesoamerica and the circum-Caribbean world, this volume also models a way of approaching a botanical subject that may be used to examine other species and to explore questions of hybrid biocultural heritage more generally (Descola 2013). It demonstrates that the collaborative synergy arising from bringing disparate fields together offers a more revealing consideration of the significant role that a particular class of plants played in Indigenous American cultures—and also gives rise to subsequent research questions on an underexplored topic. Furthermore, the contributions collected here seek to diminish the notion that plants or animals stand in isometric relationship with specific deities, or held static roles in human cultural conceptions; rather, they emphasize the dynamic relationships that existed—and continue to operate—among subsistence practices, ecological epistemologies, and ideation. In short, the contributors show that to understand one element of a system requires a detailed knowledge of the complex holistic system of which it is but one part. Finally, to these cultural questions addressed by a number of contributors, we have invited other participants who describe the history of the study of American cycads, as well as their distribution, biology, genetics, and future conservation.

Because we recognize that some readers may not be conversant with cycad biology and their natural history, in the following section of the introduction we provide a brief overview of these issues, foreshadowing the first chapter by Calonje et al., which provides a more detailed exploration of these subjects. Likewise, stemming from our aim to bring the Mesoamerican evidence into conversation with worldwide discussions of cycads, we offer a cursory review of broader patterns of human-cycad relationships across the globe, before zeroing in on our specific regional focus on the Mesoamerican and circum-Caribbean worlds. Although this may at first appear counterintuitive for a volume whose avowed focus is the Americas, we feel that exploring relations between cycads and humans in other regions helps to provide both context and a point of reference for the considerations unpacked in the individual chapters collected here. In this sense, we seek to first contextualize the Mesoamerican and Caribbean examples within broader global patterns that point to the vital place of these plants in human culture, as well as highlight the urgent need to better understand the role of such plants in both the past and present.

What are Cycads?

Cycads are seed-bearing plants that belong to the ancient order of Cycadales, of a lineage that originated ca. 300 million years ago. They are related to other gymnosperms including conifers, araucaria, and ginkgo. Like other gymnosperms, they do not produce flowers, but instead reproduce via cones or strobili (Figure 0.1) on male or female plants, and thus are dioecious. Each plant produces a morphologically distinct strobilus; male plants produce pollen cones (microstrobili), while female plants produce ovuliferous or seed cones (megastrobili) (Norstog and Nicholls 1997; Whitelock 2002; see also Calonje et al., this volume). Pollination depends mostly on specialized pollinators, usually a species of beetle (Terry, Tang, Taylor, et al. 2012) or other insects (Tang et al. 2018; Terry, Tang, and Marler 2012). Both sexes have woody, pachycaul or thick cylindrical trunks with a large rosette crown

Figure 0.1. Strobili of three Mexican cycad species: *a,* female *Ceratozamia fuscoviridis*; *b,* male *Dioon edule*; *c,* female *Zamia furfuracea* (*a,* courtesy Michael Calonje; *b,* photograph by Chip Jones, used under CC BY-NC-SA 4.0 license; *c,* photograph by Angélica Cibrián-Jaramillo).

Figure 0.2. Pachycaul stem, rosette crown, and pinnate leaves of cycads: *a, Dioon edule* (photograph by Andrew P. Vovides); *b, Ceratozamia fuscoviridis* (photograph by Michael Calonje).

of stiff evergreen pinnate leaves (Figure 0.2), giving them a hard and leathery texture. In this way, cycads superficially resemble—and are often mistaken for—palms or even ferns, given their circinate leaves, but they are not closely related to either of these groups.

The more than 350 extant species of cycads are classified into three families (Cycadaceae, Stangeriaceae, and Zamiaceae) and ten genera (Calonje et al. 2022; Christenhusz, Reveal, et al. 2011; Stevenson 1992). Species vary greatly in size, from diminutive ones with subterranean stems to arborescent examples reaching more than 10 meters in height. Like trees, they are slow-growing, long-lived plants (Octavio-Aguilar 2008), with some specimens known to be several thousand years old. Cycads are resilient plants and thrive in a variety of ecosystems throughout the tropical and subtropical regions of the world (see Figure 0.3; Norstog and Nicholls 1997), oftentimes in locations with extreme environmental conditions.

Cycad root adaptations allow them to exploit what are for other plants inhospitable or nutrient-poor ecological niches. Their complex roots possess two main structures: a contractile taproot that anchors the plant to the soil and stores starches, in some cases actively pulling the plant into the earth, and coralloid roots (Figure 0.4) that are lateral roots with specialized palisade cells that host facultative symbioses with cyanobacteria and other bacterial groups (Gutiérrez-García et al. 2019). These cyanobacteria and other bacterial groups, such as Rhizobiales and Actinobacteria, together help the plant fix nitrogen, transport other nutrients, and produce the neurotoxin

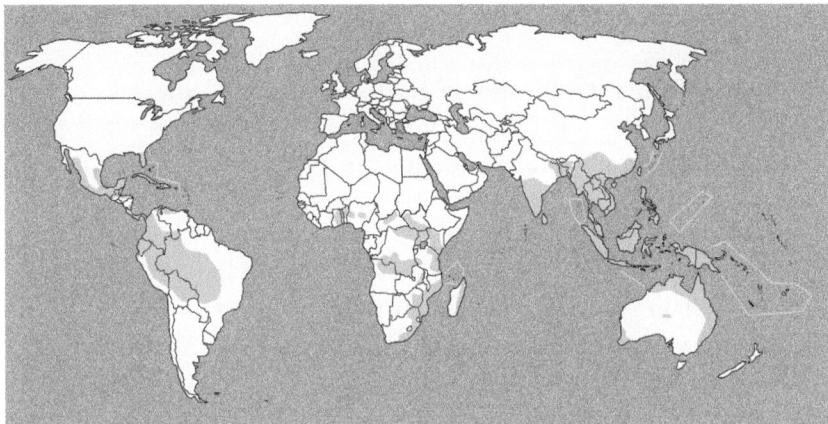

Figure 0.3. Approximate worldwide distribution of Cycadales in tropical and subtropical regions (map by Mark A. Bonta).

β-N-Methylamino-L-alanine, or BMAA (Cruz-Morales et al. 2017; Rai et al. 2000). The symbiotic relationship with these bacteria perhaps explains their perseverance through time as a million-year-old lineage (Gutiérrez-Arroyo et al. 2018; Gutiérrez-García et al. 2019). While ancient people may not have understood—in modern scientific terms—the biological mechanisms of nitrogen fixation, the fertility of cycads was likely not lost on them. Perhaps their vitality in otherwise difficult environments stood out to early people, as it continues to in a number of traditions. This also might explain why cycads were interplanted with other food crops to increase soil fertility, or for other agroecological purposes—a point discussed briefly below in the example of

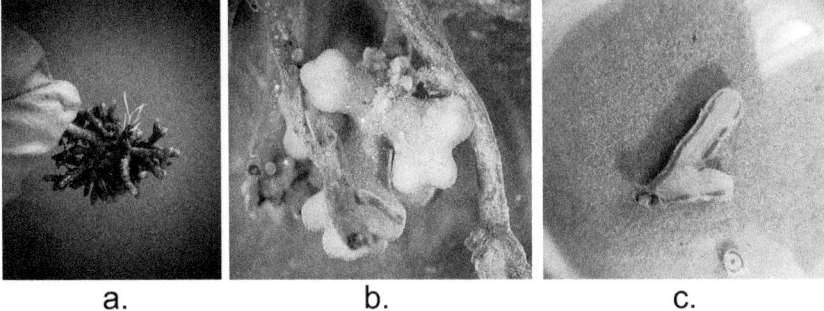

a.　　　　　　　　b.　　　　　　　　c.

Figure 0.4. Coralloid root structures in cycads: *a*, cortex of specialized *Dioon* root (photograph by Naishla Gutiérrez-Arroyo); *b*, light micrograph of a preserved specimen of coralloid roots of the genus *Cycas* (image by Curtis Clark, used under CC BY-SA 3.0 license); *c*, close-up of the palisade cells that contain cyanobacteria and other bacteria, together known as the "cyanobacterial ring" (micrograph by Angélica Cibrián-Jaramillo).

the still extant agroecological practices of the Ryukyu Islands, and more fully within Mesoamerican contexts in Amber VanDerwarker's chapter 4. It is possible that their vitality, longevity, and food potential, despite—or perhaps in conjunction with—their natural toxicity, made cycads compelling plants to think with, to adapt Claude Lévi-Strauss's (1962:132) formulation *"bonnes à penser."*

A Brief Natural History of Cycads

The 300-million-year history of cycads (Gao and Thomas 1989) has made these plants witnesses to fundamental transformations of our planet. Cycads existed on Paleozoic Pangaea, and since then they have seen oceans and mountains surge and recede, continents drift apart, and the evolutionary rise—and fall—of countless species, including our own. They have survived warming periods and ice ages, atmospheric changes, and four massive extinction events. The oldest known fossil with anatomical affinities to the modern Cycadales (e.g., monoxylic axis, girdling leaf traces [see Calonje et al., chapter 1, this volume, Fig. 1.1]) was recently found in the Irati Formation of Brazil's Paraná Basin, and dates to the Cisuralian epoch of the early Permian, ca. 300–275 mya (Spiekermann et al. 2021). Although their fossil record is generally thin, cycad diversity was likely much greater prior to the Permian-Triassic extinction event—the "Great Dying" ca. 250 mya—at which point five families of the extinct Medullosales order were extinguished. Nonetheless, cycads enjoyed their apogee during the Mesozoic Era, the time of the dinosaurs, during which they formed a major component of the world's vegetation prior to the development of angiosperms (flowering plants).

Morphologically, cycads have changed very little since their heyday in the Jurassic period, compared with some relatively major evolutionary changes in other plant orders. For example, the modern *Cycas revoluta,* likely the most closely related plant to the last common ancestor of Cycadales, ginkgo, conifers, and gnetophytes, bears megasporophylls (fertile, seed-carrying leaves) that are strikingly similar to Permian-period Crossozamia fossils (Gao and Thomas 1989). Another remarkable ancestral characteristic is cycads' multiciliated male gametes (akin to motile spermatozoa); the only other living species that conserves this trait is *Ginkgo biloba.* An additional ancestral trait is found in circinate leaf emergence in cycads, which parallels that of ferns (Norstog and Nicholls 1997). Perhaps because of these evolutionary continuities, existing cycads are often thought of as "living fossils," even if this perception is misplaced given that cycads continue to evolve, adapt, and diversify. Although the cycad lineage itself is indeed ancient, the

K-Pg Extinction at the end of the Mesozoic ca. 65 mya caused the extinction of most Cycadales and the entirety of the Cycadeoidales or Bennettitales order. The few Cycadales families that survived into the Cenozoic gave way to speciation events that resulted in the extant cycad species known today, most of which are of relatively recent evolution since the Miocene ca. 12 mya (Nagalingum et al. 2011). In fact, only one cycad genus, *Bowenia*, dates to the Cretaceous or earlier. Today, modern cycads and their evolutionary lineage hold a special place in the human social imaginary, as aliens among us, relics of a fundamentally different world that we are only just beginning to understand.

Global Cycad-Human Relationships: A Brief Overview

Since our species first encountered cycads, they have played integral roles in human cultures worldwide. Although cycads have never been domesticated, archaeological, genetic, historical, and ethnographic evidence demonstrates that cycads have held considerable alimentary and symbolic significance in many areas of the world, from at least the Pleistocene–Holocene transition to the present. Human uses of cycads range from the mundane to the sacred, from a pre-grain subsistence food, starchy staple, or famine food to the involvement of cycads in ritual and religious practices. The use of cycads as a foodstuff has been documented in a number of contexts, from the Upper Pleistocene in Australia (Beck 1992; Smith 1982)—where they have been documented in Aboriginal rock art (Figure 0.5) associated with the Madjedbebe rockshelter, dated to >50 kya (Clarkson et al. 2015; although the paintings themselves are likely of more recent origin [Roy Osborne, personal communication, 2021])—to the Caribbean ca. 7800 BP (Pagán-Jiménez et al. 2015) to more recent contexts in these areas, as well as Japan, South, East, and Southeast Asia, sub-Saharan Africa, Central and South America, and of course Mexico (see, e.g., Beck 1992; Bonta et al. 2019; Bradley 2005; Hayward and Kuwahara 2012; Khuraijam and Singh 2012; Krishnamurthy 2014; Mickleburgh and Pagán-Jiménez 2012; Patiño 1989; Thieret 1958; Veloz Maggiolo 1992, among many others). In all these cases various parts of cycads have been consumed, used as medicine, or crafted into utilitarian items. The large, starch-rich seeds are especially attractive and, in general, can be consumed once detoxified. A variety of methods have been used to accomplish this, the majority of which involve boiling or leaching, and then grinding the seeds into a flour from which a variety of preparations arise. Humans also consume cycad leaves, trunks, and other parts, mostly after detoxification, and the fleshy outer cover of some cycad seeds, the sarcotesta, can be eaten

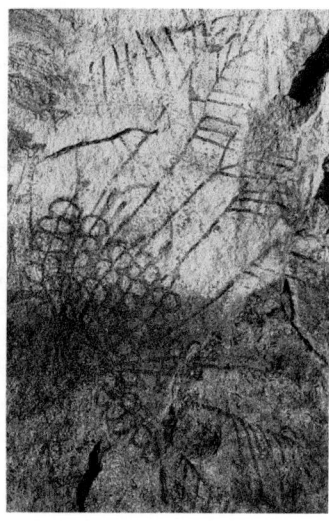

Figure 0.5. Australian Aboriginal rock art representations of cycads at the seed-producing stage, Alalak Bene, near the Madjedbebe/Malakunanja II archaeological site in the East Alligator Valley, Arnhem Land, Northern Territory (images courtesy Ian Morris).

raw (Bonta et al. 2019; see Khuraijam and Singh 2012:157, fig. 11.1j and 159, fig. 11.2h).

Beyond consumption as a foodstuff, cycads have been employed in a variety of other utilitarian preparations in many regions, from pest control to medicinal. Such practices are particularly well documented throughout India. In the northwestern states of Assam and Meghalaya, Khuraijam and Singh (2012:157–158) report that young men consume the young microsporophylls and pollen-filled microsporangia of *Cycas pectinata* to enhance sexual potency (Figure 0.6). This reflects a common association—documented in other regions—between cycads and fertility and/or longevity, likely due to the long lifespan of these plants—and perhaps also to the phallic shape of the strobili. These authors also detail that microsporangia, megasporophylls, and other parts of the plant are used to make poultices that alleviate ulcers, cramps, and other stomach ailments. Krishnamurthy (2014) documented similar human uses of *C. circinalis* in the ethnically distinct southern Indian states of Andhra Pradesh, Kerala, and Tamil Nadu. In addition, the dried piths of *C. beddomei* and *C. circinalis* are used to treat rheumatism, muscular pain, and debility (see Krishnamurthy 2014:36, figs. 3 and 4, 107–108; Reddy et al. 2006). Such practices parallel those observed in China with *C. revoluta* (Pant 1973), in southern Africa, where *Encephalartos* spp. and *Stangeria eriopus* are used for medicinal, magical, and narcotic purposes (Bonta and Bamigboye 2018; Cousins et al. 2012; Donaldson 2003; Osborne et al.

Figure 0.6. Vendor selling young microsporophylls and pollen-filled microsporangia of *Cycas pectinata* for consumption and use in medicinal preparations in Shillong, Meghalaya, India (image courtesy J. S. Khuraijam) (see also Krishnamurthy 2014:36, figs. 3 and 4; Khuraijam and Singh 2012:157, 159, figs. 11.1 and 11.2).

1994), and in Mexico (see, e.g., Bonta et al. 2019) and Central America (Taylor Blake et al. 2008:421).

In addition, there is evidence that cycads were a key component of Indigenous horticultural practices in many contexts, perhaps playing a crucial role in "non-linear transitions" to agricultural lifeways in some areas (Freeman et al. 2015; Gremillion et al. 2014). For example, in Japan's Ryukyu Islands, *sotetsu* (*Cycas revoluta*) were often deployed as wind- and sea-breaks to protect fields and other food crops (Kira and Miyoshi 2000; Tōyama and Ankei 2015; see Figure 0.7) and their cut leaves were used as green manure. Further, archaeological evidence (Pearson 1969:116; Takamiya 2013:324; Tsuji et al. 2007) has revealed cycad exploitation in the Ryukyus dating to as early as 4800 BCE, thus indicating the preagricultural importance of this plant—data which may align with genetic evidence that potentially suggests human involvement in the spread and diversification of *C. revoluta* throughout the archipelago beginning ca. 19,000 years ago (Kyoda and Setoguchi 2010; see also Chang et al. 2019; Englehardt and Carrasco 2022). Nonetheless, the specific uses or "management status" of cycads (i.e., as part of forest gardens,

Figure 0.7. Use of cycads in regional agroecology: *sotetsu* (*Cycas revoluta*) hedged fields near Kanamizaki village, Tokunoshima (photograph courtesy Sueo Kuwahara; see Hayward and Kuwahara 2012:30, fig. 2).

perhaps) in Late Pleistocene contexts remain a mystery (cf. Pearson 1969; Tawada 1975). A particularly striking image from the nineteenth-century *Nantōzatsuwa,* or *South Island Chronicle,* details the processing of cycads to make both sake and food—and more recent historical photographs suggest that the methods and techniques employed have changed little in the last 200 years (see Figure 0.8; see also Hayward and Kuwahara 2012:35, fig. 3; Kira and Miyoshi 2000). Indeed, cycads are still used as a foodstuff in the Ryukyus, and one can buy cycad miso at specialized shops on Amami Oshima (Hayward and Kuwahara 2012:41–42, figs. 6 and 7) and even through online retailers (www.amazon.co.jp/ヤマア-そてつみそ-なりみそ/dp/B00J7LF7VA, accessed July 9, 2021; see Figure 0.9). Cycads also appear to function as an identity marker in Amamian culture, in which they feature in folk songs known as *shima uta* (lit. island songs), and the leaves of *C. revoluta* form one of the main motifs in the famous *Oshima tsumugi* regional textile tradition (Linton 2020).

As the examples suggest, aside from utilitarian uses in domestic and local economies, cycads also have a number of important symbolic associations and ritual functions. Modern ethnographic evidence demonstrates that cycads have been and are used as entheogens, as narcotics, as part of ceremonies, and even as a poison (derived from the cyanobacteria in cycads' coralloid roots) in a variety of cultural practices the world over. Outside of the Mesoamerican and Caribbean contexts with which the authors are most familiar—and on which this volume is focused—Vanuatu offers a

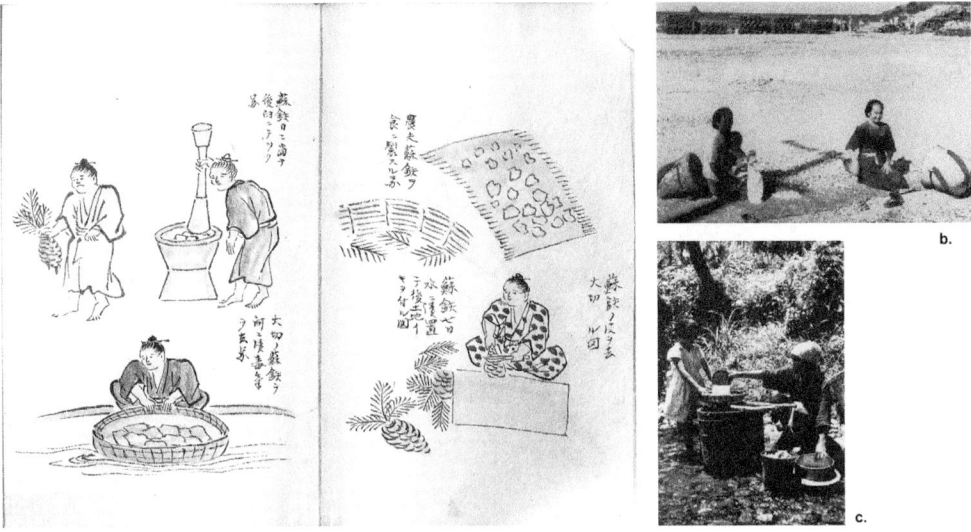

Figure 0.8. Historical cycad use in the Amami Islands, Japan: *a*, p. 27 of the 19th-century Nantōzatsuwa (Amami Museum Collection; see also Nagoya 1984); *b* and *c*, 20th-century photographs by Hideo Hakari of cycad processing in the Amami Islands.

particularly compelling example of the symbolic power of cycads. In this Pacific island nation, cycads are known as *namele* and are both symbols of traditional culture, revered as sacred ancestors, and a sign of chiefly authority (Forsyth 2009; Layard 1942). Further, the *namele* carries an extremely strong tabu: the presence of *namele* leaves in a doorway bars entrance to all

Figure 0.9. Cycad miso paste (そてつみそ, *sotetsu miso*) available in Japan (left image by authors from www.amazon.co.jp, right photograph courtesy Hiroyuki Takanashi).

Figure 0.10. Chief Seni Mau Tiruspe of the Malvatumauri Council of Chiefs, with Prince Charles in Vanuatu, dressed in chiefly regalia, including *Cycas rumphii* cycad leaves, as the Prince of Wales received the *kastom* title of Mal Menaringmanu (photograph by Dan McGarry, courtesy Vanuatu Daily Post).

those without chiefly authorization (McGarry 2018). Yevgeniiy Osiievskyi (personal communication, 2021) also reports that *namele* leaves were traditionally placed on sacrificial victims—both animal and human—as a sign of peace and prosperity. *Namele* leaves form part of Vanuatu's coat of arms and appear on its national flag and in various other official insignia. On a recent visit to the island, the British Prince Charles received a chiefly title from the people of Vanuatu; part of his ceremonial attire—and of his hosts—included a *namele* leaf (Figure 0.10).

Another striking illustration of cultural-symbolic significance of cycads comes again from Japan. On the grounds of the Buddhist temple of Myōkoku-ji in Osaka stands an ancient *C. revoluta* that, according to legend, is more than 1,000 years old. In the sixteenth century, the feudal lord Oda Nobunaga attempted to transplant the cycad to his base at Azuchi Castle some 80 km to the northeast. Once it had been installed at the castle, residents reported hearing an eerie voice pleading, "Take me back to Myōkoku-ji." This unnatural disturbance upset Nobunaga's people and, angered, he ordered the tree cut down. On attempting this, however, it is said that the tree bled and fainted from the pain, revealing itself as a *yōkai*—a kind of supernatural spirit or demon that in this case resembled a great snake. Fearful, Nobunaga sent the tree home to the temple in Osaka. The head priest took pity on the dying cycad and recited 1,000 sutras in an attempt to save it. Following this, a hybrid snake-human deity appeared in the priest's dream, thanking him for his prayers and swearing sacred oaths to ease the pain of childbirth, alleviate hardship, and bring happiness to the poor. The priest subsequently named the cycad as the guardian god of the temple and built a hall in its honor; in the nineteenth century, pilgrims to the site could purchase votive images of the cycad (Figure 0.11). In the early twentieth century, the cycad was declared a national natural monument by the government of Japan. The incident with Nobunaga was commemorated in a series of nineteenth-century *ukiyo-e* woodblock prints (Figure 0.12). In fact, cycads were featured in Edo period art, figuring prominently in works by celebrated artists such as Yoshitoshi and Hiroshige.

A stunning example of the symbolic power of cycads is that of the Mudjadji, or Rain Queen, the hereditary ruler of the Balobedu people from what is now Limpopo Province, South Africa (Krige and Krige 1943; Motshekga 2010). The Rain Queens are said to possess the ability to create rain, leading a specialized rainmaking ceremony each November. Her power as a rainmaker is reflected in the lush garden that surrounds her royal compound in an otherwise arid landscape—a garden that not coincidentally nurtures one of the world's largest cycads, *Encephalartos transvenosus* (known as the Mudjadji cycad, in honor of the queen), which grows up to 12 m in height. Cycad leaves are also used to decorate the Rain Queen's royal *kraal* (Caitlin Mapitsa, personal communication, 2021). A beautiful formal portrait of the fourth Rain Queen, Mukoma Mudjadji IV, depicts her in ceremonial attire positioned among the cycads that bear her name (Figure 0.13). The relationship between cycads and the Rain Queen appears to reflect the same conceptual association between cycads and fertility or longevity noted in the Indian and Japanese examples discussed briefly above.

A final example that illustrates this pattern, and the symbolic power of cycads, is the placement of cycad leaves as adornments or decorations at temples and shrines, as well as their incorporation in traditional and syncretic rituals. Such practices are particularly evident throughout Asia, where many species of cycads adorn the grounds of temples and shrines (Figure 0.14). In northeastern and southern India, leaves of *C. pectinata, C. orixensis,* and *C. circinalis* are used to decorate altars, temples, and houses for ceremonial purposes, often for weddings (Khuraijam and Singh 2012:158, 160, fig. 11-3; Krishnamurthy 2014:104–105, image 2), paralleling similar uses reported by Bonta et al. (2019:figs. 19, 30) in Mexico and Honduras. In the state of Assam, cycad leaf decorations are also said to prevent bad dreams and repel snakes

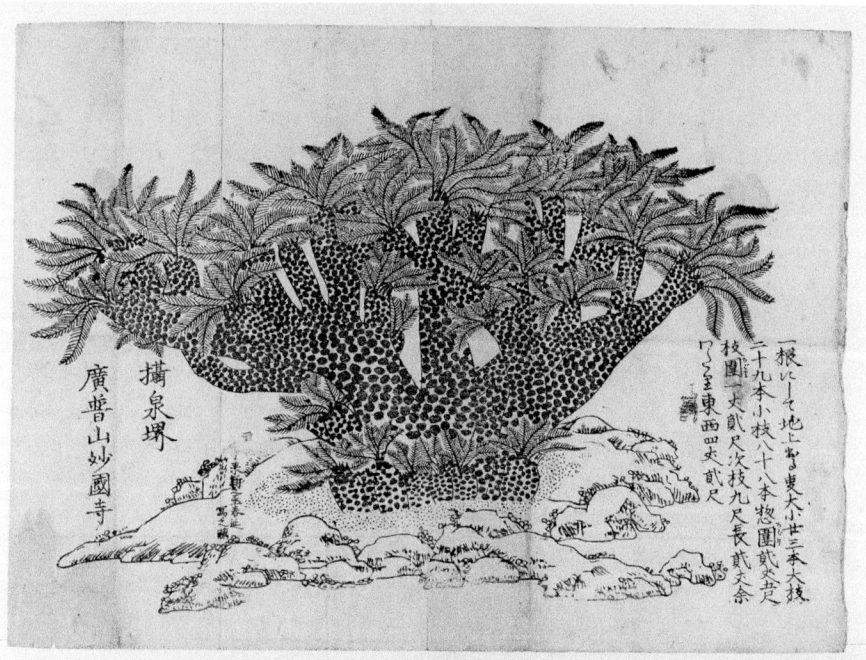

Figure 0.11. Images of the Myōkoku-ji cycad. *Top*: votive print dedicated to the Myōkoku-ji cycad (print from the collection of Michael Carrasco); *bottom*: photograph of tourists visiting the Myōkoku-ji cycad in the Edo period (public domain image from the archives of the National Diet Library, NDL-401-5).

Figure 0.12. Utagawa Yoshitsuya, *Harunaga-ko and the Angry Sotetsu from 54 Battle Stories of Hideyoshi*, 1864, detailing the legend of Oda Nobunaga and the sacred cycad of Myōkoku-ji: as Nobunaga's men attempt to cut down the cycad, it reveals itself as a *yōkai*, with snake tongues and demonic eyes as it bleeds (public domain image from the archives of the National Diet Library, NDL-434-00-023).

(Khuraijam and Singh 2012:158). In the Philippines, Gamil (2014) presents exquisite photos of the syncretic incorporation of *C. revoluta* (known locally as *oliba*) leaves in Catholic celebrations of Palm Sunday, much like the use of *espadaña* (*Dioon merolae*) cycad leaves in the festivities surrounding the Feast of the Holy Cross in Chiapas, Mexico, reported by Pérez-Farrera and Vovides (2006), or the placement of cycad design motifs on the facades of Catholic churches throughout central and western Mexico (Bonta et al.

Figure 0.13. A formal portrait of Mukoma Mudjadji IV, the fourth Rain Queen, who reigned from 1960 to 1980, in royal attire and surrounded by Mudjadji cycads (painting by Pawel Kot, image courtesy Caitlin Mapitsa, originally published in Motshekga 2010).

2019:figs. 2–7). The incorporation of cycads in ritual landscapes in Asia appears to once again reflect their conceptual association with longevity, fertility, and abundance.

The limited examples briefly discussed above illustrate the multiple ways humans have used and conceived of cycads. Together with linguistic data that reveal a large number of terms for cycads in Indigenous languages throughout the world (e.g., Bonta and Osborne 2007; Bonta et al. 2019:table S1), these data illuminate how humans have perceived cycads and incorporated them into their quotidian cultural practices and belief systems throughout history. More recent ethnographic data from these contexts indicate that many of these cultural practices and beliefs continue in the present, along with more recently incorporated uses of cycads, such as for ornamental plants in landscaping.

Figure 0.14. The incorporation of cycads in Asian temples and shrines: *a*, *C. rumphii* group cycad within the grounds of the Gunung Kawi temple, Ubud, Bali (photograph courtesy Greg Holzman); *b*, *C. elongata* flanking entrance to Dinh Tien Hoang Temple complex, Hoa Lư District, Ninh Bình Province, Vietnam (photograph by Michael Carrasco); *c*, *C. orixensis* leaves adorning an altar to the Hindu god Ganesh, Assam, India (photograph by Rita Singh, courtesy J. S. Khuraijam).

Cycads in the Indigenous Mesoamerican and Caribbean Worlds

Like the traditions reviewed briefly above, the historical records of Mesoamerica and the Caribbean reveal similar cultural practices. As a botanical resource that occurs in diverse ecosystems, cycads have been—and continue to be—used in a variety of ways within the Americas. Although individual ethnobotanical studies have noted their significance in regional foodways and mythologies, for the first time this volume treats these plants within larger cultural systems, their symbolic relationships with other plants and concepts, and as an index of Archaic and Formative period interregional interaction and exchange systems. Because such uses, and their histories, are the focus of this book, we will not present them in great detail here. Nonetheless, it is useful to contextualize the individual contributions within a larger picture of the place of cycads in Mesoamerican and Caribbean epistemologies and ethnoecological systems.

As Calonje et al. detail in chapter 1, in Mexico, three genera and 70 species of cycad have been documented, representing 58% of all Neotropical cycad species (see also Calonje et al. 2022; Gutiérrez-Ortega et al. 2018; Nicolalde-Morejón et al. 2013; Vázquez Torres 1990). In fact, Mexico is second in the world in terms of cycad diversity. Cycads of the genera *Dioon* (17 species) and *Ceratozamia* (35 species) are endemic to the country. Eighteen species of *Zamia* are also present in Mexico, and 26 species of *Zamia* have been documented in the Caribbean region, which is also home to the monotypic genus *Microcycas* (see also Pagán-Jiménez's chapter 3). As the population genetics data presented by Cibrián-Jaramillo and colleagues in chapter 2 reveal, surprising genomic relationships are evident among various distinct genera and populations in these regions.

In Mesoamerican and Caribbean cultural systems, and in both ancient and modern contexts, cycads play a significant role in agroecology, Indigenous foodways and domestic economies, and diverse cultural practices and beliefs (see, e.g., Alcorn 1984; Bonta 2010a, 2010b; Bonta et al. 2019; Pérez-Farrera and Vovides 2006; Sifuentes de Ortiz 1983; Smith 1951; Tristán 2012; Tristán Martínez et al. 2020; Valdez 2009; Vite-Reyes et al. 2010), just as they did in other areas of the world, such as those discussed above. Of particular note is the temporal depth of evidence for cycad use in the Mesoamerican archaeological record, as Englehardt et al.'s chapter 5 details—echoing the indirect evidence for ancient cycad-human relationships suggested by the population genetics data (Cibrián-Jaramillo et al., chapter 2). In many ways, the most intriguing aspects of the Mesoamerican and Caribbean evidence are those that attest to the conceptual associations between cycads and other staple food crops, especially maize. Significantly, as Bonta's and Carrasco's contributions detail, cycads in Mesoamerica also occupied an important cultural place, in which they often possess a close relationship with maize (see also Bonta 2010b; Carrasco 2012, 2015; Diego-Vargas 2017).

This association is manifested in a variety of linguistic, artistic, and other cultural expressions, as many of the chapters in this volume attest. Exploring the conceptual relationships between cycads and maize, as well as the potential role of cycads in Mesoamerican domestication processes, was—and remains—a primary motivation that drove the initial collaborations of many of the contributors to this volume. And while, as Bonta details in his two chapters, this issue remains salient and a potentially fruitful avenue of research, our collective investigations have given rise to a host of other critical questions that are equally as intriguing and whose continued investigation is just as potentially productive. For example, in chapter 4 VanDerwarker asks, what roles could cycads have played in pre-grain subsistence strategies in these regions? Responding to such a query has the potential to contribute to wider anthropological debate on the origins of agriculture and the role of other wild, managed foodstuffs in pre-agricultural societies—a severely understudied subject that is often disregarded in favor of research that emphasizes the origins of domesticates, both in the specific contexts of Mesoamerica and the Caribbean, and more broadly.

Yet, as Englehardt et al. (chapter 5) note, despite the potential and richness of the evidentiary record in Mesoamerica and the Caribbean, in many instances we lack the methodological tools that would allow researchers to more fully explore these possibilities. In offering some preliminary thoughts on new methodologies and techniques, these authors seek to build on the pioneering work of Pagán-Jiménez (chapter 3; see also Mickleburgh and Pagán-Jiménez 2012; Pagán-Jiménez et al. 2015) to expand our knowledge of cycad cultures and their histories. The Mesoamerican and Caribbean worlds provide a particularly fertile data set—given the tremendous biological diversity, temporal depth, and sustained cultural use of cycads—on which to base future research along the lines suggested by VanDerwarker in chapter 4.

Each of the chapters in this volume seeks to more fully unpack one component of this broader data set, focusing on a specific aspect related to cycad-human relationships in discrete and at times overlapping temporal, spatial, and disciplinary contexts. This is possible since the place of cycads in Mesoamerican and Caribbean cultural and agroecological systems essentially reflects the same key patterns noted in other regions of the world. Precisely because the significance of cycads in both ancient and modern contexts in these regions is manifested in a variety of data, by collating a collection that integrates cycad biology, population genetics, archaeological and ancient art historical evidence, and more recent ethnographic and human geographical data regarding the cultural uses of these plants in Indigenous foodways, epistemologies, mythologies, and representational systems, this volume seeks to create a positive feedback loop and interdisciplinary template that can drive research on these and other questions, and in these and other contexts, and to promote cycad conservation, for years to come.

Scope and Aims of This Volume

A fundamental impetus for this volume was the lack of attention to cycads in the academic literature, particularly in the social sciences. To address the gaps in our understanding of the social and environmental roles of cycads in ancient and modern societies, the contributions in this collection speak to this underappreciated and critically endangered plant from a critical heritage perspective. Such a perspective permits the fuller consideration of a series of key issues that span multiple disciplines, including archaeology, ethnography, human geography, and art history, as well as genomics and evolutionary biology (Bonta et al. 2019; Englehardt et al. 2020). These issues include the cultural uses and symbolic significance of cycads, as well as the role of cycads in Indigenous agroecological systems, foodways, epistemologies, and cultural practices, particularly as they relate to local conceptions and performances of sociocultural identity.

This volume critically examines these and other issues from an interdisciplinary perspective via a range of discrete yet interrelated data sets. Our aim is to place the Mesoamerican and Caribbean data in conversation with larger, global discourses on cycad cultures, ethnobotany, and conservation. In addition to presenting new and valuable comparative data, many chapters offer novel strategies and methods by which to approach the study of cycads and cycad cultures. In doing so, a secondary goal of this collection is to stimulate further discussion on the significance of cycads in Mesoamerican and Caribbean beliefs, foodways, and ecologies. Although each of the contributors approaches the topics at the core of this volume through specific disciplinary lenses, we all agree that a closer examination of cycad-human relationships has the potential not only to highlight the environmental and cultural role of cycads in human societies—as well as the significance of ancient and modern cultural practices that incorporated this natural-cultural hybrid—but also to motivate conservation of these endangered plants in ways that engage local communities and celebrate the diversity of practices that involve this ancient plant lineage.

In this sense, this volume has several potentially wider impacts. First, it draws attention to a form of biocultural heritage in urgent need of conservation and protection by international law. Second, it highlights ancestral cultural traditions in danger of being lost and largely ignored in received scholarship. Third, comparative data presented in the contributions will complement investigations of the use of cycads or other hybrid heritage in discrete contexts. Thus, this work contributes to scholarly discourse on the management of wild foods in pre-grain subsistence strategies, a critical question to which we alluded above. Finally, the volume seeks to promote engagement with local communities to incentivize holistic preservation of the cultural practices, histories, and values related to cycad heritage, thereby supporting the protection of both tangible and intangible patrimony.

Because we conceive of this volume as the beginning of a wider interdisciplinary discussion, we do not intend it as a unified, standalone statement, per se. As such, we have consciously chosen to exclude an explicit conclusion or concluding chapter. Rather, the volume closes with a brief epilogue by renowned cycad expert Dennis Stevenson, who offers some preliminary thoughts on the present state of cycad studies, the contributions to this volume, and potential future directions.

The Urgency of Cycad Studies in These and Other Contexts

Despite the evident significance of cycads in a variety of Indigenous subsistence strategies, agroecological systems, and cultural practices, they remain relatively underappreciated and surprisingly understudied. Although a handful of sources review the use of cycads as a foodstuff in many contexts (see, e.g., Beck 1992; Bradley 2005; Hayward and Kuwahara 2012; Khuraijam and Singh 2012; Mickleburgh and Pagán-Jiménez 2012; Thieret 1958; Tōyama and Ankei 2015; Veloz Maggiolo 1992), as well as other cultural uses (Bonta and Bamigboye 2018; Bonta et al. 2006; Kira and Miyoshi 2000; Patiño 1989; Pérez-Farrera and Vovides 2006), the broader features of intriguing regional cycad cultures and their archaeological histories remain virtually unknown outside a small circle of academic specialists and cycad enthusiasts and collectors. Even those focused on Mesoamerican foodways and archaeology often know little to nothing about the significance of these plants. In this sense, cycad cultures have contributed very little to broader scholarly discourse on a number of critical topics. Yet more detailed studies of the incorporation of cycads into Indigenous horticultural practices and economies may offer insights that contribute to a better understanding of the transition to agriculture and the development of social complexity (Ford 1985; Freeman et al. 2015; Gremillion et al. 2014; Terrell et al. 2003; Winterhalder and Kennett 2006). The lack of scholarly focus on those in Mesoamerica appears to stem from their toxicity (Bonta et al. 2019; Cox and Sacks 2002; Whiting 1963, 1989), which, while requiring similar processing to that of manioc (*Manihot esculenta* Crantz), another ancient, starchy food source that also requires extensive detoxification, seemingly diverted attention away from cycads.

Moreover, and in spite of cycads' antiquity and their centrality in human life for millennia, they are in critical danger. Indeed, cycads are among the world's most threatened land plants (Calonje et al. 2022). Of the 80 recognized native species in Mexico and northern Central America, 21 are critically endangered, 23 endangered, and the rest in lesser risk categories, with only one categorized as a species of least concern (Bonta et al. 2019; Calonje et al. 2022; Donaldson et al. 2003). Worldwide, there are two primary causes of the demise of cycad populations. First, they are threatened by habitat destruction due to a variety of human activities (e.g., agriculture, agroforestry, pasture expansion) and natural factors (e.g., climate change) (see Vovides 1989). Second, increasing illegal extraction of plants from the wild by collectors in the nursery trade, for use in horticulture and landscaping, has severely impacted cycad populations, almost to extinction in some cases (for example, *Encephalartos woodii*; Donaldson 2003; Donaldson et al. 2003). Although cycads are protected on paper by the International Convention on the Trade of Endangered Species, as well as by many country-specific laws and conservation frameworks, it is clear that much work remains to be done. In some cases, governments, academic institutions, conservation NGOs, and local communities have engaged in productive dialogue and developed innovative strategies to protect cycads (see, e.g., Ruiz-Mallén et al. 2015; Vovides 1989; Vovides and Iglesias 1994; Vovides et al. 2010). Nonetheless, the results of such efforts have been mixed, at best. For example, as Bonta and colleagues (2019) argue in the case of Mexico, " . . . [e]x situ efforts such as community nurseries incentivizing habitat conservation through seed harvest, propagation, legal plant sales, and reintroduction have been marginally successful due to bureaucratic insufficiency and lack of marketing and follow-up studies." We concur with their conclusion that additional on-the-ground measures are necessary to fully ensure cycad preservation. Indeed, the best protectors of cycads in many cases have been the Indigenous and local communities that rely on them, as documented by Bonta et al. (2019) in Honduras, but also an increased sense of meaning and ecological relevance among collectors and the community in general.

These situations pose immediate challenges for cycad studies. The lack of scholarly interest in the topic, combined with increasing threats to habitats and cycads in the wild, may lead to irreparable losses, particularly for those cultures in which cycads continue to play significant economic and symbolic roles. This situation is particularly notable in the Amami Islands of Japan's Ryukyu archipelago. There, despite a millenary history and a relatively greater degree of academic and community attention, traditional uses of cycads and the agroecological heritage landscapes formed by the plants are in peril as habitats succumb to development and as the economic and symbolic significance of cycad use fades from social memory or is actively erased due to historical associations with poverty or otherness—cycads were and remain associated with a period of historical deprivation in the region that was known as *sotetsu jigoku,* or "cycad hell" (Hayward and Kuwahara 2012; Kuwahara 2013; Tōyama and Ankei 2015). Sadly, similar dynamics are playing out in many other areas of the world. For example, in South Africa, many cycad species have been collected almost to the point of extinction; in the Caribbean, foodstuffs associated with cycads are fading in traditional culture; and in Mexico an expanding livestock industry is degrading cycad populations, as ranchers eradicate the plants, which can be poisonous to cattle.

For these reasons, the immediate documentation and investigation of cycad cultures is critical, both to conserve these ancient plants and for the scientific value inherent in the study of this relic of pre-grain subsistence strategies. Research on the historical and contemporary roles of cycads in human societies assists in crafting deeper anthropological understandings of the role of cycads in human societies, and elucidates aspects of cycad ethnobotany that can aid conservation strategies, such as those discussed by Calonje et al. in chapter 1.

Final Thoughts

Cycads continue to hold a fascination for a variety of scholarly and lay audiences. Our current knowledge of these plants and their cultural uses has been informed by contributions from numerous professional researchers and amateur enthusiasts in multiple disciplines since the nineteenth century. As we have argued elsewhere and throughout this introduction, the environmental and cultural role of cycads reveals multiple convergences, as well as the significance of ancient and modern cultural practices that incorporated this natural-cultural hybrid to preserve alimentary history and encode ecological knowledge in higher order mythologies, ideological systems, and epistemologies (Bonta et al. 2019; Carrasco 2020; Englehardt et al. 2020). In sum, available data suggest the existence of long and complex cycad-human relationships.

The limited glimpse provided by this volume reveals the potential richness of the subject. As we have recently suggested elsewhere (e.g., Bonta et al. 2019; Englehardt et al. 2020), a closer examination of cycad-human relationships stimulates further discussion on the significance of cycads in dynamic cultural practices, foodways, and ecologies worldwide, particularly when conceived of as a collaborative process between traditional knowledge holders and outside experts—precisely as Bonta et al. (2019) exhort. We would hope that future researchers give adequate recognition to these issues, thereby enhancing interpretations of cycad ethnobotany and traditional knowledge systems associated with this plant, as well as to promote community-based biocultural conservation initiatives. It is our hope that further study will advance our understanding of this plant, the complex Indigenous ecologies in which it figures, and the communities that valued—and continue to value—cycads as an integral component of their sense of place and cultural identity.

1

An Overview of Cycadales in Mesoamerica and the Caribbean

Biology, Distribution, and Conservation

MICHAEL CALONJE, ANDREW P. VOVIDES,
AND JOSÉ SAID GUTIÉRREZ-ORTEGA

Cycads, belonging to the order Cycadales, are long-lived, slow-growing, dioecious, pachycaul plants with starch-rich stems bearing terminal crowns of pinnate leaves (although some species of the genus *Cycas* present bipinnate or multipinnate leaves). Cycads superficially resemble palms or even ferns, although they are biologically unrelated to either group. They are distinguished from other plants by a combination of morphological traits, including a lack of axillary buds, the production of specialized roots hosting nitrogen-fixing cyanobacteria (coralloid roots; see Figure 1.1; see also Figures 0.2, 0.4, Introduction, this volume), the presence of unique chemical compounds (e.g., cycasins, macrozamins, and BMAA [the amino acid β-N-methylamino-L-alanine]; see Brenner, Stevenson, and Twigg 2003; Suárez-Moo et al. 2019), and rudimentary ecological interactions with symbiotic cyanobacteria and insect pollinators (Norstog and Fawcett 1989; Tang 1987). All cycad species are tropical or subtropical, most commonly occurring at low altitude. Populations are typically small, isolated, and restricted to narrow geographic areas. They are considered the world's most threatened land plants and are highly imperiled by human activities, especially habitat destruction and the extraction of plants from the wild for horticultural uses. Cycads consist of 10 genera and a total of 367 currently accepted species (Calonje et al. 2022). The genus *Cycas* is the sole representative of the family Cycadaceae, whereas the other nine genera belong to the Zamiaceae (Christenhusz, Reveal, et al. 2011). Four endemic genera occur in the New World: *Ceratozamia, Dioon, Microcycas,* and *Zamia*. Below we provide additional

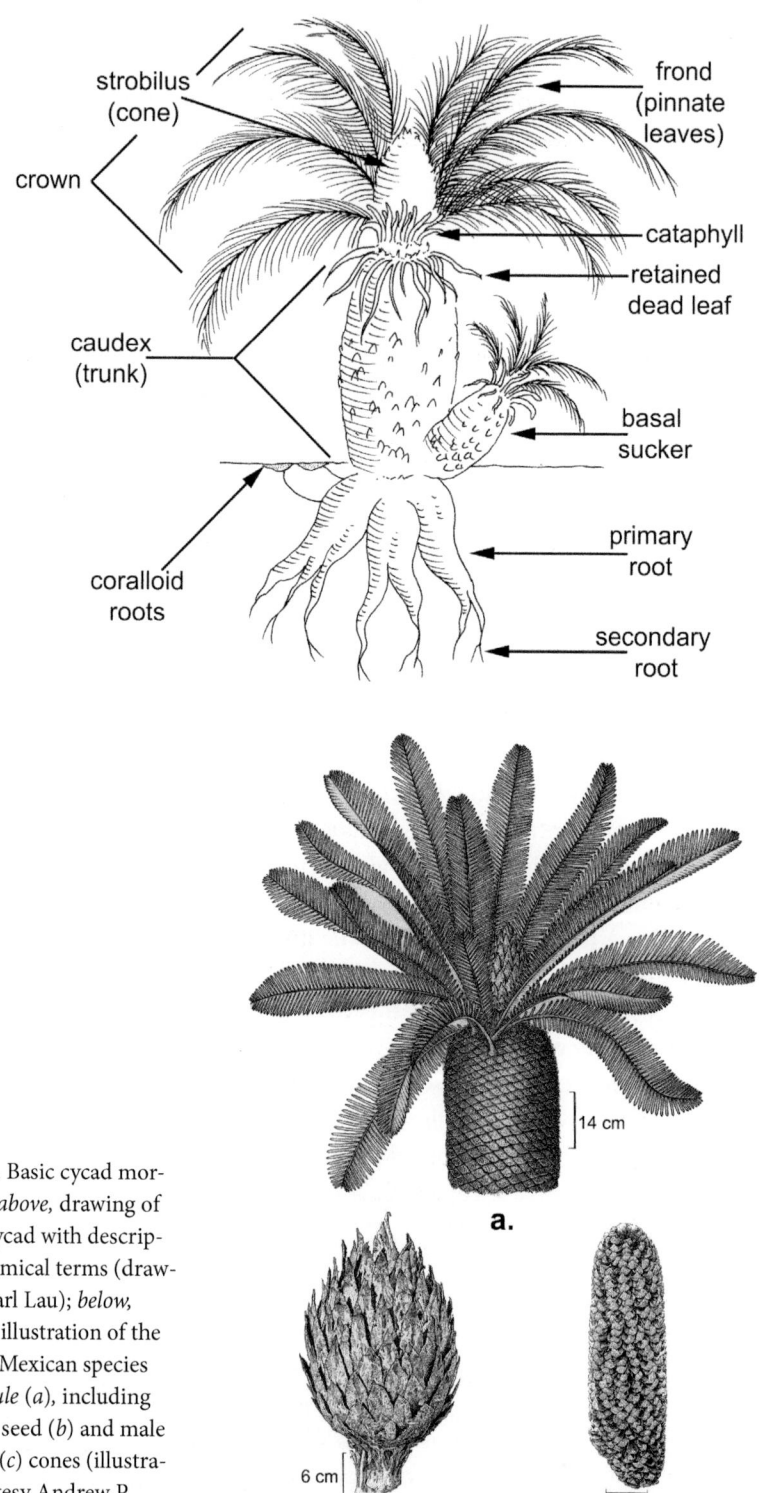

Figure 1.1. Basic cycad morphology: *above,* drawing of generic cycad with descriptive anatomical terms (drawing by Pearl Lau); *below,* botanical illustration of the common Mexican species *Dioon edule* (*a*), including female or seed (*b*) and male or pollen (*c*) cones (illustration courtesy Andrew P. Vovides).

details about the biology, diversity, and distribution of cycads, focusing primarily on those occurring in Mesoamerica and the Caribbean region.

General Characteristics of Cycads

Cycad Morphology and Anatomy

Cycad stems are pachycaulous (thick-stemmed) and manoxylic, meaning they possess very little woody tissue (Chamberlain 1911; Stevenson 1990). They contain a small amount of vascular tissue and a relatively massive pith and cortex composed mostly of starch-rich parenchyma cells (Greguss 1968). Unique to cycad stem anatomy are the encircling or girdling leaf traces (Figure 1.2) where a vascular bundle originating at the central stele encircles the cortex to exit at a leaf petiole emerging from the opposite side. Although a leaf receives a number of traces that have separate cauline origins, some leaf traces originate close to a petiole attachment, while others completely

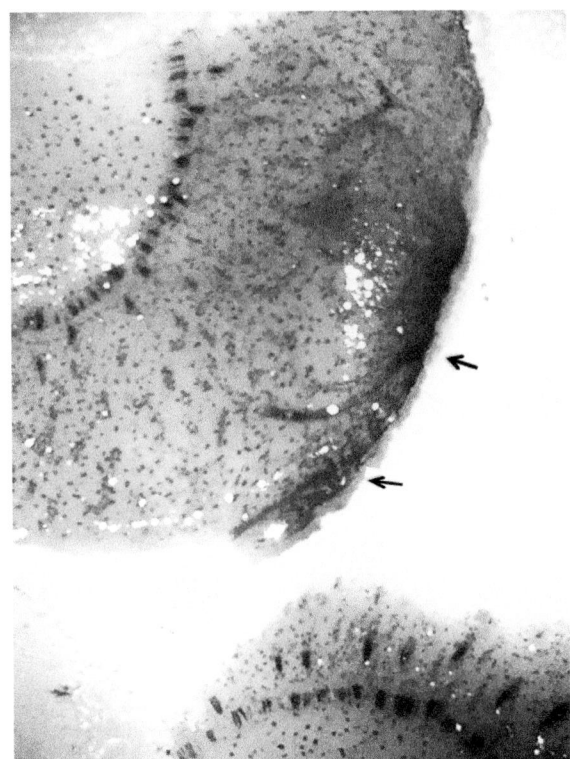

Figure 1.2. Encircling leaf traces in cortex of *Zamia* stem (arrows) (photograph by Andrew P. Vovides).

girdle the stem cortex (Bierhorst 1971). The function of these traces is open to speculation, but they may help strengthen parts of the stem in which they occur (Norstog and Fawcett 1993).

Cycad stems can be arborescent or subterranean and, in the latter case, are often tuberous. *Microcycas* and *Dioon* are exclusively arborescent, whereas *Zamia* and *Ceratozamia* have both subterranean-stemmed and arborescent-stemmed species. Cycad stems range from solitary to highly branched, the latter more typical of subterranean-stemmed species. Cycads lack axillary buds, so all branching is either dichotomous (arising from equal division of the apical bud) or arising from adventitious buds from undifferentiated cortical tissues produced in association with leaf bases (Norstog and Nicholls 1997). The stems of species of *Ceratozamia* and *Dioon* are protected by a dense armor of persistent leaf bases and cataphylls (protective scale-leaves), which may offer structural protection from mechanical damage or fire (Norstog and Nicholls 1997), whereas in *Zamia* the stems are bare and protected only by a thin layer of periderm. In *Microcycas,* leaf bases cover younger stems or the upper sections of older stems, whereas the lower sections of old stems are covered by a thick, corky bark. Cycad leaves are pinnate (bipinnate in *Bowenia* and bipinnate to multipinnate in some *Cycas*), spirally arranged at the apex of each stem, and interspersed with cataphylls (see Figure 0.2, Introduction, this volume).

Cycads produce a root system consisting of normally functioning roots, contractile roots, and apogeotropic roots bearing coralloid masses at the soil surface (Figure 1.1). Contractile roots (and stems) can pull the cycad's growing apex below the soil surface, offering protection from drought and fire conditions that are common in some habitats (Stevenson 1980). Coralloid roots are coral-like upright growing roots that host a diverse array of bacterial endophytes, including symbiotic cyanobacteria (Gutiérrez-García et al. 2019; Suárez-Moo et al. 2019). The cyanobacteria, readily observable in root cross-sections as distinct bluish-green bands, can fix atmospheric nitrogen and convert it to nitrogenous compounds that benefit cycads (see VanDerwarker, this volume). This symbiotic relationship is thought to help cycads survive in the nutrient-poor environments in which they are commonly found.

Cycad Reproduction

Cycads are strictly dioecious, with pollen and seed bearing sporophylls borne on separate male and female plants (Figure 1.3). These sporophylls are spirally arranged in determinate structures known as *cones,* except in

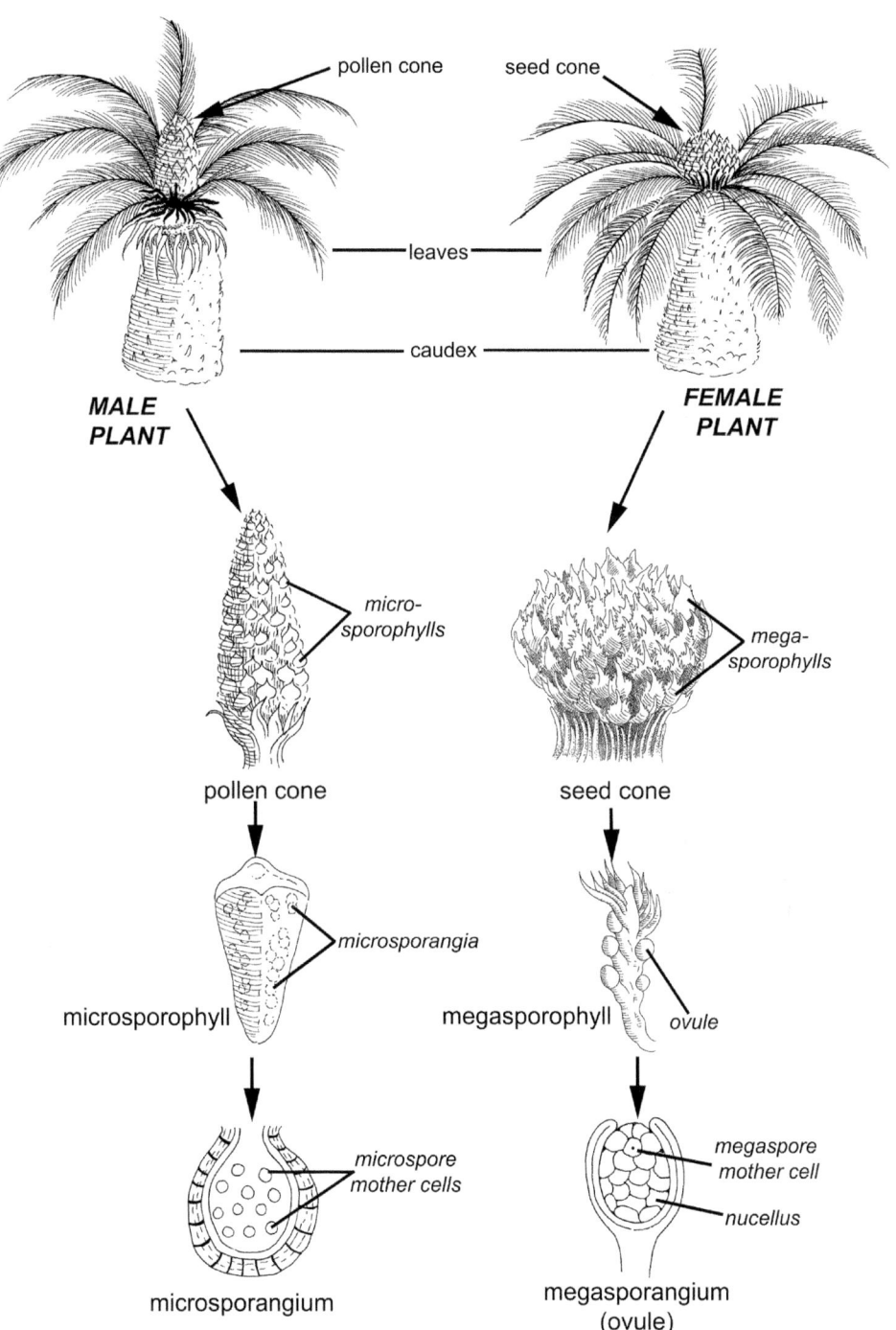

Figure 1.3. Basic cycad reproductive anatomy (drawing by Pearl Lau).

the case of female plants of the genus *Cycas,* in which case they are produced on multiple individual leaflike sporophylls that are initially aggregated into an open conelike structure known as a *pseudocone* (Niklas and Norstog 1984). Thus, cycads, like conifers, are gymnosperms, or seed plants that produce unprotected (or "naked") seeds inside cones; however, cycad cones are simple and do not have bract scales associated with the fertile seed-bearing scales. In contrast, the cones of conifers, such as pines, are compound, having a single bract scale associated with each fertile scale (Foster and Gifford 1974). Cycads and *Ginkgo* (another gymnosperm) are the only two living seed plant groups bearing ciliate antherozoids and an ancestral ovule structure, and share similar stem anatomy (Norstog and Nicholls 1997).

Cycads are known to be insect-pollinated, mostly by beetles (Terry, Tang, Taylor, et al. 2012; Vovides 1991b). Some species of *Macrozamia* are pollinated by thrips (Terry 2001), and microlepidopteran moths are suspected to play a role in the pollination of *Cycas micronesica,* some Caribbean *Zamia* species, and some *Macrozamia* species (Terry, Tang, and Marler 2012). In the New World, there are two major lineages of beetles inhabiting cycad cones: beetles in the family Erotylidae, which are found in all New World cycad genera, and weevils in the subtribe Allocorynina, which are restricted to *Dioon* and *Zamia* (O'Brien and Tang 2015; Tang et al. 2018). After pollination, fertilization is carried out by ciliate antherozoids introduced by the pollen tube and released into an archegonial chamber of the ovule where they swim and eventually enter an archegonium to fertilize an egg cell. This mechanism in cycads represents an ancestral connection between the evolution of nonseed plants, such as the ferns, where dependence on free water is essential to fertilization, and seed plants, where there is independence of free water (Brenner, Stevenson, and Twigg 2003; Norstog and Nicholls 1997).

Cycad Evolutionary History and Distribution

Cycads have an ancient evolutionary history tracing back to at least the Lower Permian, some 270–280 mya (million years ago) (Gao and Thomas 1989). They are considered among the most primitive living seed plants (Brenner, Stevenson, McCombie, et al. 2003) and are often referred to as "living fossils" due to their long fossil record and retention of primitive seed plant characters such as motile sperm cells (Nishida et al. 2004). In fact, cycads exemplify the early stages in the evolution of reproductive mechanisms of vascular plants, such as the production of seeds and pollen, or the appearance of insect-pollination systems (Brenner, Stevenson, and Twigg 2003; Norstog and Nicholls 1997). However, the term "living fossils" is somewhat inadequate, as it suggests that cycads have remained essentially unchanged throughout their evolutionary history. The cycad lineage itself is indeed quite ancient, yet living cycad species are believed to have evolved as recently as ca. 12 mya, with most diversification occurring in the Miocene epoch (Crisp and Cook 2011; Nagalingum et al. 2011). Moreover, while cycads have retained some primitive morphological seed-plant characteristics, they also have many derived traits. In a similar manner to their popular but inaccurate depiction as "living fossils," cycads are often mistakenly portrayed as a fossil-rich group reaching their peak in abundance and diversity during the Mesozoic (Harris 1961), a time period alternatively referred to as the "age of dinosaurs" or "age of cycads." This misconception may be due in part to the fact that cycads are often considered part of an informal group called the "cycadophytes," which include several seed plant groups with compound leaves (e.g., Cycadales, Cycadeoidales [sometimes referred to as Bennettitales], Nilssoniales), whose members superficially resemble each other but are not closely related (Coiro and Pott 2017). Although the Cycadeoidales and Nilssoniales were dominant plant groups in Mesozoic flora, the Cycadales constituted only a minor portion of the vegetation. The fossil record for extant cycad genera is quite poor, extending only to the Cenozoic (Coiro and Pott 2017). The fossil *Crossozamia* from the lower Permian of Taiyuan, China is a seed-bearing scale in which its apex resembles those of the living cycad genus *Cycas*, especially that of *Cycas revoluta* (Gao and Thomas 1989; see Figure 1.4). Cycads are now restricted to the tropics and subtropics of the globe (see Figure 0.3, Introduction, this volume) but had a broader latitudinal range in the milder climate of the Paleogene as evidenced by fossils found in Alaska, Argentina, Russia, Britain, the United States, and elsewhere. However, because of the sparse fossil record, it remains unclear when they may have been most abundant or species-rich (Mario Coiro, personal communication, 2020).

Knut Norstog, an eminent cycad biologist, compared the information stored within the cycads to the Rosetta Stone. Just as the Greek translation made possible the decipherment of the Ancient Egyptian text that revealed previously unknown aspects of Ancient Egyptian culture, cycads store important information that can advance our understanding of the early evolution of plants. For example, the epidermal anatomy of cycads, especially *Dioon* species, reflects the climate in which the different species are found (Gutierrez-Ortega, Yamamoto, et al. 2018a; Vovides et al. 2018). Cycads that grow in more mesic and humid climates with a relatively high annual rainfall evenly spread throughout the year exhibit a less sclerified epidermis, usually an absence of a hypodermis, and a shallower, wider, epistomatal chamber lacking papillae projecting into the epistomatal chamber than those species

Figure 1.4. Apex of the Permian fossil *Crossozamia* female scale compared with the living *Cycas revoluta* scale apex (photograph by Andrew P. Vovides).

that grow in harsher and more xeric environments. Thus, over long periods of time, plants may reflect historic climate change in their morphology, and such aridification can lead, in turn, to speciation (Gutiérrez-Ortega, Yamamoto, et al. 2018a; see also Cibrián-Jaramillo et al., this volume). It is in this sense that Norstog considered cycads "the Rosetta Stone of plant biology" (Norstog and Nicholls 1997).

New World Cycad Genera: Characteristics and Geographic Distribution

In the mid-1980s, about 130 cycad species were known worldwide. At present, there are 367 described species, and 70 are known in Mexico, placing the country second worldwide for cycad diversity—Australia occupies the first place with about 79 species and six infraspecific taxa, and South Africa has 39 species (Calonje et al. 2022; Vovides 2000).

The New World hosts 138 endemic cycad species belonging to four genera within the family Zamiaceae (Calonje et al. 2022), representing 37% of the species richness and over 40% of the generic richness in the Cycadales. The four New World endemic genera are *Ceratozamia* (36 species), *Dioon* (18 species), *Microcycas* (1 species), and *Zamia* (83 species). Although a few New World cycad species have relatively broad geographic distributions (e.g., *C. robusta, D. tomasellii, Z. prasina*), many species have very restricted geographic distributions and consist of only a few populations (e.g., *C. alvarezii, D. caputoi, Z. inermis, Z. grijalvensis*).

The distribution of New World cycads is clearly separated into three distinct geographic regions: a Caribbean region, including several islands in the Greater Antilles and Florida; a Mesoamerican region including Mexico, Guatemala, Belize, Honduras, and El Salvador; and a region encompassing the Central American Isthmus and South America, referred to here as the Isthmian-South American region (Figure 1.5a). Each region has its unique cycad flora, with no species being shared between these regions. The Mesoamerican region hosts three genera classified in 80 species (36 *Ceratozamia*, 26 *Zamia*, and 18 *Dioon*), and has the highest generic and species richness in the New World, alone representing almost one-third of the generic diversity and 20% of the species diversity within the Cycadales. The Caribbean region hosts the monotypic genus *Microcycas*, and 8 species of *Zamia*, all belonging to a monophyletic assemblage of subterranean-stemmed small plants known as the Caribbean *Zamia* clade (Calonje et al. 2019; Meerow, Salas-Leiva, Francisco-Ortega, et al. 2018). Finally, the Isthmian-South American region exclusively hosts the genus *Zamia* and is the region with the highest species richness and morphological diversity in *Zamia*, holding 49 species belonging to the most species-rich and morphologically diverse clade in the genus (Calonje et al. 2019).

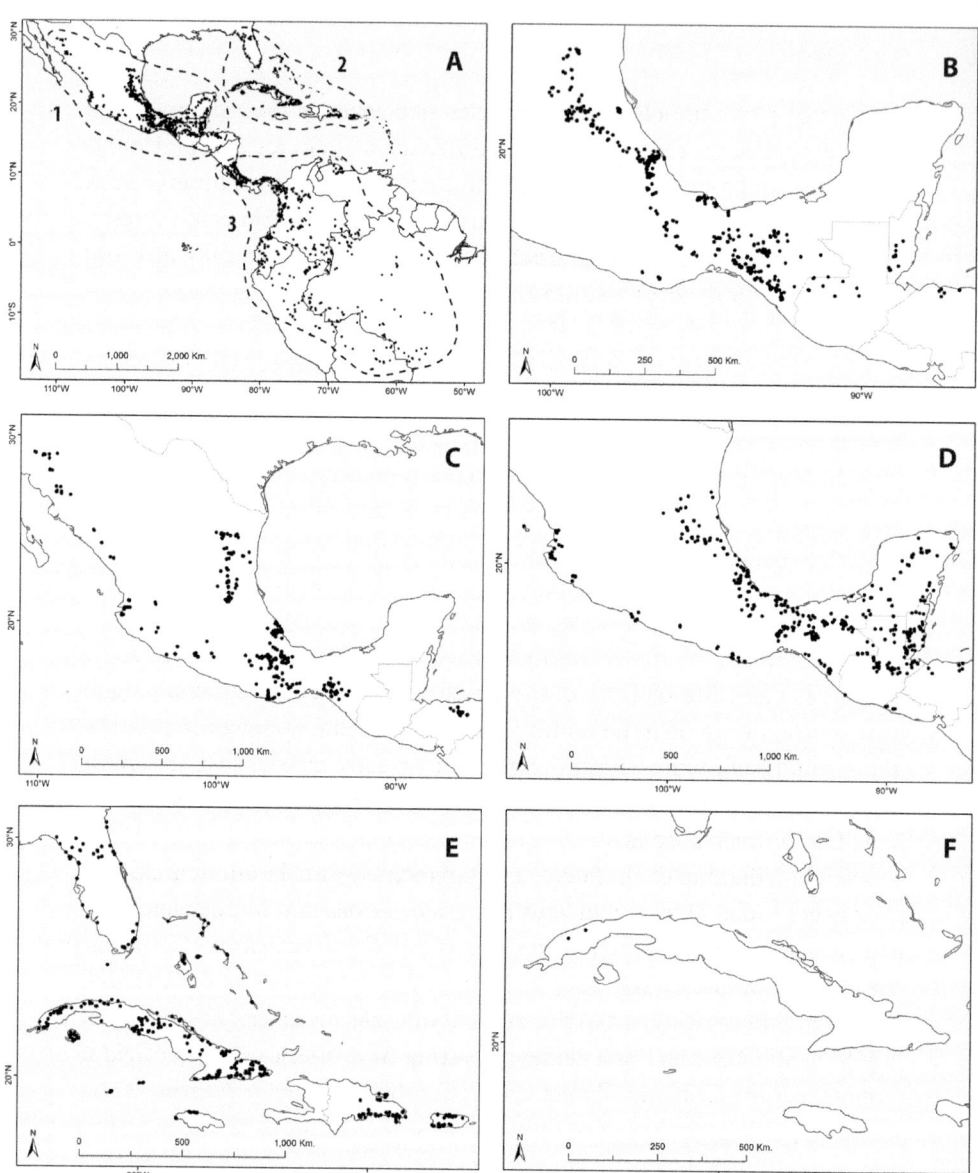

Figure 1.5. Distribution of New World cycads. Dots represent population occurrence records. *A,* Three separate biogeographic areas of distribution can be distinguished, corresponding to 1 = Mesoamerica, 2 = the Caribbean, 3 = South America. Mesoamerica is home of three cycad genera: *B, Ceratozamia, C, Dioon,* and *D, Zamia,* whereas the Caribbean harbors two genera: *E, Zamia* and *F, Microcycas* (figure by the authors).

Mesoamerican Cycads

Ceratozamia

The genus *Ceratozamia* is the most species-rich genus in the Mesoamerican region. It is mostly distributed in Mexico, sharing few species with Belize and Guatemala, and a single species distributed exclusively in Honduras (Figure 1.5b). *Ceratozamia* is distributed primarily in high-elevation cloud forests, pine-oak forests, and elevated areas of evergreen forest, typically growing under moist understory conditions (Martínez-Domínguez et al. 2018; Vovides, González, et al. 2004).

A few species of *Ceratozamia* have subterranean stems (e.g., *C. hildae, C. euryphyllidia*), but most species are arborescent (e.g., *C. robusta, C. subroseophylla, C. vovidesii*) albeit typically with relatively short trunks rarely exceeding 1 m length. *Ceratozamia* species can be confused with *Zamia*, but they can be distinguished by the presence of persistent leaf bases on the upper portions of their trunks, and by the unique pair of hornlike projections found in both their seed-bearing and pollen-bearing sporophylls. Also, the leaflet margins on *Ceratozamia* are always entire, whereas those in *Zamia* are either toothed or entire. *Ceratozamia* grows in similar moist understory habitats as many Mesoamerican species of *Zamia*, and species belonging to the two genera can often be found growing in sympatry. However, the geographic range of *Zamia* within Mesoamerica greatly exceeds that of *Ceratozamia*, with a few species able to exploit much drier and open habitats than most *Ceratozamia* species.

Medina-Villarreal et al. (2019), using markers for nrITS and five chloroplast genes separate the genus into two major clades: the Miqueliana clade with 10 species (*C. becerrae, C. chimalapensis, C. euryphyllidia, C. hondurensis, C. matudae, C. miqueliana, C. mirandae, C. santillanii, C. vovidesii, C. whitelockiana*) restricted to the south of the Trans-Mexican Volcanic Belt (TMVB), and the Mexicana clade, consisting of 18 species (*C. fuscoviridis, C. huastecorum, C. hildae, C. decumbens, C. latifolia, C. tenuis, C. robusta, C. sabatoi, C. mexicana, C. kuesteriana, C. zoquorum, C. zaragozae, C. morettii, C. mixeorum, C. norstogii, C. alvarezii*, and two yet undescribed species of *Ceratozamia*) occurring on both sides of the TMVB. The Miqueliana clade was also treated previously by Martínez-Domínguez et al. (2017) and Pérez-Farrera et al. (2009), and while these treatments and the recent phylogenetic work by Medina-Villarreal et al. (2019) agree on the membership of some species belonging to this group (e.g., *C. miqueliana, C. euryphyllidia, C. santillanii*), there is considerable disagreement about the membership of several other species. Future phylogenetic work including additional independent molecular markers and anatomy may help clarify the membership of this group and other relationships in *Ceratozamia* (see Cibrián-Jaramillo et al. and Englehardt et al., this volume).

Dioon

Species of *Dioon* are easily distinguishable from the other two genera in Mesoamerica by their distinctive egg-shaped seed cones with dense tomentum and flattened sporophylls, and by their non-articulated leaflets. All *Dioon* species have aboveground trunks, with some species having relatively short trunks rarely exceeding 1 meter tall (e.g., Tomasellii clade species: *D. stevensonii, D. tomasellii, D. sonorense,* and *D. vovidesii*). An interesting fact about *Dioon* is that species are extremely long-lived, with some species believed to live for more than two millennia (Vovides 1990). Also, the most massive cycads in the New World belong to this genus: the "tree dioons," a monophyletic group classified within the Spinulosum clade. The species of this group can reach up to 5 m tall in the case of *D. rzedowskii* (Whitelock 2002), 10 m tall in *D. mejiae* (Haynes and Bonta 2007), and more than 16 m in *D. spinulosum* (Chamberlain n.d.). These three large species also produce massive seed cones. The largest cone recorded for any gymnosperm may be that of a cultivated *D. spinulosum* which produced a cone 96 cm long and 65 kg in weight that produced more than 800 seeds (Calonje et al. 2016).

The genus *Dioon* is almost exclusive to Mexico, except for the single species *D. mejiae* from Honduras (Figure 1.5c). Most species are restricted to steep, rocky hillsides with poor soils (Sabato and De Luca 1985), often in pine-oak or tropical deciduous forests. The three Spinulosum clade species occur in humid rainforests. The remaining species occupy a mosaic of habitats ranging from high-rainfall mesic zones to arid zones with extreme variation in temperature and precipitation (Gutiérrez-Ortega, Yamamoto, et al. 2018a). Excluding the Spinulosum clade species, *Dioon* species are quite tolerant of water stress, having adapted a variety of epidermal traits such as deeper and narrower epistomatal chambers, and more protected stomata (Barone Lumaga et al. 2015; Vovides et al. 2018). The ability of several *Dioon* species to survive in fairly dry, rocky environments has resulted in the genus being able to persist in habitats to which the other two genera were not able to adapt. Consequently, despite having the lowest species richness among Mesoamerican cycad genera, *Dioon* has the broadest geographic distribution within Mesoamerica (from Sonora to Honduras), exceeding that of both *Zamia* and *Ceratozamia*.

Recent phylogenetic studies agree on the separation of *Dioon* into four major clades (Dorsey et al. 2018; Gutiérrez-Ortega et al. 2018): the Spinulosum clade ("tree dioons") is sister to the rest of the genus. The Edule clade, consisting of *D. edule* and *D. angustifolium*, occurs in the Gulf of Mexico coastal lowlands and the Sierra Madre Oriental, and is sister to a clade consisting of two sister groups: the Purpusii clade and the Tomasellii clade. The Purpusii clade consists of nine species (*D. argenteum, D. califanoi, D. caputoi, D. holmgrenii, D. merolae, D. oaxacensis, D. planifolium, D. salas-moralesiae,* and *D. purpusii*), occurring in the eastern region of the Sierra Madre del Sur, with *D. merolae* also occurring east of the Isthmus of Tehuantepec in the Sierra Madre de Chiapas. The Tomasellii clade comprises four species (*D. stevensonii, D. tomasellii, D. sonorense,* and *D. vovidesii*), occurring in the Rio Balsas Basin, the westernmost area of the Trans-Mexican Volcanic Belt, and the Sierra Madre Occidental. The genus *Dioon* represents the most ancient cycad lineage in the New World, with most recent phylogenetic studies placing it as sister of the rest of the family Zamiaceae (Condamine et al. 2015; Nagalingum et al. 2011; Salas-Leiva et al. 2013). Published estimates for when extant *Dioon* species began to diversify (i.e., the crown age) vary widely depending on molecular markers, methods used for node calibrations, and/or the parameters utilized in the analyses, with point estimates for the crown age of *Dioon* ranging from 7.77 mya to 55.92 mya (Condamine et al. 2015; Crisp and Cook 2011; Dorsey et al. 2018; Nagalingum et al. 2011; Gutierrez-Ortega, Yamamoto, et al. 2018a; Salas-Leiva et al. 2013). Lineage splitting in *Dioon* is variously attributed to processes and conditions of the Pleistocene (Dorsey et al. 2018) or climate change and orogeny in the Miocene (Gutierrez-Ortega et al. 2018; Gutierrez-Ortega, Yamamoto, et al. 2018a; Nagalingum et al. 2011). Further studies at the interpopulation level have suggested that aridification and the initial difficulty of *Dioon* lineages to explore the new arid niches promote geographic isolation among populations leading to speciation (Gutiérrez-Ortega et al. 2020; Gutiérrez-Ortega et al. 2021).

Mesoamerican *Zamia*

The Mesoamerican region hosts 26 species of *Zamia*, representing approximately 30% of the species richness in the genus. Most Mesoamerican *Zamia* species were treated in great detail by Nicolalde-Morejón et al. (2009), although a few significant changes have occurred since then, including the publication of new species (e.g., Calonje 2009; Calonje et al. 2009; Pérez-Farrera et al. 2012), as well as papers proposing additional taxonomic and/or nomenclatural changes (Calonje and Meerman 2009; Pérez-Farrera et al. 2016). The highest concentrations of species richness in Mesoamerican *Zamia* are found in the Isthmus of Tehuantepec region in Mexico, and in an area encompassing parts of southern Belize, western Guatemala, and northern Honduras (Calonje et al. 2019; see Figure 1.5d). The majority of *Zamia* species occurring in Mesoamerica are small and subterranean-stemmed, with only six species developing considerable aboveground stems (Calonje et al. 2019). The tallest species are *Z. tuerckheimii*, with stems up to 3 meters tall (Whitelock 2002), and *Z. onan-reyesii*, with stems up to 2 meters tall (Schutzman et al. 2008). Most *Zamia* species in Mesoamerica (except *Z. inermis* and *Z. fischeri*) have prickles on their petioles, and most have prominently toothed leaflet margins (except *Z. inermis, Z. decumbens, Z. meermanii, Z. soconuscensis,* and *Z. tuerckheimii*). Mesoamerican zamias occur in a wide range of different forest types, including pine-oak forests, oak forests, tropical dry forests, evergreen tropical rainforest, sub-deciduous tropical forests, and tropical deciduous forests (Nicolalde-Morejón et al. 2009). Recent phylogenetic work in the genus *Zamia* (Calonje et al. 2019) has helped clarify the relationships between Mexican species of *Zamia*, with four major groups evident:

1. The Fischeri clade, sister to the rest of mainland American species (excluding *Z. integrifolia* from Southeast USA, which belongs in the Caribbean *Zamia* clade) and consisting of the three species from northeastern Mexico: *Z. fischeri, Z. vazquezii,* and *Z. inermis;*
2. The Furfuracea subclade, a broadly distributed group (Mexico, Guatemala, Belize, and Honduras) consisting of seven species (*Z. herrerae, Z. paucijuga, Z. variegata, Z. spartea, Z. furfuracea,* and *Z. loddigesii*);
3. The Purpurea clade, consisting of six Mexican endemic species (*Z. lacandona, Z. grijalvensis, Z. purpurea, Z. cremnophila, Z. splendens*) associated mostly with the Northern Area of the Chiapas Highlands biogeographic province (*sensu* Morrone et al. 2017); and
4. The Tuerckheimii subclade consists of six species (*Z. decumbens, Z. onan-reyesii, Z. tuerckheimii, Z. meermanii, Z. sandovalii, Z. standleyi*) restricted to evergreen tropical rainforests in Guatemala, Belize, and Honduras. Calonje et al. (2019) did not include *Z. katzeriana* or *Z. oreillyi* in the molecular phylogenetic analysis, but the former is believed to represent a species of hybrid origin (Pérez-Farrera et al. 2016), and the latter based on its morphological characteristics and geographical distribution is likely in the Tuerckheimii subclade.

Caribbean Cycads

All *Zamia* populations in the Caribbean region belong to a single monophyletic assemblage known as the Caribbean *Zamia* clade (Calonje et al.

2019; Meerow, Salas-Leiva, Francisco-Ortega, et al. 2018; see Figure 1.5e). Caribbean zamias are relatively small plants with subterranean stems and unarmed petioles, and all species have the same $2n=16$ karyotype (Moretti, Caputo, and Cozzolino 1993). Caribbean cycads grow mostly at low elevations ranging from sea level to 700 m in a variety of habitats, including coastal scrubs, savannas, pine forests, and hardwood hammocks. There are currently eight accepted species of Caribbean zamias (Calonje et al. 2022), but their taxonomy remains controversial (Eckenwalder 1980; Stevenson 1987) in part due to the over-reliance on using vegetative characters such as leaflet width for species delimitation. This is problematic because leaflet width can vary greatly between plants within a single population depending on solar exposure (Newell 1985), making this character inefficient for diagnosis. In addition, similar leaflet morphotypes appear on different islands and are currently treated as conspecific even in cases where the islands have never had a land connection. The convergent evolution of morphological character states appears to be common among different Caribbean Islands in response to variable environmental factors and stochastic processes (Meerow, Salas-Leiva, Calonje, et al. 2018), so the current taxonomic delimitation of Caribbean zamias likely requires substantial revision.

Microcycas calocoma, the sole representative of the monotypic genus *Microcycas* and the sister genus of *Zamia* (Calonje et al. 2019; Salas-Leiva et al. 2013), is endemic to the Caribbean region (Figure 1.5f). It is known from only a few locations in the Pinar del Río province of western Cuba, where it occurs in scrubs, pine-oak forests, and semideciduous forests on stony-sandy yellowish quartz-allitic soils or rocky limestone hills (Chaves and Ferrer 2007). *Microcycas calocoma* is a tall cycad bearing a compact crown of leaves. It may develop trunks more than 10 m tall (Chaves and Ferrer 2007), the tops of which are protected by a crown of old leaf bases and cataphylls that eventually shed and are replaced with a corklike bark. The leaves of *Microcycas* are truncate and rather short for such a large cycad, reaching only up to 1 m long and carrying narrow, strongly deflexed leaflets.

Hot Spots of Cycad Species Richness in Mesoamerica

The highest cycad species richness in Mesoamerica is found in southeastern Mexico, with notable hot spots in the Isthmus of Tehuantepec and northwestern Chiapas for *Zamia* and *Ceratozamia* (Figure 1.6), and in the Papaloapan/Santo Domingo River basins for *Dioon*. These areas represent centers of cycad diversification and are not only rich in cycad species, but also harbor some of the highest species diversity of seed plants in Mexico (Rzedowski 1991). There are some ideas that may explain the high species

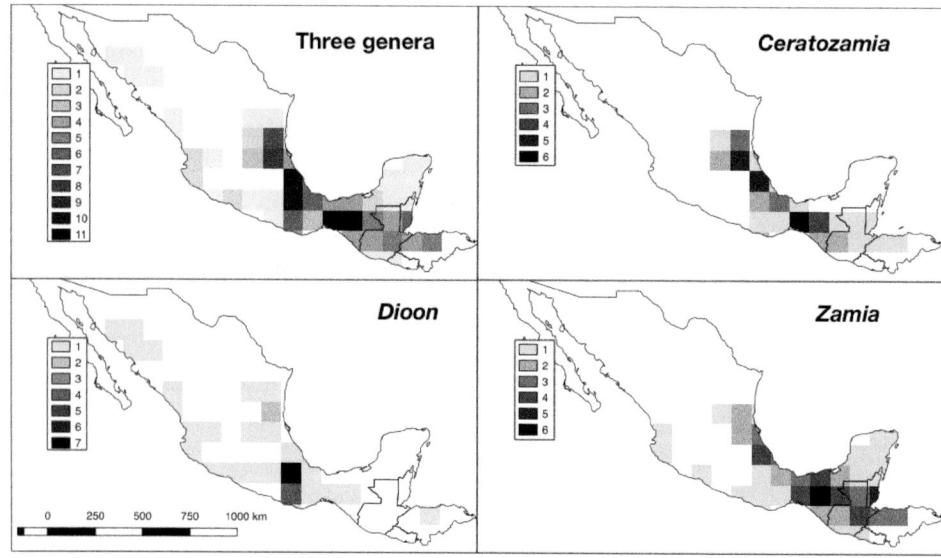

Figure 1.6. Species richness maps for three Mesoamerican genera, combined and for each genus individually. Grid cells are 1.5 × 1.5 degrees, with darker colors representing areas of maximum species counts (figure by the authors).

diversity of cycads (and other plant groups) in Southeastern Mexico. In the case of *Zamia* and *Ceratozamia*, the evergreen forests of Chiapas, Oaxaca, and Veracruz have maintained appropriate habitats for these genera since the Miocene (Rzedowski 1991; Wendt 1987), favoring their persistence in habitats that might have served as floristic refugia during cool past conditions. Some of these areas are the Baizabal, Los Tuxtlas, and Soconusco Cenozoic floristic refuges identified by Wendt (1987).

Consequently, demographic events, geographic isolation among ancestral populations, local adaptation, or secondary contact might have influenced their diversification (González and Vovides 2002; Pérez-Farrera et al. 2014). The case of *Dioon* seems to be different. The hot spot of *Dioon* species richness is located between Puebla and Oaxaca, in an area that encompasses extreme habitat variation, from the humid evergreen forests of the Papaloapan River, which host *D. spinulosum* and *D. rzedowskii*, and the deserts and xeric scrublands of the Tehuacán-Cuicatlán Desert and surrounding areas, which host *D. caputoi*, *D. califanoi*, *D. purpusii*, *D. argenteum*, and *D. oaxacensis*. These two groups with contrasting habitats belong to different clades, suggesting that the richness of *Dioon* in this area is due to both phylogenetic and historical factors (Dorsey et al. 2018; Gutierrez-Ortega, Yamamoto, et al. 2018a, 2018b).

The areas with the lowest cycad species richness in Mesoamerica may represent sites where dispersal from the centers of diversification occurred at more recent times. Thus, the elapsed time since dispersal has not been enough for speciation and the distinction of new taxa. Moreover, some areas need further investigation, as it is possible that new species have not yet been detected. We believe that the number of cycad species in Mesoamerica will increase when additional field surveys are undertaken in under-explored areas and future research further clarifies the species delimitation within the three genera.

A Brief History of Cycad Research in Mexico

The first American cycad to be reported cultivated in Europe was probably that of Giovanni Lerio, a *Zamia* from Brazil in 1576 (De Luca 1990). *Zamia furfuracea* (Figure 1.7) was the first Mexican cycad known to be cultivated in Europe. It was reported to be cultivated in England in 1691 at the Royal Garden at Hampton Court Palace (Plukenet 1691; Vovides 1983) and was the first Mexican species described under the modern system of binomial nomenclature (Linnaeus 1789). John Lindley in 1843 described *Dion edule* (Figure 1.8) from a plant and female cone brought to the Royal Horticultural Society, England by a Mrs. Lavater.

Figure 1.7. *Zamia furfuracea* (photograph by Andrew P. Vovides).

Figure 1.8. *Dioon edule* in tropical seasonally dry forest, Veracruz (photograph by Andrew P. Vovides).

Friedrich Anton Wilhelm Miquel, a physician and professor of botany in Amsterdam and Utrecht (1846–1871), was also director of the Rijksherbarium, Leiden, Belgium, and was well versed in the Classics, Latin, and Greek. Miquel suffered from ill health and never traveled outside Europe. He described many cycad species from Mexico from material collected in Veracruz by Schiede and Deppe and sent on to Schlechtendal. Schlechtendal and Miquel were in close correspondence; thus material was made available to Miquel (Stafleu 1966, 1970). A few Mexican species described by Miquel include *Ceratozamia latifolia* and *Zamia loddigesii*. The following three species were later placed into synonymy under *C. mexicana* by Thiselton-Dyer (1884): *C. brevifrons, C. robusta,* and *C. longifolia* (Miquel 1848). *Ceratozamia brevifrons* (Figure 1.9) was later revised by Vovides et al. (2012) from a morphological and anatomical viewpoint, enabling its reestablishment as a valid species. Verschaffelt spoke eloquently of Miquel's "infatigable [*sic*] zeal in the elucidation of genera and species from the already confusing literature brought about by the great variability of species" (Lemaire and Verschaffelt 1869). Miquel also corrected the spelling of the genus name *Dion* to *Dioön* on etymological grounds (Miquel 1848). However, De Luca et al. (1982) adopted the spelling *Dion,* and later, Vovides and Moreno (1983) proposed conservation of the spelling *Dioon* against *Dion*. De Luca, Sabato, and Stevenson (1984) opposed the conservation of the spelling *Dioon* in favor of

Figure 1.9. *Ceratozamia brevifrons* (photograph by Andrew P. Vovides).

Dion, which was rejected by the Committee for Spermatophyta (Brummitt 1987), and thus the spelling *Dioon* has been conserved on the grounds of usage. Vovides et al. (1983) published the *Flora de Veracruz* fascicle on Zamiaceae with the then known nine species for Veracruz, including two new species: *Zamia inermis* (Figure 1.10) and *Z. purpurea*; at the time of writing this chapter, there are 21 described species for Veracruz (Calonje et al. 2022).

In 1854, Wendland described *C. miqueliana* (Figure 1.11) from a sterile specimen from an unknown locality. Fortunately, Thiselton-Dyer's (1884) excellent illustration showing leaflet and strobili details enabled its rediscovery in southern Veracruz (Vovides et al. 1983). Since then, further localities of this species have been found in Tabasco and northern Chiapas (Pérez-Farrera et al. 2009).

In 1878, Moore described "tentatively" *Ceratozamia fuscoviridis* and mentioned that it had been brought to Ireland from Havana, probably collected in the Real del Monte, a mining region in the state of Hidalgo, by Thomas Coulter (1793–1843). Coulter was an Irish physician who served in Mexico from 1824 to 1829 at the Real del Monte Mining Company. Coulter sent living plants, mainly cacti, to the Trinity College Botanical Garden in Dublin, Ireland, and he also sent seeds to Moore, director of the National Botanic Gardens, Glasnevin, in Dublin, via Cuba. For many years, this species existed in obscurity because its original locality was unknown, coupled with

Figure 1.10. *Zamia inermis* (photograph by Andrew P. Vovides).

undetected herbarium specimens, until a voucher at Kew (K) was found annotated as "*Ceratozamia fusca-viridis,* Hort. Bot. Glasnevin 1881," and was assigned as the holotype of *C. fuscoviridis* by Osborne et al. (2006) in an attempted validation of the species name. However, the authors were unaware that the species name had been previously validly published in 1878 in William Bull's horticultural catalogue. To maintain nomenclatural stability, Calonje and Sennikov (2017) designated this same specimen as the neotype for Bull's *C. fuscoviridis*.

Finally, Martínez-Domínguez et al. (2018) designated a specimen from the Molango locality as an epitype serving as an interpretive type for the neotype. Living specimens of this taxon are still extant at the Botanic Gardens in Ireland, and duplicates are also in the Botanical Garden of the University of Florence, Italy (Osborne et al. 2006). Though originally known as *Ceratozamia* "Molango" from the Molango, Hidalgo region, its range has since been expanded with a few new localities found (Martínez-Domínguez et al. 2018; Pulido-Silva et al. 2015).

Another case involving the horticultural catalogues of William Bull is about a *Zamia* species. Populations of *Zamia* occurring within the Yucatan

Figure 1.11. *C. miqueliana* (photograph by Andrew P. Vovides).

peninsula were previously circumscribed under *Z. loddigesii* when their karyology was studied by Vovides and Olivares (1996); however, Stevenson et al. (1998) found morphological differences and segregated these populations and those occurring in adjacent Belize into the new species *Z. polymorpha*. Calonje and Meerman (2009) examined the protologue description of *Zamia prasina* published in William Bull's 1881 horticultural catalogue (Bull 1881), as well as a specimen deposited at Kew collected by Bull, and realized the name was a validly published name for the same taxon, having nomenclatural priority over *Z. polymorpha*.

The early twentieth century saw research done by Sister Alice Lamb (1923) on cycad anatomy, and Charles J. Chamberlain, a plant morphologist based at the University of Chicago. Chamberlain managed to rediscover the type locality of *Dioon edule* at Chavarrillo with the help of the then governor of Veracruz Theodore Dehesa. Chamberlain, with the aid of Prof. Luis Murillo, a botany teacher at the teachers' training college, Xalapa, calculated the age of a 5-foot-tall *D. edule* growing in Luis Murillo's garden. Murillo sent leaf production data of this plant to Chamberlain over a period of 11 years. With these data and the number of leaf petiole scars counted on the cycad's trunk, Chamberlain estimated the cycad's age at about 950 years (Chamberlain 1919). This *Dioon* specimen is still extant at the Quinta Las Rosas, now a retirement home. *Dioon edule* ages have been estimated using Chamberlain's method on size classes of *D. edule* individuals in their natural habitat where plants with trunks more than 2.5 meters tall were estimated to be more than 2,000 years old (Vovides 1990).

Chamberlain (1935) published his monumental *Gymnosperms: Structure and Evolution,* in which he described in detail the structure of the male and female gametophytes of cycads and erroneously claimed that cycads are wind pollinated. Anemophily in cycads has since been refuted by the work of Norstog and Stevenson (1980), Tang (1987), and Norstog and Fawcett (1989). Since then, much work has been done on the cycad pollination syndrome (Hall et al. 2004; Terry, Tang, Taylor, et al. 2012; Terry et al. 2007; Vovides 1991a, 1991b; Vovides et al. 1993).

When Professor Marcos E. Becerra (1870–1940), a writer, poet, politician, and ethnographer, carried out a floristic study in the mountainous zone of southern Tabasco and northern Chiapas during 1914–1915, he discovered and collected an interesting *Ceratozamia,* which he identified as *C. miqueliana*. Upon further examination of Becerra's voucher (s/n) and comparison with *C. miqueliana,* morphological differences were found, especially in leaflet pairs and texture, with greater distances between the leaflet articulations than in *C. miqueliana*. The new species *C. becerrae* was named in honor of Professor Becerra (Vovides, Pérez-Farrera, Schutzman, et al. 2004) but was later synonymized with *C. zoquorum* (Martínez-Domínguez et al. 2017). However, molecular evidence appears to contradict this synonymization (Medina-Villarreal et al. 2019), and a rebuttal with additional lines of leaflet anatomical evidence has been presented (Vovides et al. 2020).

Eizi Matuda made a comprehensive study of the vegetation of the Soconusco region in Chiapas and in June 1938 found a sterile specimen of a *Zamia* sp. sympatric with *Ceratozamia matudae*. This *Zamia* species is morphologically similar enough in vegetative characteristics to cause confusion, largely due to its entire leaflet margins typical of *Ceratozamia* spp. and was therefore labeled as such by Matuda. During a visit to the Soconusco locality in January 1985, fertile specimens were found of this intriguing *Zamia* and described as *Zamia soconuscensis* (Schutzman et al. 1988).

With a renewed interest in cycad taxonomy beginning in the late 1970s and continuing to date, researchers were hampered by the still confusing literature due to high variability within species as mentioned by Lemaire and Verschaffelt (1869). Stevenson and Sabato (1986a, 1986b) took on the task of putting order in the literature by identifying all valid nomenclatural combinations in New World Cycad genera, identifying their types, and typifying those names requiring it. This important nomenclatural research established a precedent for further research. For example, there has been some confusion on the identity of *Ceratozamia mexicana*. Chamberlain (1919) recognized a population near Xalapa as such, but this population differs from that of El Mirador, Veracruz. The plants from El Mirador correspond to the *C. mexicana* holotype at Paris (P) Brongniart s.n., collected by Ghiesbrecht, and an illustration in Brongniart (1846). In 1999, Stevenson discovered oversized sheets of cycads at the Kew Herbarium that were far removed from the normal-sized collections. In those collections are three sheets of a single leaf of a specimen labeled *C. mexicana* var. *tenuis* collected from a cultivated plant at Kew (no longer alive) in 1881 and annotated by Thiselton-Dyer that has been designated as the lectotype for that name (Vovides et al. 2016). Further morphological and anatomical examination on living collections and in the field has enabled the establishment of *Ceratozamia tenuis* (Dyer) D. W. Stev. & Vovides corresponding to the population near Xalapa (Vovides et al. 2016).

Aldo Merola, director of the Horto Botanico in Naples, Italy, began research on Mexican cycads around 1960. More recently, his students, led by Paolo De Luca, Sergio Sabato, and Aldo Moretti, collected extensively in Mexico in collaboration with Mario Vázquez Torres of the Universidad Veracruzana during the 1980s. The Italian team with Vázquez Torres as a field guide discovered and described a number of *Dioon* species from the Pacific seaboard of Mexico and Oaxaca, namely *D. califanoi*, *D. rzedowskii*, *D. caputoi*, *D. merolae*, *D. holmgrenii*, and *D. tomasellii* (De Luca, Sabato, and Vázquez-Torres 1980; De Luca and Sabato 1979; De Luca et al. 1981a, 1981b; De Luca, Sabato, and Vásquez-Torres 1984). Moretti did cytotaxonomic studies on *Zamia paucijuga* Wieland of the Pacific seaboard and reported chromosomal fissions (Moretti and Sabato 1984). Later, Vovides and Olivares (1996) also found chromosomal fissions in *Zamia loddigesii* Miq. (now *Z. prasina*) of the Yucatan peninsula. Caputo et al. (1990) published the first molecular phylogeny for the genus *Dioon* during the Second International Conference on Cycad Biology, followed by that of Moretti et al. (1993).

Figure 1.12. National Cycad Collection, Clavijero Botanic Garden, Xalapa, Veracruz (photograph by Andrew P. Vovides).

From 1979 to 1986, Vovides was establishing the Clavijero Botanic Garden and the National Cycad Collection at Xalapa (Figure 1.12) while working on the Flora de Veracruz for Zamiaceae (Vovides et al. 1983) in collaboration with John D. Rees, a geographer with knowledge of Mexican cycads and localities. Field expeditions to cycad localities with Rees, and later with Bart Schutzman, enabled the establishment and construction of the National Collection. This documented living collection has aided in cycad research in the fields of cycad taxonomy and cytotaxonomy (Schutzman et al. 1988; Vovides 1983, 1985; Vovides and Olivares 1996), molecular phylogenetics (Dorsey et al. 2018; González and Vovides 2002; González et al. 2008; Gutierrez-Ortega, Yamamoto, et al. 2018a; Medina-Villarreal et al. 2019), and anatomy (Pérez-Farrera et al. 2014, 2016; Vovides and Galicia 2016; Vovides et al. 2012, 2016, 2018). This recent research based on living collections has confirmed the importance of documented living collections in botanic gardens and has been reinforced by Blackmore et al. (2011) and Vovides et al. (2013).

Traditional Use and Toxicity of Mexican Cycads

Cycad Consumption

Cycad seeds contain a large starchy female gametophyte and, along with stems, are harvested as a carbohydrate-rich starch source. In Mesoamerica, cycad starch, obtained mostly from cycad seeds of species in the genus *Dioon* and *Ceratozamia*, and to a lesser extent, from stems of species of *Zamia*, has been used as food for millennia and is still used mainly in Mexico and Honduras (Bonta et al. 2019). In the Caribbean region, stems of *Zamia* have been exploited as a starch source for millennia (Chinique de Armas et al. 2015; Mickleburgh and Pagán-Jiménez 2012), and were even once extracted for commercial starch production at a grand scale in Florida (Burkhardt 1952). Cycad starch is known to be poisonous (Whiting 1963), containing toxic compounds including methylazoxymethanol azoxyglycosides such as cycasin and macrozamin (Moretti et al. 1983), as well as BMAA. These compounds do not appear to cause long-term neurological damage, but their effects on health are yet not fully understood, although BMAA is thought to cause amyotrophic lateral sclerosis (ALS). But toxicity in humans through ingesting food prepared from cycad tissues appears to be restricted to the Western Pacific region, especially Guam, but not reported in other regions where cycads are consumed on a regular basis such as in Mexico or Honduras. This is further discussed in detail by Bonta et al. (2019).

Therefore, for human consumption, cycad starches must first be detoxified by several complex processes, usually involving combinations of cutting, boiling, water leaching, and drying, to remove cycasins, macrozamins and the nerve toxin BMAA stored in the tissues (Bonta et al. 2019; Norstog and Nicholls 1997; Vovides et al. 1993; see also Englehardt et al., chapter 5, this volume; Cibrián-Jaramillo et al., chapter 2, this volume, Figure 2.5). However, some parts are eaten without removing toxins, while seed and stem starches are detoxified via several complicated multi-day processes such as leaching in water and *nixtamalization,* a process of detoxifying by adding wood ash (*Quercus* or *Pseudobombax*) or lime (Bonta 2010b) to the water. The cycad starch thus processed is used to make tortillas, tamales, and other dishes considered delicacies in the Pamería district of northeastern Mexico—although the neighboring Teenek consider cycads a famine food, associated with negative memories of hard times (Bonta et al. 2019).

Nonetheless, in some Indigenous communities, cycads are an important food as well as a religio-magic resource, and the inhabitants conserve their habitat. In folklore, cycads are related to maize (teosinte) or a maize forerunner, though there are no botanical affinities between the two (Bonta 2010b; Bonta et al. 2019; Osborne 1989; see also the chapters by Bonta and Carrasco, this volume). Carbohydrate-rich cycad seeds, and to a lesser extent stems, have been both famine foods and staples for at least six millennia across the Mesoamerican region and are still consumed in Mexico and Honduras.

Symbolic and Other Utilitarian Uses of Cycads in the Mesoamerican Region

As the Introduction to this volume details, in Mexico, as in many areas of the world, cycads are incorporated into syncretic Roman Catholic–Mesoamerican religious ceremonies for the decoration of altars and church entrances, such as during the Day of the Dead and pilgrimages during or near Easter Week. In the communities of the Suchiapa region, in the central depression of Chiapas, leaves of *espadaña* (*Dioon merolae*) are harvested from the holy mountain Nambiyugua during late April, culminating on the Día de la Santa Cruz (Day of the Holy Cross) festival on the third of May. Drumming, burning of incense, chanting of prayers in the Chiapanec language, and drinking of *pozol* (a fermented maize beverage) at certain stations on the way, usually next to large trees, is performed during the leaf harvesting pilgrimage. The leaves are used to adorn altars during this Catholic religious festival every May. It appears to be a pre-Hispanic Chiapanec tradition that has undergone syncretism with Catholic practice during colonial times (Pérez-Farrera and Vovides 2006).

Leaves of *Ceratozamia fuscoviridis* are used to decorate church facades and entrances to villages in the municipalities of Tlanchinol, La Misión, and Chapulhuacán in the state of Hidalgo (see also Bonta et al. 2019:figs. 2–7). Past and current uses of cycads in this region include food, recreation, medicine, poison, construction for roofing that lasts around 15 to 30 years, and as ornamental plants in gardens. The most frequent use nowadays is the elaboration of arches to decorate churches and schools (Vite-Reyes 2010). Cycads are often perceived as ancestors and protectors of maize, revealing a close relationship between both groups; in this region, cycads have common names such as *chamal, chamal chico, chamalillo, teocintle,* and *tzompollo,* which vary according to the species and place of occurrence (see Bonta, chapter 8, this volume). The term *teocintle* comes from the Nahuatl -*téotl* or Téutle, meaning God, and *cintli,* corn (God of corn), relating to the cycad cone that is similar to an ear of corn (Vite-Reyes 2010). Traditional beliefs and practices afford cycads important roles in cultural heritage and sense of place, and, as detailed above and in the Introduction, cycads are still consumed in many Mesoamerican regions today. However, because most traditional knowledge and uses are rapidly disappearing, new community-based biocultural conservation efforts need to be undertaken. These should incorporate tradition where possible and seek inspiration from existing successful cases in countries such as Honduras and Mexico (and possibly Colombia, where *Zamia chigua* is also consumed).

There are 213 cycad terms across 14 languages among 20 ethnic Indigenous communities within Mexico, and some of these rural communities contribute to the cycads' in situ conservation since the plants are greatly appreciated. Seed remains of *Dioon* dating back to 4000 BCE have been found in rockshelters, and cycads were a predominant food among hunter-gatherers in the Sierra de Tamaulipas from 4000 BCE through the 1700s CE (Bonta 2010b; Bonta et al. 2019; see also Englehardt et al., chapter 5, this volume). In northern Oaxaca, seeds from the *palma de chicalite* (*Dioon spinulosum*) are used to make bullroarers called *zumbadores* that make an eerie noise when spun rapidly on the end of a length of string. Ground *Ceratozamia* seeds mixed with sugar or jam are used as an insecticide in central Veracruz (Vovides et al. 1983).

Cycad Endangerment, Sustainability, and Conservation

Cycads are the world's most threatened plant group and are listed as such in the IUCN Red List of Threatened Species, as well as being protected by international laws such as CITES (Convention on International Trade in Endangered Species of Wild Fauna and Flora) and local laws (Brummitt et al. 2015; Fragnière et al. 2015; Marler and Marler 2015). While some species may be naturally declining, human activities are undoubtedly affecting cycad populations far more than natural processes (Donaldson 2003). Like the rest of earth's biodiversity, cycads are experiencing accelerated human-induced species losses, leading many to believe humans are causing and experiencing the sixth mass extinction on earth (Ceballos et al. 2015). The main threats to cycads—and primary causes of cycad population demise today—are undoubtedly the alteration and destruction of their native habitats as well as the extraction of plants from the wild for landscaping, commercial collecting for plant collections (Donaldson 2003), and traditional uses (Cousins et al. 2012). Because cycads are very charismatic and beautiful, they are appreciated and highly sought after as ornamentals among landscape architects, gardeners, cycad enthusiasts, and collectors. Their unquestionable ornamental appeal, combined with their slow growth, has unfortunately made many species the target for illegal and destructive harvesting of entire plants from their native habitats.

The effects of introduced alien species on cycads, although still poorly understood and requiring further study, also can be potentially devastating. The clearest example of this is invasion of the scale insect *Aulacaspis yasumatsui* into the island of Guam in 2003, which has decimated native populations of *Cycas micronesica*. The scale insect, known as Cycas Asian Scale (CAS), feeds on fluids from phloem or parenchyma cells and eventually kills the plant due to stress-induced carbohydrate depletion (Marler and Cascasan 2018). Additionally, CAS has virtually eliminated seedling recruitment in *Cycas micronesica* in Guam by increasing seedling mortality (Marler and Lawrence 2012). Consequently, the species went from being the most abundant tree in Guam (Donnegan et al. 2004) to being unable to reproduce (Marler and Terry 2011) and listed as Endangered (EN) in the IUCN's Red List.

The threats to native cycads vary in their impact and nature by geographic region (Donaldson et al. 2003). In the New World, the major cause of decline appears to be the destruction and transformation of habitats, followed by the collection of plants from the wild, the latter most evident in Mexico, where populations of several species have been decimated by collectors (Stevenson et al. 2003)

Nursery Strategies for Cycad Conservation in Mexico

In spite of national and international legislation aimed at protecting threatened and endangered species, poaching and illegal international traffic still occur. In Mexico, poaching of *Dioon edule* crowns for sale by street peddlers in large cities leads to low recruitment in natural populations, since the cycad stump left behind takes a long time to bud off a side shoot, and this shoot takes even longer to grow to a suitable cone-bearing girth (Octavio-Aguilar et al. 2008; Vovides 1990; Vovides et al. 2010). Another cycad, *Zamia furfuracea* or cardboard palm, so named owing to its very coriaceous leaflets, was extracted from the coasts of southern Veracruz at a rate of 40 tons/month during the 1980s for the landscaping industry in the USA and elsewhere (Vovides, Pérez-Farrera, Vázquez Torres, et al. 2002).

Conservation strategies have been implemented to address the issues of habitat loss through agricultural expansion and commercial collecting, two of the major factors responsible for the demise of cycads in the wild. Conservation through sustainable utilization is a concept that has been highlighted by many conservationists and plays a major role in the framework of the

Convention on Biological Diversity (CBD) (see Article 10, Appendix 1). It has been postulated by some (and questioned by others) whether putting an economic value to populations can create an incentive for their conservation. Under this concept, sustainably managed cycad nurseries theoretically have potential, in which conservation through propagation and utilization can take place. The prospect of being able to sell propagated plants creates an incentive for farmers to protect wild cycad populations that have been over-collected or heavily damaged by agricultural expansion.

Botanic gardens are playing an ever-increasing role in the conservation, monitoring, and collaboration with authorities for ex situ conservation of threatened species, and their contribution cannot be understated (Akeroyd et al. 1994; Blackmore et al. 2011; Vovides et al. 2013). A case study is that of *Dioon edule;* the survivorship curve of an ecological study of more than four years at a *D. edule* habitat showed a high seedling mortality (Vovides 1990), and it was assumed that seeds could be harvested from the habitat for propagation purposes, under the condition of reintroduction back into the habitat of nursery-grown plants to compensate for seed harvest. To address this issue, the first sustainable management nursery for cycads was created in 1990 with the collaboration of local farmers who live near the *D. edule* habitat at Monte Oscuro, Veracruz in conjunction with personnel of the Clavijero Botanic Garden of INECOL, Xalapa (Figures 1.13 and 1.14).

Here, subsistence farmers (campesinos) were clearing sections of cycad habitat and allowing outsiders to lop the crowns off large cycads which they would later sell as apparently well-established large plants. In Monte Oscuro, the community agreed to set up a nursery using wild-collected seed on the understanding that farmers would (1) conserve the natural habitat as a seed source, and (2) carry out reintroduction of nursery produced plants to compensate for seed removal. The Botanic Garden gave training courses in basic horticulture and assessment with the premise of encouraging habitat conservation through seed harvesting, artificial propagation, and plant sales followed by reintroduction of nursery-produced plants back into the habitat (Vovides and Iglesias 1994). The result has been that about 80 hectares of tropical thorn-forest habitat has been conserved, not only for the cycads but also for the harvesting of emergent palm fronds (*Brahea dulcis*) for the elaboration of palm ornaments for Roman Catholic Palm Sunday rituals as well as other forestry products for household use.

Most wild harvesting is restricted to seed collecting for nurseries, and is controlled by authorities governing the use of wild fauna and flora in Mexico under the General Law of Ecological Balance and Protection of the Environment passed by Congress in 1986. Threatened species are also included in an

Figure 1.13 Monte Oscuro nursery with a *Dioon edule* crop (photograph by Andrew P. Vovides).

Figure 1.14. Nursery-produced *Dioon edule, Beaucarnea recurvata,* and *Zamia furfuracea* plants delivered from the Monte Oscuro nursery for sale at the Clavijero Botanic Garden shop by the producer, Mr. Concepción Villa Díaz and daughter Iany (photograph by Andrew P. Vovides).

act passed in 1994 (NOM-059-ECOL-1994), and updated in 2010 (NOM-059-ECOL-2010), that makes provision for artificial propagation in registered sustainable management nurseries known as UMAs (Unidades de Manejo para la Conservación de la Vida Silvestre). This system is a wildlife management unit for plants overseen by the country's environment secretariat implemented in the year 2000 and in particular for cycads (SEMARNAT 2000). The nursery at Monte Oscuro was the forerunner of the official UMA system.

The role of UMA nurseries is in accord with Article 9 of the CBD on ex situ conservation which states that *"ex situ conservation should preferably be in the country of origin as this facilitates research, rescue and propagation of threatened species; regulates and names ex situ collections that do not threaten ecosystems and in situ populations"* (Vovides, Pérez-Farrera, Vázquez Torres, et al. 2002). The Monte Oscuro nursery has since expanded to growing the coastal *Zamia furfuracea* and *Beaucarnea recurvata* (pony tail palm), the latter native to the area. The nursery began with 24 members and by 1992 only 5 members remained. Only the family heads (males) ran the nursery on a weekly rotating basis in which any one member attends the nursery once every five weeks. Coordination of intensive work involved all five. Supervision by technical assessors was on average two visits per month in order to coordinate more intensive work. Approximately 15,000 plants of *Dioon edule* were produced during the period 1990–1998, resulting in plants with ages varying from one to nine years. Conservation of an 80 ha cycad habitat in tropical deciduous forest has been attained (Vovides et al. 2010).

Two further nurseries have been established in Veracruz State under the leadership of researchers from the Universidad Veracruzana. The first was established in 1992 at Ciénega del Sur, on the southern coast of Veracruz near Alvarado, for *Z. furfuracea,* in response to the 1980s over-extraction mentioned above. This cycad grows very well on the stable coastal sand dunes in this region. The second project was initiated with a campesino community at Tlachinola, near Xalapa, for the management of *Ceratozamia tenuis*. Sadly, neither of these two nurseries exists today. However, four additional nurseries were established in the early to mid-1990s in the state of Chiapas for the management of *Dioon merolae* and *Ceratozamia mirandae,* situated in the buffer zone of La Sepultura Biosphere Reserve, and *C. matudae* and *Zamia soconuscensis,* located in the buffer zone of El Triunfo Biosphere Reserve. Three of the nurseries are still active to date, but the El Triunfo nursery is not.

A certain level of success has been achieved through plant sales that encouraged the growers to continue the project for more than 25 years. Nurseries in Chiapas focused on the propagation of four cycad species: *Ceratozamia mirandae, C. matudae, Dioon merolae,* and *Zamia soconuscensis,* as well as local *Chamaedorea* palms under a similar model; however, owing to the remoteness of some of these nurseries in Chiapas and the lack of efficient marketing mechanisms, it has been difficult to connect the growers with markets. Only the Monte Oscuro nursery had certain success with this because the one remaining grower owns a pickup truck and therefore enjoys access to markets and town fairs. These producers are peasant farmers who are hampered by the bureaucratic hurdles of registry and annual reporting of their activities to the licensing authorities, and often need outside assessment that can be costly and at times disillusions the producers. Compounding these difficulties, sales and exports have occurred only on a limited scale (Vovides et al. 2010).

An experimental reintroduction of nursery-produced *D. edule* was done in September 1997, in which three class ranges of plants (2, 5, and 7 years) were reintroduced into the habitat. This was done in order to verify the earliest age at which the plants can be returned to nature with minimum loss through transplant and environmental factors encountered in the wild such as transplant adaptation and survival through the dry season. This called for monitoring for a period of a few years following reintroduction, the first year being the most critical. Parallel studies on naturally dispersed seed showed that during the first year there is a pre-germination mortality through predation and seed loss and to a lesser extent a post-germination mortality through dehydration during the dry season (Vovides et al. 2002).

Over the years, the lesson learned by the team of technical assessors is that each community has idiosyncrasies in the way members work together (or not) and must be respected. In some communities in Chiapas, the members prefer to have their own family mini-nursery in their backyard, making assessment by technicians inefficient and time consuming, while others in Veracruz (Monte Oscuro) agreed on a single communal nursery run by the men. On the southern Veracruz coast, the communal nursery was run only by the women and children, since men "had more important activities" (Vovides et al. 2002). Unfortunately, as mentioned above, the latter coastal nursery is now abandoned.

Cycad Sales

Sales and marketing had always been a problem since no previous marketing study to address the issue had been conducted by experts in this field. There was only the assumption that a demand existed for cycads because there was an illegal trade to satisfy this demand. An attempt to introduce the *Dioon edule* plants grown at Monte Oscuro to the German horticultural trade

Figure 1.15. Nursery-produced *Dioon edule* plant adorning urban gardens in the city of Xalapa, Veracruz (photograph by Andrew P. Vovides).

through a GTZ-Germany funded project in the mid-1990s was not successful, since the *Dioons* could not compete with the already established market for the Japanese *Cycas revoluta*. Such is the world of commercial ornamental horticulture. The domestic market was the only option for *D. edule*, though there was a onetime export of seedlings to the United States. Local sporadic plant sales have occurred, including a sale of 250 dioons to the city of Xalapa's parks and gardens department; these now adorn the city's urban roads and public gardens (Figure 1.15). Likewise, *Z. furfuracea* was used similarly in the port of Coatzacoalcos; its high-wind and sea-salt resistance made it an ideal plant for exposed coastal promenades. Small-scale sales of the plants also occur at the Botanic Garden shop (Figure 1.14). The Monte Oscuro nursery had average annual sales of US$4,015 over a 19-year period, and the Chiapas nurseries US$1,368 over an eight-year period (Vovides et al. 2010).

Dioon edule plants are extremely slow growing in their natural habitat, and large adult plants have been estimated to be more than 2,000 years old (Vovides 1990); however, this growth rate is up to five times faster under nursery conditions. In a reintroduction experiment, a cohort of seven-year-old seedlings were reintroduced into the habitat, and some kept at the nursery as a control. At the time of reintroduction, the plants measured around

4 cm stem diameter with three to four leaves. After more than 10 years, those in habitat remained the same size with no measurable difference in stem diameter and leaf production, whereas those in the nursery grew to a size of more than 25 cm stem diameter with a full leaf crown. By 15 years, the first male plants produced cones, and after 17 years, the female plants coned (Vovides et al. 2010). Because of heavy seedling mortality under natural conditions, it was deemed not feasible to reintroduce seedlings into habitat. A demographic study by Octavio-Aguilar et al. (2008) using elasticity matrices showed that the best conservation strategy was to conserve adult reproductive plants and to reintroduce very few large, nursery-grown sexed plants, more females than males, instead of many seedlings, with most doomed to perish during the dry season. This study has shown a more feasible conservation strategy for plant reintroduction back into habitat and underlines the importance of demographic studies of a species to point the way to an efficient and reliable conservation strategy. The reintroduction of large nursery plants has not yet been done owing to lack of funding.

Conclusions

Our current knowledge of the systematics of Neotropical cycads is due to the contributions of numerous researchers and enthusiasts in multiple fields of study since the nineteenth century. The taxonomy and geographic distribution patterns presented here, emphasizing Mesoamerican and Caribbean cycads, represent a snapshot in time that will surely change as researchers continue to describe novel taxa, clarify taxonomic delimitations, and test evolutionary scenarios to understand the origins of extant species. We hope that upcoming researchers pay adequate attention to the characteristics that define species, utilize an appropriate combination of methods currently at their disposal in future cycad systematics studies (i.e., morphological, anatomical, and molecular), and use previously constructed knowledge to formulate effective measurements for conservation. We also expect that this summary of the history of cycad research in Mexico and the stories about attempts to create sustainable methods for cycad horticulture and the conservation of local traditions can stimulate further interest in the propagation and conservation of cycad species.

Such conservation is crucial, not only since cycads are among the most endangered seed plants on the planet, but also because their study may reveal data that impact research on, for example, climate change and the evolution of plant life on earth. Cycads come from a Mesozoic ancestral stock that survived the last mass extinction at the end of the Cretaceous, yet the question remains: will they survive the rampant habitat destruction of the Anthropocene? The botanical gardens and sustainable nurseries discussed in the latter half of this chapter offer one potential strategy to ameliorate some threats. Given the specific biological circumstances of Cycadales (e.g., long generation times, limited number of taxa, small population sizes, in situ conservation threats, etc.), such conservation resources are of increasing importance. Many large conservation centers, such as the Montgomery Botanical Center in Miami, Florida, the Francisco Clavijero Botanic Garden in Xalapa, Mexico, the New York Botanical Garden, and the Horto Botanico, Naples, Italy—among others—hold comprehensive regional and historical cycad collections for purposes of ex situ conservation and systematic horticultural or taxonomic research. Nonetheless, efforts to conserve cycads whether in situ or ex situ will be futile if their habitats and whole ecosystems—including insect pollinators—are not also protected, and if local stakeholders cannot be convinced of the value of conservation. In situ sustainable management nurseries in Africa, Australia, Asia, and Latin America may offer a viable long-term policy for cycad conservation, but experience has shown that impediments—bureaucratic and otherwise—to their successful and efficient operation still remain and must be overcome.

2

Cycad Population Genetics in Mexico and the Caribbean

Angélica Cibrián-Jaramillo, Francisco Pérez-Zavala, Naishla M. Gutiérrez-Arroyo, Dánae Cabrera-Toledo, Jorge González-Astorga, and Joshua D. Englehardt

The study of genetic variation understood as variable expressions of DNA at the level of populations—or population genetics—allows us to trace the natural history of organisms over time (Hamilton 2009). Population genetics aims to describe and explain genetic variation within and among groups, and to estimate how individuals are related to each other within the time span of a few recent generations, which typically encompass hundreds or thousands of years (Hamilton 2009). The intersection of population genetics and social sciences such as anthropology and archaeology permits a deeper understanding of the role of humans in the evolutionary history of plants, specifically in terms of their potential impact on plant genetic variation. Humans move seeds and whole plants of their interest, primarily as food sources, but also for other reasons, such as ritual purposes, for raw materials, etc. (Banks et al. 2015; Kantar et al. 2019). Such human-induced dispersal can affect the genetic structure of these species and may be reflected in their biogeographic patterns, in some cases to a degree so extreme that the evolutionary history of the plant itself is obscured. For example, the buoyancy of coconuts allows their dispersal via ocean currents, a fact that, coupled with the multiple introductions of these palms throughout the world by humans, makes it nearly impossible to know where coconuts originated and in which places they were first introduced (Gross and Zhao 2014; Summerhayes 2018). Likewise, the intricate and complex history of most domesticated plants is a result of human-induced movement as well as artificial selection for desirable traits (Song et al. 2018). Rice, for example, displays a complex pattern of genetic variation that has led to decades of debate in the biological literature regarding the domestication of this crop (Gross and Zhao 2014). Archaeological data alone may reveal the general routes of rice movements throughout history, but aspects such as the timing of such movements, the detailed relationships between wild rice populations, or the role of the environment in the dispersal patterns are discernible only by integrating population genetic studies with archaeological evidence. To date, most scholars agree on the origin of domesticated rice in East Asia about 9 kya, but the minutiae of the dispersal and diversification of both japonica and indica varieties over the last five millennia (Fuller et al. 2016) remains unknown—and will continue to be explored by the joint effort of population geneticists and social scientists. Such convergent interdisciplinary efforts hold great potential for elucidating questions regarding the evolutionary history of many plant species that continue to elude adequate explanation.

In this chapter, we explore patterns in cycad population genetics. We begin with an introduction to the field of population genetics. We then provide a brief review of the literature on cycad population genetics globally, followed by a detailed discussion of genetic patterns in species from Mexico and the Caribbean. This discussion centers on these regions due primarily to their rich species diversity. We include the Caribbean studies of *Zamia*, given the high species diversity of this genus in this relatively small geographic region (see Calonje et al., this volume; Pagán-Jiménez, this volume), which makes it all the more interesting for population-level processes associated with species diversification. Likewise, Mexico ranks second worldwide in cycad diversity, represented by 70 species distributed in three genera of Zamiaceae family; *Dioon, Zamia*, and *Ceratozamia* with 17, 18, and 35 species, respectively, 65 of which are endemic to Mexico (Calonje et al. 2022; Medina-Villarreal et al. 2019; see also Calonje et al., this volume).

Subsequently, we present our own previously unpublished data on the population genetic structure of *Dioon* species in Mexico and Honduras. This evidence suggests that in some instances genetic patterns are best explained by human-induced gene flow. Finally, we discuss how the increasing evidence for the prevalence of cycads as important elements—and a frequent food source—in many cultures, raises questions about the role of humans in their dispersal and potential human impact on cycads' genetic structure (Bonta et al. 2019; see also the chapters by Bonta, this volume). In Mexico and Central America, there are at least 235 terms among 28 ethnic groups used to describe and classify 57 cycad species in 19 languages (Bonta et al. 2019). It is also clear that cycads served as a foodstuff for at least six millennia across the Mesoamerican region and continue to be consumed in the present as both famine food and starchy staple, due to their carbohydrate-rich seeds and stems (Bonta et al. 2019; see also Englehardt et al., this volume). Certain parts such as the seed covers (sarcotestas) can be eaten without removing toxins, while whole seed and stem starches must be detoxified via several complex processes (Bonta et al. 2019). Further, cycad leaves are often incorporated into syncretic Roman Catholic–Indigenous religious ceremonies such as pilgrimages, Easter Week, and the Day of the Dead (Pérez-Farrera and Vovides 2006; see also the chapters by Bonta, this volume).

Our study of *Dioon* presents one of the few case studies that considers human-induced dispersal as a possible mechanism to explain the observed variation in population genetic patterns. In sum, our chapter provides one of the first reviews of population genetic patterns in this plant group within the contexts of ancient Mesoamerica (but see also Gutiérrez-Ortega et al. 2018) and explores the potential of population genetics methods to generate a thorough understanding of the role of humans in cycad evolutionary history.

A Brief Introduction to Population Genetics

We provide here a brief introduction to the field of population genetics for the nonspecialist. Population genetics centers on describing the distribution of patterns of genetic variation within and among populations, and explaining the various biological, geographic, and evolutionary factors that lead to a given observed distribution (Nielsen and Slatkin 2013). Population genetics is the study of alleles, or variants of a gene, and how they change due to barriers to migration or dispersal. In the presence of such barriers, a population structure (or population subdivision) emerges as a result of restricted gene flow, so not all individuals are freely connected. This structure can be minimal, with minor genetic separations between groups of individuals—or subpopulations—or it can be extreme, with subpopulations having no connection at all, no migration or genetic flow between them, a situation termed "complete population structure."

In addition to gene flow, population genetic patterns also depend on other evolutionary factors, namely mutation or random changes in the DNA; natural selection; and genetic drift, which is the random changes in allele frequencies that affect their movement across the landscape (Nielsen and Slatkin 2013). Biological traits and environmental characteristics, such as distribution range, seed survivability, seed and pollen dispersal mechanisms, habit, and reproductive strategies (selfers or outcrossing) also underlie the resulting genetic variation we observe, especially in long-lived plants such as cycads (Duminil et al. 2007; Fernández and Sork 2007).

Population genetics can be used to evaluate the ability of populations to cope with environmental changes, as well as to estimate the degree to which populations are genetically connected. Further, such data may be correlated with ecological and biological patterns from pollinators or seed dispersers to infer specific sources of dispersal. Landscape features (e.g., the width of a river, the topology of a mountain, or the existence of a highway) may also be explicitly considered in such inferences, as barriers to gene flow that directly impact the genetic structure of a population (Holderegger and Wagner 2006). Incorporating such geographic data in analyses of population genetics may thus also provide valuable information to guide conservation decisions (see Calonje et al., this volume).

Models and estimates of population genetic patterns span dozens or a few hundred plant generations and thus speak to the relatively recent evolutionary history of a species, on the order of hundreds to a few thousands of years. Data derived from population genetic studies may be evaluated against evidence related to human genetic groups within a region or a country (Avila-Arcos et al. 2019). Since human genetic structure also speaks to processes such as demographic movements across the landscape, comparing these data sets allows for the correlation of the human cultural record with the evolutionary history of plants and their genetic structure. Such comparative evaluation, in turn, makes it possible to assess the role and impact of humans as barriers to or facilitators of plant gene flow (Gutaker et al. 2019).

Molecular and Genomic Markers

In order to model population structure, it is necessary to identify markers that can be extracted from DNA or proteins by which we can measure genetic variation and diversity. Early population genetics studies (Hamrick and Godt 1996) focused on allelic variation at loci coding for soluble proteins known as allozymes. These are variable molecular forms of enzymes that have identical or similar functions in the same individual and are codominant, which allows for the identification of homozygous (same inherited alleles) and heterozygous genotypes (different inherited alleles) (Hamrick and Godt 1996). A limiting factor of allozyme variation is their slow mutation, resulting in less statistical resolution in the determination of population structure compared with other markers (Frankham and Briscoe 2004). After allozymes, various molecular markers became more common in population genetics studies. Inter Simple Sequence Repeats (ISSRs), for instance, are markers with relatively faster mutation rates that can reveal polymorphisms (variable phenotypes derived from genetic variation) and have greater reproducibility than allozymes. These characteristics of ISSRs allow these markers to provide a comparatively clearer picture of population structure. The main disadvantage of ISSR markers, however, is that they are dominant, which permits only detection of homozygotes. Therefore, their applicability to wider questions that require higher resolution data at the individual level, such as family pedigrees, is reduced (Rentería 2007).

In contrast, Microsatellites or Simple Sequence Repeats (SSRs), or tandem repeats of short DNA sequences (known as motifs) scattered throughout the genome, are used because their extremely rapid mutation rate results in greater variability within populations and the fact that they are codominant, permitting identification of both homo- and heterozygosity. SSRs are found in almost all animal and plant species, allowing for comparative studies across genera, facilitating the reconstruction of family pedigrees (Frankham and Briscoe 2004; Rentería 2007).

Advances in next-generation sequencing (NGS) technologies have allowed for the simultaneous generation of a large number of data (Liu et al. 2017). NGS technologies are a relatively cost-effective tool for sequencing genetic markers such as SSRs, as well as single nucleotide polymorphisms (SNPs), which recently have become the preferred molecular marker in population genetics analyses (Ganal et al. 2009; Xu et al. 2017). Compared with SSRs, SNPs are more suitable for high-throughput automated genotyping assays, allowing samples to be genotyped faster, more conveniently, and more economically (Fischer et al. 2017; Hurley et al. 2004). SNPs are widespread among individuals of populations and constitute the most abundant type of molecular variation in the genomes of plants and animals (Ganal et al. 2009; Wu et al. 2019). Moreover, SNPs identified by sequencing the expressed portion of the genome (transcriptomes) offer a particularly robust marker to assess genetic diversity within a species (see, e.g., recent studies of *Ginkgo biloba* [Wu et al. 2019] and *Dendrobium officinale* [Xu et al. 2017]). Each molecular and genomic marker offers specific advantages and disadvantages, and the selection of specific markers is contingent upon particular research questions.

Methods for Estimating and Modeling Population Genetic Structure

A wide array of methods are available for measuring population genetic structure, and new methods are constantly being devised. Analytic methods to estimate population structure from molecular or genomic markers were first established by Wright (1943, 1951), which, with subsequent modifications, continue to form the basis of the methodologies most frequently employed in current research (Nei and Chesser 1983). The most common statistical measures developed by Wright include expected and observed heterozygosity (*He*, *Ho*), measured via the extent of genetic variation (where values closer to 1 mean that all individuals are heterozygotes), and genetic differentiation, typically measured by the 0–1 *F*st scale, in which values that are closer to zero indicate a greater degree of gene flow and, conversely, values closer to 1 indicate less gene flow (Nei and Chesser 1983). It is important to note that *F*st values are relative to specific molecular or genomic markers, and results derived from analyses of distinct markers are not directly comparable.

Wright's methods thus include estimates and parameters that are capable of measuring the distribution of genetic variation among and within populations. Nonetheless, Wright's proposals required an arbitrary *a priori* definition of the "population," usually based on sampling sites (Excoffier et al. 1992; Meirmans 2006). In contrast, more recently developed methods to infer population structure utilize individual genotypes to estimate how many subpopulations exist. Once that baseline estimate is established using various likelihood models, individuals are each assigned to one or several of these predicted subpopulations. Building on Wright's initial proposals, this type of method can detect admixed individuals and describe hybrid zones, as well as identify the origin and ancestry of previously unknown individuals (Alexander et al. 2009; Pritchard et al. 2000), thereby offering less biased estimates of population structure.

Issues of scale are also crucial in population genetics. Most population genetic studies take place among populations that are separated by a few dozen to a few hundred kilometers, yet a portion of studies focus on very small geographic scales, on the order of meters to only a few kilometers at most. This scale of work is termed Fine Scale Genetic Structure (FSGS). FSGS is useful in plants with limited dispersion of seeds and pollen, such as cycads. Previous evidence of genetic structure at this fine scale exists for the genera *Dioon* and *Zamia* in New World contexts (Cabrera-Toledo et al. 2012, 2019; López-Gallego and O'Neil 2010; Octavio-Aguilar et al. 2017). In FSGS patterns, the greater the linear distance, the lower the pairwise relatedness between individual plants. In other words, these variables are autocorrelated: greater distance indicates less genetic relationship. In such cases, it is expected that genetic "neighborhoods" emerge (Vekemans and Hardy 2004). The statistically significant extent of this autocorrelation delimits the genetic neighborhood size (Smouse and Peakall 1999). Regardless of the specific marker employed, FSGS provides finer-grain detail at smaller scales and complements estimations of population structure at larger spatial scales.

For optimal resolution at larger scales, population genetics methods can incorporate tools and concepts derived from landscape ecology, which is the study of ecological processes and their impact on spatial patterns and features in the landscape. Considering features of the landscape in tandem with population genetic models and methods significantly augments our understanding of the evolutionary processes that shape population structure (Turner and Gardner 2001). Methods that incorporate landscape ecology are subsumed under the wider category of landscape genetics, which seeks to describe and explain the correlation between population structure and the environment. For example, if there is an unexpectedly low or null genetic flow between two very close populations, landscape genetic studies could explore ecological and geographical traits such as altitude, temperature, type of vegetation, etc., to identify factors that potentially influence observed genetic discontinuities (Manel et al. 2003; Storfer et al. 2010). It is also possible to incorporate ethnographic, ethnobotanical, or archaeological data in considerations of the landscape, which can be further correlated with genetic data. Additionally, landscape genetics can define populations as evolutionarily significant units, management units, or conservation units; such studies therefore may provide vital data for natural-resource managers.

Estimating the overall gene flow patterns among populations over large geographic distances, and associating these patterns with other types of connections, such as those that result from human or animal movements, is also a critical consideration in population genetics. Population graphs are a common method used to carry out such overlapping connectivity estimates and assist in visualizing results. Generally speaking, a graph is made up of vertices (commonly called nodes) that are connected by edges (also called links) (Minor and Urban 2007). A population graph is constructed by creating an initial graph in which all populations are connected to all other populations (all nodes connected), and the connection degree between two nodes is calculated based on the pairwise genetic distance between populations, as identified via analysis of selected markers. The statistical significance of each link or edge between two populations is tested, and only significant edges are maintained in the final graph (Dyer and Nason 2004). Once this final graph is generated it is possible to test various hypotheses by analyzing the general topology or specific features of the graph itself (Dyer and Nason 2004), for example to identify key patches that are very important to habitat connectivity and long-term population persistence across the landscape. Population graphs lack any initial assumptions of how populations are distributed in geographic space, which makes them less sensitive to biases based on defining barriers to gene flow *a priori*, a potential difficulty with Wright's initial methods, and also allows for correlation with multiple independent data sets. For these reasons, we employ population graphs as the method of choice to analyze the new data presented on *Dioon* in the final section of this chapter.

Finally, in most population-level studies it is very important to estimate the effective population size (N_e). The formal definition of this term is "the census sizes of a real population into the size of an idealized population showing the same rate of loss of genetic diversity as the real population under study" (Husemann et al. 2016). In essence, N_e refers to the number of individuals within a given population that contribute with progeny to the next generation. In practice, effective population size is often estimated via theoretical approximations using models. The mathematical theory of the coalescent (Kingman 1982) is used to conceptualize how sampled gene variants within a population may have originated from a common ancestor, assuming no recombination, no natural selection, no gene flow, and no population structure. The theory forms a strong base by which N_e may be estimated and modeled. This approach uses the genetic variation of the present populations to reconstruct the genetic history, or genealogy, of ancestral populations (Kingman 1982). Coalescent theory-based models allow for the estimation of effective population size, as well as potential variability and population change over time (Hedges and Kumar 2009). These theoretical genealogies may then be evaluated against observed data to identify population genetic parameters, for example, migration rates, population sizes, and recombination between populations, in both the present and the past. Subsequently, it becomes possible to infer the demographic history of a population, for instance, when and how major events such as expansions, bottlenecks, colonization of new areas, population splits, and others, took place (Rosenberg and Nordborg 2002). Such models also provide data and hypotheses that are frequently employed to evaluate the potential effects of humans on gene flow (as barriers or facilitators, for example; see Griffith et al. 2022 for an excellent example), and to characterize major changes in human populations themselves, such as migrations or demographic shifts (DeGiorgio et al. 2009).

In sum, the field of population genetics encompasses a variety of tools and methods that provide a wealth of valuable data regarding the processes that take place during the most recent generations of a group of individuals. These genetic data may then be correlated with other interdisciplinary evidence to provide a more complete understanding of the various factors that may influence the evolutionary history of a given population or species. In the specific contexts of this chapter, population genetics offers many avenues to facilitate the exploration of the intersection of patterns of genetic variation and human cultural practices. Such convergent considerations offer a more detailed view of the evolutionary history of plants, in this case, cycads, and the potential impacts of humans on that history. Likewise, integrative analyses provide critical information that can be applied to the development of management or conservation strategies for threatened or highly endangered groups such as cycads.

An Overview of the Population Genetics and Genomic Patterns in Cycads Worldwide

Before turning to our own analyses, we now briefly review specific population genetics studies in cycads, providing a short synopsis of previous research worldwide. This review is intended to offer context for our own research on New World populations, and it is by no means exhaustive. A great variety of studies examine cycads from the perspective of population genetics. Most initial studies focused on describing the levels of genetic variation and how this variation is structured in the regional landscape. Notably, most of the early research focused on the use of allozymes and ISSRs, which as mentioned above, present limitations in extent of variation they can detect due to their lower mutation rates. Among the main insights gleaned from these earlier studies is the notion that cycads are plants with low levels of genetic variation and low levels of gene flow. Indeed, several estimates suggest that the dispersal of pollen and seeds ranges from 2 to 10 km, for example, in species such as the Australian *Macrozamia heteromera* (Sharma et al. 1999), the Chinese *Cycas guizhouensis* (Xiao et al. 2004), Chinese and Vietnamese *Cycas multipinnata* (Gong et al. 2015), and Taiwanese *Cycas taitungensis* (Huang et al. 2004; Lin et al. 2000) (see Table 2.1). In terms of biological factors in cycads, low gene flow and low genetic variation have been explained by these authors in part by the slow reproduction rate characteristic of the Cycadales order and a strictly dioecious reproduction mode. For instance, although most cycads reproduce earlier, the average minimum age at which wild female plants of *D. edule* have been inferred to begin to reproduce is about 500 years (Vovides 1990). This slow reproductive rate, coupled with the intermittent appearance of a few male or female megastrobili in a population (see Calonje et al., this volume, for a brief explanation of cycad reproduction), would decrease the effective population size over time, thus reducing the amount of genetic diversity and increasing population structure.

With respect to geographic factors, low gene flow could be the result of a dramatic reduction of geographic ranges. This reduction could also decrease

Table 2.1. Summary of genetic diversity and structure of cycad species

No.	Species	H_O	H_E	F_{ST}/G_{ST}	M	Site of study	Reference
1	D. mejiae	0.346	0.375	0.000	SSR	Olancho, Honduras	This chapter
2	D. spinulosum	0.350	0.399	0.000	SSR	Soyaltepec, Oaxaca	This chapter
3	D. rzedowskii	0.251	0.341	0.206	SSR	San Bartolomé, Oax.	This chapter
4	D. califanoi	0.173	0.250	0.153	SSR	Huautla, Oax.	This chapter
5	D. purpusii	0.298	0.311	0.000	SSR	Coyula, Oax.	This chapter
6	D. caputoi	0.078	0.079	0.621	SSR	Majada Garambuyo, Puebla	This chapter
7	D. caputoi	0.133	0.145	0.349	SSR	Coatillo, Puebla	This chapter
8	D. merolae	0.064	0.067	0.688	SSR	Lachiguiri, Oax.	This chapter
9	D. merolae	0.258	0.196	0.000	SSR	Loma Colorada, Oax.	This chapter
10	D. merolae	0.196	0.177	0.041	SSR	Río Flores, Chiapas	This chapter
11	D. merolae	0.180	0.125	0.122	SSR	Jiquipilas 1, Chiapas	This chapter
12	D. merolae	0.209	0.222	0.000	SSR	Jiquipilas 2, Chiapas	This chapter
13	D. merolae	0.093	0.145	0.546	SSR	Agua Prieta, Chiapas	This chapter
14	D. merolae	0.130	0.169	0.363	SSR	Las Minas, Chiapas	This chapter
15	D. holmgrenii	0.439	0.369	0.000	SSR	Río Limón, Oax.	This chapter
16	D. holmgrenii	0.408	0.503	0.000	SSR	Loxicha, Oax.	This chapter
17	D. edule	0.249	0.216	0.000	SSR	El Farallón, Veracruz	This chapter
18	D. edule	0.333	0.383	0.000	SSR	Soldad de Guadalupe, Qro.	This chapter
19	D. edule	0.137	0.424	0.429	SSR	Valle Verde, Qro.	This chapter
20	D. stevensonii	0.239	0.245	0.000	SSR	Altamirano, Guerrero	This chapter
21	D. stevensonii	0.287	0.284	0.000	SSR	Higueral, Michoacán	This chapter
22	D. tomasellii	0.200	0.271	0.127	SSR	El Tuito, Jalisco	This chapter
23	D. tomasellii	0.156	0.162	0.321	SSR	Compostela, Nayarit	This chapter
24	D. tomasellii	0.239	0.268	0.000	SSR	Panuco, Sinaloa	This chapter
25	D. sonorense	0.136	0.229	0.407	SSR	Nuri, Sonora	This chapter

(continued)

Table 2.1—Continued

No.	Species	H_O	H_E	F_{ST}/G_{ST}	M	Site of study	Reference
26	D. vovidesii	0.279	0.296	0.000	SSR	Mazatlán, Sonora	This chapter
27	D. vovidesii	0.296	0.286	0.000	SSR	Presa Novillo, Sonora	This chapter
	Avg. ± SD	0.228 ± 0.098	0.257 ± 0.108	0.162 ± 0.219			
28	D. angustifolium	0.215	0.218	0.167	A	NE Mexico	González-Astorga et al. 2008
29	D. holmgrenii	0.204	0.17	0.069	A	Oaxaca, Mexico	Ibid.
30	D. tomasellii	0.309	0.295	0.145	A	Western Mexico	Ibid.
31	D. sonorense	0.33	0.314	0.151	A	Sonoran coast, Mexico	Ibid.
32	D. edule	0.323	0.386	0.148	A	Veracruz, Mexico	Octavio-Aguilar et al. 2009
33	D. caputoi	0.522	0.358	0.06	A	Central Mexico	Cabrera-Toledo et al. 2010
34	D. merolae	0.713	0.446	0.07	A	SE Mexico	Ibid.
	Avg. ± SD	0.37 ± 0.18	0.31 ± 0.09	0.12 ± 0.04			
35	Ceratozamia mexicana	—	0.235	0.185	ISSR	Veracruz, Mexico	Rivera-Fernández 2012
36	C. zaragozae	0.8	0.6	0.102	SSR	San Luis Potosí, Mexico	Gutiérrez-Arroyo et al. 2018
37	C. kuesteriana	0.66	0.53	0.25	SSR	Tamaulipas, Mexico	Rubio-Tobón and Octavio-Aguilar 2019
38	C. fuscoviridis	0.77	0.74	0.127	SSR	Hidalgo, Mexico	García-Montes et al. 2020
	Avg. ± SD	0.74 ± 0.28	0.53 ± 0.22	0.17 ± 0.04			
39	Microcycas calocoma	0.2	0.17	0.337	A/I	Cuba	Pinares et al. 2009
40	Zamia loddigesii	0.263	0.266	0.179	A/I	Gulf Coast Mexico	González-Astorga et al. 2006
41	Z. lacandona	—	0.216	0.108	A/I	Mexico	Ibid.
42	Z. katzeriana	—	0.298	0.191	A/I	Mexico	Ibid.
43	Z. variegata	—	0.355	0.085	A/I	Mexico	Ibid.
44	Z. cremnophila	—	0.429	0.174	A/I	Mexico	Ibid.
45	Z. purpurea	—	0.485	0.025	A/I	Mexico	Ibid.
46	Z. lucayana	0.483	0.49	0.067	SSR	Bahamas	Calonje et al. 2013

No.	Species	H_O	H_E	F_{ST}/G_{ST}	M	Site of study	Reference
47	Z. inermis	0.152	0.348	0.734	SSR	Mexico	Iglesias-Andreu et al. 2017
48	Zamia pumila complex	0.538	0.542	0.129	SSR	Puerto Rico	Meerow et al. 2012
49		0.458	0.452	0.196	SSR	Cayman Islands	Ibid.
50		0.424	0.436	0.055	SSR	Dominican Republic	Ibid.
51		0.398	0.385	0.204	SSR	Jamaica	Ibid.
52		0.408	0.411	0.132	SCNG	Cayman Islands	Ibid.
53		0.321	0.351	0.066	SCNG	Dominican Republic	Ibid.
54		0.413	0.444	0.313	SCNG	Jamaica	Ibid.
55		0.276	0.28	—	SCNG	Puerto Rico	Ibid.
	Avg. ± SD	**0.376 ± 0.20**	**0.387 ± 0.09**	**0.177 ± 0.17**			
56	Macrozamia communis	—	0.045	0.27	A/I	Australia	Ellstrand et al. 1990
57	M. riedlei	—	0.263	0.274	A/I	Australia	Byrne and James 1991
58	M. parcifolia	0.02	0.037	0.09	A/I	Australia	Sharma et al. 1998
59	M. pauli-guilielmi	0.043	0.081	0.03	A/I	Australia	Ibid.
60	M. heteromera	0.049	0.061	0.1	A/I	Australia	Ibid.
61	M. plurinervia	0.08	0.111	0.35	A/I	Australia	Ibid.
	Avg. ± SD	**0.048 ± 0.03**	**0.10 ± 0.08**	**0.19 ± 0.13**			
62	Cycas pectinata	0.066	0.076	0.387	A/I	China, India, SE Asia	Yang and Meerow 1996
63	C. siamensis	0.114	0.134	0.291	A/I	China, India, SE Asia	Ibid.
64	C. seemannii	0.047	0.057	0.594	A/I	Vanuatu, New Caledonia, Fiji, Tonga	Keppel 2002
65	C. taitungensis	0.021	0.039	0.051	A/I	SE Taiwan	Huang et al. 2004
66	C. guizhouensis	—	0.059	0.139	ISSR	SW China	Xiao et al. 2004
67	C. parvula	—	0.053	0.097	ISSR	SW China	Ibid.
68	C. balansae	—	0.13	0.4	ISSR	SW China	Ibid.
69	C. micronesica	0.349	0.545	0.323	SSR	Guam	Cibrián-Jaramillo, Daly, Brenner, et al. 2010

(continued)

Table 2.1—Continued

No.	Species	H_O	H_E	F_{ST}/G_{ST}	M	Site of study	Reference
70	C. simplicipinna	0.394	0.447	0.261	SSR	SW China, Laos	Feng et al. 2014
71	C. multipinnata	0.387	0.497	0.295	SSR	SW China, N Vietnam	Gong et al. 2015
72	C. debaoensis	0.389	0.484	0.114	SSR	Yunnan, Guangxi China	Gong and Gong 2016
73	C. panzhihuensis	0.189	0.328	0.331	SSR	Yunnan, Sichuan, China	Xiao et al. 2019
	Avg. ± SD	0.21 ± 0.17	0.24 ± 0.2	0.27 ± 0.16			

Note: H_O: heterozygosity observed; H_E: expected heterozygosity; F_{ST}/G_{ST}: index of genetic differentiation between populations; M: type of molecular marker used.

genetic diversity, the result primarily of population bottlenecks and founder effects (migration of only a few individuals), at least during the initial stages of these events. The genetic variation in *Cycas revoluta* in the Ryukyu Islands, for instance, was estimated to be very low in terms of haplotype diversity in comparison to its relative in Taiwan, *Cycas taitungensis,* according to Kyoda and Setoguchi (2010). A reasonable explanation for this pattern is the presence of severe bottleneck effects that resulted from the submersion of low islands and the subsequent diminished total landmass of the Ryukyus beginning in the interglacial age of the Quaternary Period (Kyoda and Setoguchi 2010).

More recent studies using SSRs as molecular markers have detected that certain groups, such as several species of *Cycas* distributed in the Pacific, have a greater degree of gene flow among populations than most other cycads. The increased flow in this genus is likely a result of their seeds' ability to float with ocean currents, which would result in more migrants exchanged among sites (Cibrián-Jaramillo, Daly, et al. 2010). Similarly, in *Cycas debaoensis,* a critically endangered species endemic to China, there is low genetic differentiation among populations, probably due to long-distance pollen dispersal (Gong and Gong 2016).

Other recent studies suggest that some cycad species have complex population genetic patterns that have resulted from a combination of recent population changes and older biogeographic events affecting their genetic structure (Table 2.1). For instance, *Encephalartos barteri* ssp. *barteri,* an endemic West African species, showed high levels of genetic diversity, but also high differentiation among populations, potentially due to geographic isolation, as well as to limited seed dispersal and pollen transfer (Ekué et al. 2008). Further, recent research on species appearance and diversification (in the range spanning ca. 19–9 mya) for several genera, including *Ceratozamia, Dioon,* and *Zamia* in the New World, African *Encephalartos,* and Asian-Australian *Cycas* and *Macrozamia* (Condamine et al. 2015; González and Vovides 2002, 2012; Medina-Villarreal et al. 2019), suggests that demographic changes resulting from speciation or divergence, as well as concomitant bottlenecks, may affect population structure and lead to lower genetic diversity.

Research has also revealed surprising population genetic patterns. For example, some species with extremely small population sizes and distribution ranges, such as some *Dioon* species (Cabrera-Toledo et al. 2008; González-Astorga et al. 2003), *Ceratozamia kuesteriana* (Rubio-Tobón and Octavio-Aguilar 2019), and *Ceratozamia zaragozae* (Gutiérrez-Arroyo et al. 2018) in the Americas, present unexpectedly high levels of heterozygosity and moderate gene flow. These data may be explained in part by the life stages in which these species were sampled. In long-lived species, different life stages may reveal the occurrence of specific demographic events that took place over many years. For example, major population reductions that took place hundreds of years ago will still be evident in adults, while the impact of very recent environmental conditions or changes in reproduction patterns in the last decades would be detected only when sampling sapling and juvenile plants. Some *Dioon* species can live 2,000 years or more (Cabrera-Toledo et al. 2019; Vovides 1990; Vovides and Peters 1987), and thus sampling either adults or juveniles could bias measurements of genetic variation and structure (Octavio-Aguilar et al. 2009). Additionally, the variation found in early life stages (saplings or juveniles) could provide insight on trends of loss, gain, or stability of genetic variation in future generations (Gutiérrez-Arroyo et al. 2018). In the case of the New World *Dioon* species, for example, long-lived populations that display an unusually high number of heterozygotes are primarily adult plants (Cabrera-Toledo et al. 2008; González-Astorga et al. 2003).

The global patterns presented above are but brief summaries. As should be evident even from such a cursory review, patterns revealed through population genetics of cycads are as diverse as the phenotypes and individual histories of the individual plant species themselves, and we cannot assume that all species will conform to the same evolutionary pattern. In that sense, and given the paucity of research on the subject (but cf. Calonje et al., this volume; see also Englehardt et al., this volume), any attempt to generalize the population genetic patterns in the Cycadales order would be premature. Rather, each species may display discrete patterns depending on its biology and the variable idiosyncrasies of its natural habitats and environments. Finally, such patterns will also vary depending on the markers of choice. As detailed above, markers are not always directly comparable. Thus, we must exercise caution when comparing patterns suggested by population genetics studies carried out with distinct molecular markers. The selection of markers is therefore a critical factor to consider not only when comparing results across studies but also when planning new research.

To address these lacunae in our understanding, and to provide additional comparative data, in the remaining sections of this chapter, we provide a more detailed discussion of genetic patterns among cycad species in Mexico, Honduras, and the Caribbean. To complement the other chapters in this collection, we also seek to emphasize the role of humans in the observed genetic diversity of cycads in this region, as revealed through our own previously unpublished data.

Genetic Patterns in Cycads from Mexico, Honduras, and the Caribbean

In what follows, we provide more detailed data derived from population genetics studies of species whose distribution is restricted to Mexico, Honduras, and the Caribbean, the geographical focus of this chapter. Most previous genetic research on cycads in this region (Cabrera-Toledo et al. 2010; González-Astorga et al. 2006; Octavio-Aguilar et al. 2009; Pinares et al. 2009) employed allozymes as molecular markers, which present some of the inferential difficulties described above. More recently, SSR and SNP markers have been increasingly employed (Gutiérrez-Ortega et al. 2018; Meerow, Salas-Leiva, Calonje, et al. 2018; Meerow, Salas-Leiva, Francisco-Ortega, et al. 2018) since they offer the previously detailed analytic advantages. Additionally in some of these genera, single-copy nuclear genes (SCNGs) have been used to complement SSRs. SCNGs provide a phylogenetic view that recapitulates deeper time scales compared with SSRs, as a result of their slower mutation rates (Meerow, Salas-Leiva, Calonje, et al. 2018). Below, we briefly detail the genetic patterns for the *Zamia* and *Ceratozamia* genera, as revealed in prior studies, before presenting our own research on *Dioon* population genetics.

Zamia

The genus *Zamia* L. (Zamiaceae), consisting of 83 species, is the most species-rich and widely distributed cycad genus in the Caribbean and the Americas, and arguably the most morphologically and ecologically diverse genus of cycads (see Calonje et al., this volume). A phylogenetic tree of the genus *Zamia* revealed five major clades that are geographically delimited: (1)

the insular Caribbean clade, (2) the Fischeri clade, with only three species, all endemic to Mexico, (3) the Mesoamerica clade, (4) the Isthmus clade, with mainly Panamanian and Costa Rican species, and (5) the South American clade (see Calonje et al. 2019:fig. 4), with a recent age of 10 mya of origin within the Mesoamerican region, and an eventual southward migration into the Isthmus of Panama and South America (Calonje et al. 2019; Condamine et al. 2015).

The diversification of the genus is characterized by gradual speciation and low extinction rates, resulting in many species. This pattern is manifested in its population genetics, such as those observed in the *Zamia pumila* species complex (Meerow, Salas-Leiva, Francisco-Ortega, et al. 2018; see also Table 2.1). Most of the genetic studies of the genus *Zamia* are of the Caribbean island clade (Meerow et al. 2012; Meerow, Salas-Leiva, Calonje, et al. 2018; Meerow, Salas-Leiva, Francisco-Ortega, et al. 2018). Most *Zamia* populations in this clade, including populations from Puerto Rico, the Cayman Islands, Dominican Republic, and Jamaica, appear to have little genetic flow between them and in some cases, almost no connections outside the islands according to SSRs and haplotypes from 10 single-copy nuclear genes (SCNGs) (Meerow, Salas-Leiva, Calonje, et al. 2018; Meerow, Salas-Leiva, Francisco-Ortega, et al. 2018). This is similar to the patterns observed in *Zamia lucayana* in the Bahamas (Calonje et al. 2013), and, to some extent, to those found in *Cycas micronesica* (Cibrián-Jaramillo, Daly, et al. 2010) (cf. Table 2.1).

Overall, the population genetics data from the Caribbean region indicate that each island has a different diversification history, although most of them likely originated in Cuba (Meerow, Salas-Leiva, Calonje, et al. 2018). For example, populations from the Dominican Republic have high genetic flow, while those in Puerto Rico seem to be vicariant (separated by geographic barriers), suggesting that observed genetic diversity is the result of multiple introductions. Northern and western populations in Jamaica constitute a separate introduction to the island with a subsequent decrease in genetic flow among them, while the Cayman Island populations display admixture with those from Jamaica, suggesting migration between these populations (Meerow et al. 2012; Meerow, Salas-Leiva, Calonje, et al. 2018). According to these authors, the complex pattern of genetic differentiation between and within islands is congruent with the precolumbian anthropogenic activity that has affected the Caribbean *Zamia* distribution, although they did not offer specifics on this hypothesis.

Jaime Pagán-Jiménez (Mickleburgh and Pagán-Jiménez 2012; Pagán-Jiménez 2013; Pagán-Jiménez et al. 2015; see also Pagán-Jiménez, this volume) has expanded on this line of research; his results suggest a great deal of potential human agency in the spread and diversification of *Zamia* throughout the Caribbean. Significantly, Pagán-Jiménez details considerable archaeological evidence—based primarily on starch grains—that suggests the importance of *Zamia* as a starch source for the native Amerindian societies of the Caribbean. These reports include areas where extant *Zamia* populations have never been documented, suggesting that the seeds and/or plants were carried with humans when they were migrating. We suggest that a direct correspondence between human demographic movements and the evident genetic flow among cycad populations from the north of Jamaica to the Cayman Islands is the most parsimonious explanation for these data. Interestingly, Simms (2014) has detected *Zamia* starches in pre-Hispanic tools at an archaeological site in Mexico's Yucatán Peninsula, which may indicate a more extensive movement of *Zamia* throughout the wider circum-Caribbean region. Further genetic studies of these populations, however, are needed to evaluate this potentiality.

Ceratozamia

Ceratozamia contains a diverse group of 36 recognized species, distributed from Honduras to Mexico and found in a wide range of habitats, from rainforests to deserts and at varying elevations. *Ceratozamia* has several morphological variations both within and between species; for instance, leaflets range from less than 3 mm wide in *C. zaragozae* to 163 mm in *C. euryphyllidia*. Likewise, leaves range in length from 60 cm to 3.6 m, and stems show similar variation, from 7.5 cm to 2 m in length (Calonje et al. 2019; Medina-Villarreal and González-Astorga 2016; Whitelock 2002). Despite this ample morphological diversity, studies based on common intergenic nuclear and chloroplast sequences have revealed only a few genetic differences among species within the genus (González and Vovides 2012). Such low differentiation could be a result of their relatively recent divergence, approximately 19.2 mya (Condamine et al. 2015; Medina-Villarreal et al. 2019).

Previous phylogenetic analyses (González and Vovides 2012; Medina-Villarreal et al. 2019) have provided insights into the biogeography of the *Ceratozamia* genus and supported a southern/southeastern Mexican origin for this genus. These models suggest that most of these original populations survived the Pleistocene climatic changes, whereas some species (e.g., *C. morettii, C. mexicana, C. brevifrons, C. microstrobila, C. fuscoviridis, C. kuesteriana, C. hildae, C. zaragozae, C. sabatoi,* and *C. decumbens*) occurring to the north and northeast of the Neovolcanic range (see González and Vovides 2002:659, fig. 5, 2012:934, fig. 1) may be the result of more recent diversification, probably closer to the Pleistocene–Holocene transition. The low molecular differentiation could also be explained by the comparatively long life cycle of some *Ceratozamia* species (González and Vovides 2002). Generation times from seed germination to reproductive stages under optimal cultivation conditions span at least 15 years, and it is at least double in the wild. This would result in only about 300 generations of *Ceratozamia* species occurring since the end of the Pleistocene ca. 12 kya (González and Vovides 2002, 2012).

Despite their ample distribution and high species diversity, surprisingly few population genetic studies have been carried out on cycads of the *Ceratozamia* genus. Pérez-Farrera et al. (2017) found low genetic diversity and genetic differentiation between species of the *C. norstogii* complex from southern Mexico, suggesting that the complex is currently undergoing a divergence process, probably via genetic drift and/or founder effects. Genetic analysis using SSRs in three *Ceratozamia* species (*C. zaragozae, C. fuscoviridis,* and *C. kuesteriana;* Table 2.1; Figure 2.1) showed overall high heterozygosity (He) and low genetic differentiation (Fst), indicating high gene flow among populations. *Ceratozamia zaragozae* showed the highest levels of heterozygosity ($Ho = 0.805$; $He = 0.600$) and the lowest genetic differentiation between populations ($Fst = 0.102$) (Gutiérrez-Arroyo et al. 2018). These authors suggested that traits like dioecism, demography (adults had the highest survival rate [lx] at $63.02 \pm 54.79\%$) and saplings the lowest ($5.34 \pm 23.58\%$) (Castillo-Lara et al. 2018), and longevity could explain these results. Furthermore, this species, along with *C. fuscoviridis* and *C. kuesteriana,* is situated in a clade with a more northerly distribution, which is assumed to have a recent origin (González and Vovides 2002, 2012). Gutiérrez-Arroyo et al. (2018) also found that the two most geographically distant populations (50 km) of *C. zaragozae* are the most genetically similar. A previous study on this species reported limited seed dispersal, facilitated mainly by gravity (Castillo-Lara et al. 2018), and gene flow via pollen dispersal was estimated to occur between 2 and 10 km in other cycad genera (e.g., *Cycas pectinata* [Yang and Meerow 1996]; *Zamia furfuracea* [Norstog and Nicholls 1997]). Therefore, natural gene flow between these populations seems less likely to occur given the large geographic distances. Indeed, some researchers have suggested that these results may be explained by anthropogenic activities, such as relocation of adults, saplings, and/or seeds. For example, a great number of *C. zaragozae* plants have been extracted for illegal trade, and their final destination is unknown (Castillo-Lara et al. 2018), which may explain observed genetic diversity in the present and may parallel the effects of human action at deeper time scales.

Figure 2.1. *Ceratozamia* species analyzed using SSRs: *a*, Adult and *b*, female cone of *C. zaragozae* (photographs by Arturo de Nova-Vázquez); *c*, Adult and *d*, female cone of *C. kuesteriana* (photographs by Sergio Medellín [*c*] and Ken Hill [*d*]); *e*, *C. fuscoviridis* saplings, which show an example of seed dispersal by gravity; *f*, male cones and *g*, adult of *C. fuscoviridis* (photographs by Naishla Gutiérrez-Arroyo; all photos not by the authors used under terms of CC BY 4.0 license).

Similarly, García-Montes et al. (2020), using ISSR and SSR markers, evaluated the genetic variation in *C. fuscoviridis* populations with different degrees of anthropogenic disturbance. Their results show human and cattle-raising activity negatively affected effective population size and that genetic variation was greater in conserved sites compared with disturbed ones. Tristán (2012) suggests an inverse relationship between the practice of seed collection and the availability of seeds in *D. edule*; that is, the greater the collection, the smaller the population size, and vice versa: the smaller the collection, the greater the population size. García-Montes et al. (2020) also found that genetic structure estimated using SSRs did not correspond with isolation by distance (IBD) models, in which geographic distance correlates with genetic distance. Such a pattern has been observed in some cycad genera and species (e.g., *Cycas debaoensis* [Gong and Gong 2016]; *Ceratozamia kuesteriana* [Rubio-Tobón and Octavio-Aguilar 2019]), but in the case of the *C. fuscoviridis* populations studied by García-Montes et al. (2020), genetic variation corresponded to the degree of habitat disturbance level of each population. Studies such as this illustrate the effects that humans may have on cycad population genetics in both past and present.

Dioon

In addition to providing an overview of the population genetic patterns of *Dioon*, we also present new results from our own research. Our study employed SSRs as genetic markers and was based on sampling throughout the entire range of *Dioon* distribution in Mexico and Honduras (see Calonje et al., this volume, Figure 1.5). Our work is distinct from most previous research insofar as there are very few previous studies of cycad population genetics, in this or other regions, that include most of the species within a genus in their sampling (but cf. Gutiérrez-Ortega et al. [2018 and 2020] for a similarly comprehensive study of *Dioon* via SNPs and chloroplast DNA [cpDNA]). These authors included most *Dioon* species in their sampling; in the 2018 study they focused primarily on reconstructing the biogeographic events of vicariance and dispersal of this genus across the Neotropical biogeographic provinces, which takes place at a much deeper timescale (on the scale of millions of years), rather than the more recent population-level patterns that we present here—and which may be correlated with archaeological evidence of human activity in the recent past. In the 2020 research, Gutiérrez-Ortega et al. concluded that lineage divergence is associated with barriers to gene flow and increased genetic drift, which resulted in isolated populations that adapted to their own local niche. Our work complements these results by providing a more detailed view of genetic connections among populations

Figure 2.2. *Dioon* cycads in habitat: *a, Dioon edule* in habitat, San Luis Potosí (photograph by Francisco Barona-Gómez); *b, D. edule* adult plant with a new leaf flush (ex situ collection, photograph by Angélica Cibrián-Jaramillo); *c, D. merolae* in habitat in Chiapas (photograph by Fret Cervantes-Díaz); *d, D. caputoi* in its natural habitat in Puebla (photograph courtesy World Cycad List, used under terms of CC BY 4.0 license).

across their distribution in more recent times, and highlights the importance of human dispersal after populations diverged during speciation.

Dioon (Figure 2.2) is the cycad genus with the fourth broadest distribution in the world, and the only one spanning all the Mexican transition ecological zones, from the Nearctic to the Neotropical biotas (Gutiérrez-Ortega et al. 2018). *Dioon* species are distributed from northern Mexico to northeastern Honduras (with the exception of the Yucatán Peninsula; see Hernández-Ramírez et al. 2020; see also Calonje et al., this volume, Figure 1.5c). The morphology of *Dioon* species varies widely in terms of seed size (from 3.5 cc to 40 cc) and color (white, yellow, or red), seedlings (types I, II, and III) and leaflets (from 5 mm to 20 mm), and cone size (male: from 20 cm to 50 cm, female: from 30 cm to 90 cm) (Sabato and De Luca 1985). Phenotypic traits in *Dioon* such as thick cuticles, deep and narrow stomatal chambers, and the presence of furrows and papillae are considered adaptations to xeric environments and can even be beneficial in areas with volcanic activity where *Dioon* cycads are also present (Barone Lumaga et al. 2014). This observation is congruent with the hypothesis of the fragmented distribution of modern populations as the result of adaptation to aridification during the expansion of arid zones in the Miocene (Gutiérrez-Ortega et al. 2018).

Pollination of *Dioon* is usually carried out by weevils and thrips (Thysanoptera), as with most cycads (Vovides 1990). The pollen dispersal capacities of both pollinators are still unknown, but studies of palms report that weevils can fly up to 50 km in 24 hours, with an average of 10 km per day (Hoddle et al. 2015). Thrips, on the other hand, are mostly short-distance dispersers. The dispersal of *Dioon* seeds is mainly by squirrels *(Sciurus* spp.), the Mexican deer mouse *(Peromyscus mexicanus)*, and the white-throated magpie-jay *(Calocitta formosa)*. Other birds and rodents may also be seed dispersers, as the sarcotesta is edible and does not contain the toxic compounds normally found in the seeds. Additionally, birds and some butterflies and rodents are immune to the megagametophyte toxic compounds, so the sarcotesta with small amounts of seed tissue could be a reward to the animals that disperse the cycad seeds (González Christen 1990; Ruíz-García et al. 2015; Vovides 1990).

Various phylogenetic studies concur that the *Dioon* genus is divided into four clades: Spinulosum, Edule, Purpusii, and Tomasellii (González-Astorga et al. 2008; Gutiérrez-Ortega et al. 2018; Moretti et al. 1993). The Spinulosum clade encompasses *D. spinulosum, D. rzedowskii,* and *D. mejiae*. The latter is the only species found outside of Mexico and is approximately 700 km away from the known populations of other species. All the populations of *D. rzedowskii* and the majority of *D. spinulosum* populations are concentrated in northern Oaxaca, with some reports of *D. spinulosum* in Veracruz (Sabato and De Luca 1985). This clade contains the tallest *Dioon* species, with reported heights of ~16 m (Sabato and De Luca 1985). There are no specific genetic studies for this clade to our knowledge.

The Edule clade includes *D. edule* and *D. angustifolium*. These species can grow up to ~5 m, which is the average for most *Dioon* species. *Dioon* can be found in various habitats, from sandy beaches and dunes to the pine forests of the Sierra Gorda in Querétaro. *D. angustifolium* populations have suffered bottlenecks but maintain high levels of genetic diversity (González-Astorga et al. 2005). *D. edule* also presents high diversity, an excess of heterozygotes, and high gene flow between populations (González-Astorga et al. 2003; Mora et al. 2013).

The Purpusii clade includes *D. argenteum, D. califanoi, D. purpusii, D. holmgrenii, D. caputoi, D. oaxacensis, D. planifolium,* and *D. merolae*. Cabrera-Toledo et al. (2010) reported the highest observed heterozygosity (0.52 for *D. caputoi* and 0.71 for *D. merolae*) and a relatively low *Fst* value for both species (0.06 for *D. caputoi* and 0.07 for *D. merolae*). These authors suggest that such diversity might be the result of dissimilar allele frequencies between male and female plants. These results may also be a consequence of climate change dynamics such as the Pleistocene refugia, during which population expansions and contractions took place (Hewitt 1996), which could have given rise to small, isolated populations in specific geographic areas that were capable of surviving these changes. *D. caputoi* showed strong signs of bottlenecks and population reduction, and it has high genetic diversity, but relatively less compared with *D. merolae*. The diversity of *D. caputoi* is probably maintained by adult individuals with minimal recruitment of new individuals. Its sister species, *D. merolae*, has higher genetic diversity that is explained in part by a larger geographical distribution (Cabrera-Toledo et al. 2008, 2012). *D. holmgrenii* also displays high heterozygosity (Ho = 0.2 vs. He = 0.17) and high genetic diversity (mean alleles = 1.71), despite its somewhat limited geographic distribution.

Cabrera-Toledo et al., using allozymes (2012) and ISSRs (2019), found fine-scale genetic flow (genetic connections within a few meters) within two populations of *D. caputoi* adults in Puebla (the Tehuacán-Cuicatlán Valley). Interestingly, they found that plants were more related to each other, even between adult plants from 400 to 2,000 years of age, and at a relatively long distance of 35 meters. These authors mention that this pattern is unusual for what is normally expected for cycads at this spatial scale, given their locally restricted seed and pollen dispersal (e.g., *Dioon merolae* and *D. planifolium,* Cabrera-Toledo et al. 2012, 2019, respectively; *Zamia furfuracea,* Octavio-Aguilar et al. 2017; *Zamia fairchildiana,* López-Gallego and O'Neil 2010), in which relatedness is expected from at least 5 meters. According to these authors, this pattern suggests that *D. caputoi* maintains relict dynamics with increasing loss of reproductive fitness in younger generations. Also, in general, adults tend to be less genetically related in many plants—not only cycads—due to a population "thinning process" (Luna et al. 2005; Zhou and Chen 2010), in which the high mortality of earlier life classes (saplings and juveniles) exacerbates genetic differences among adults. The longer distance relatedness in *D. caputoi* was similar to seedlings of *D. planifolium* growing in the same habitat within the Tehuacán Valley (Cabrera-Toledo et al. 2019).

As an aside, it is potentially significant that the Tehuacán Valley, the primary study area of Cabrera-Toledo et al. (2019), also contains a wealth of archaeological materials, including cycad remains (see Englehardt et al., this volume). Smith (1967:235) concludes on the basis of recovered archaeobotanical evidence that *D. edule* "had been fairly persistently used over some thousands of years" in this region. This region is also remarkable for its high biocultural diversity and for the strong and ancient human-plant interactions that have been documented there (Blancas et al. 2013; Casas et al. 2007; MacNeish 1992; Paredes-Flores et al. 2007). It is therefore tempting to hypothesize that populations of at least some plants were themselves planted by the precolumbian groups that inhabited this territory, especially given the ample ethnographic evidence for long-term human cycad use among Indigenous communities in Mexico (see, e.g., Bonta et al. 2019; Carbajal-Esquivel et al. 2012; Pérez-Farrera and Vovides 2006; Sifuentes de Ortiz 1983; Vite-Reyes et al. 2010; see also the chapters by Bonta and Calonje et al. in this volume) and worldwide (e.g., Beck 1992; Hayward and Kuwahara 2012; Kira and Miyoshi 2000; Smith 1982; Thieret 1958).

The Tomasellii clade, which contains four species, *D. stevensonii, D. tomasellii, D. sonorense,* and *D. vovidesii,* is the northernmost clade. All four have relatively high genetic diversity but are more structured (less gene flow), which may be the result of their harsh local habitats that limit seed germination or even dispersal, and also due to whole-plant extraction and notable habitat destruction (González-Astorga et al. 2008). The populations of *D. stevensonii* seem genetically related to *D. holmgrenii* from the Purpusii clade even when there are geographically closer *D. tomasellii* populations—but they are separated by the Trans-Mexican Volcanic Belt, an important barrier to seeds and pollen (González-Astorga et al. 2008).

New Genetic Data on *Dioon* at More Recent Time Scales

To complement and extend these previous genetic studies, we also undertook new analyses of *Dioon* populations in Mexico and Honduras, thereby adding a new data set, which we report on here. We were interested in understanding the extent of gene flow throughout the entire range of this genus within the region. We used population graphs to explore genetic interactions between populations of *Dioon,* and also to consider the possible movement of seeds by humans. As mentioned above, population graphs are somewhat distinct from most "traditional" approaches in population genetics, in that

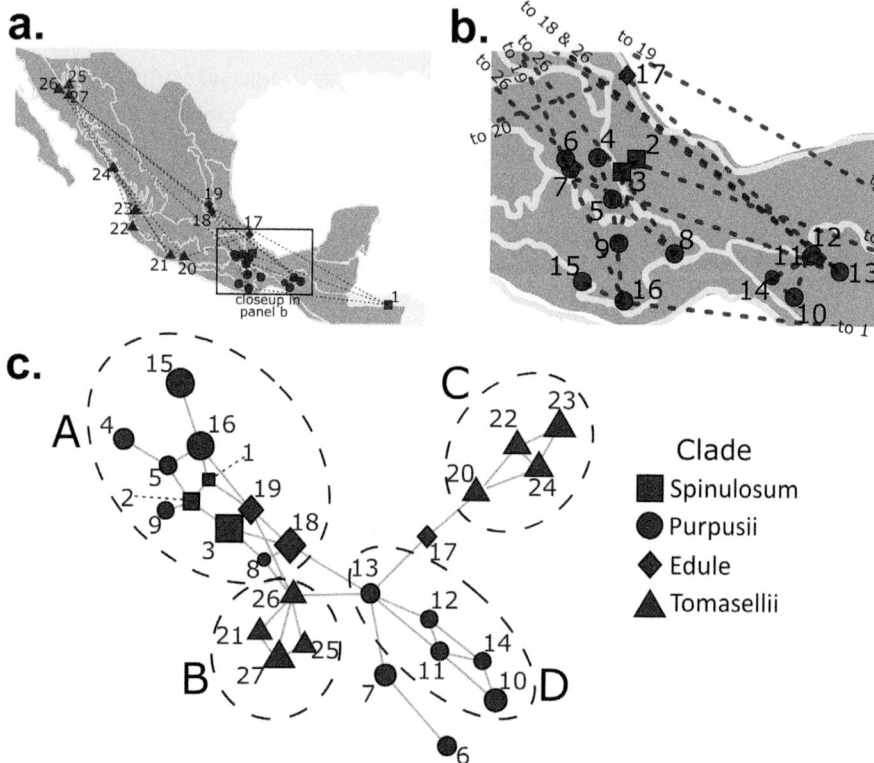

Figure 2.3. SSR analyses of *Dioon* cycads from throughout their range in Mexico and Honduras: *a* and *b*, map of the sampling localities (the dotted lines represent genetic connections based on the population graph, independent of the strength of the connection); *c*, population graph depicting genetic relationships among populations. Node size is proportional to the genetic diversity within the population (i.e., bigger nodes reflect more genetic diversity); edge length represents the genetic variation between populations: shorter edges indicate less variation among populations. The graph projection is based only on genetic data, so the position of each node does not reflect geographical distribution.

the extent of gene flow is measured in the context of the graph. This makes it possible to determine and visualize which populations are connected by migrants and which are not. We sampled and genotyped a total of 27 *Dioon* populations (345 total individuals; see Figure 2.3a and b) using 12 SSR markers. We include representatives from the entire range of distribution of *Dioon*, from 13 of the 17 recognized Mexican species. *D. planifolium* (clade undefined), *D. argenteum* (Purpusii clade), and *D. angustifolium* (Edule clade) are not present in this analysis due to issues with sampling limitations and/or inconsistency in the SSRs variation patterns, or its recent discovery in the case of *D. planifolium*.

We found that most individuals group into four subgraphs (Figure 2.3c): A: Oaxaca and Northeast, B: North *D. tomasellii*, C: South *D. tomasellii*, D: Chiapas. Many of the genetic connections are congruent with the distribution and the capacities of seed dispersers and pollinators mentioned previously, yet some are unexpected, such as the *D. edule* and *D. mejiae* connection that we discuss briefly below. All the subgraphs have a similar number of connections (ANOVA test, $p = .926$, no significant differences in the number of connections), which indicates that populations have similar levels of genetic flow within each subgraph. Given that most subgraphs have one or two connections from one subgraph to another, we can assume that gene flow is limited from one subgraph to another but is maintained within subgraphs. This pattern is congruent with what is known from previously reported data on most New World cycads (see, e.g., *D. merolae, D. caputoi, D. planifolium* [Cabrera-Toledo et al. 2012, 2019], *Zamia furfuracea* [Octavio-Aguilar et al. 2017], *Zamia fairchildiana* [López-Gallego and O'Neil 2010]).

The Oaxaca and Northeast (A) is the biggest subgraph with 11 populations, including five from the Spinulosum clade, two populations of the Edule clade, and all of the Purpusii clade except for the *D. merolae* populations from Chiapas and the two populations of *D. caputoi*. Two *D. caputoi* populations (6, 7), with only two genetic connections, and one population of *D. edule* from Veracruz (17) are isolated in the graph. These results are congruent with the genetic patterns of other *D. caputoi* populations reported by Cabrera-Toledo et al. (2008). However, it is possible that with a larger sample including populations from Veracruz of *D. edule* and other populations of *D. caputoi* in Puebla, these outliers would become part of other subgraphs or become new subgraphs.

The Chiapas subgraph (D) contains only populations of *D. merolae*. These populations are subject to relatively constant environmental pressure, such as fire and the exploitation of their leaves by humans (Pérez-Farrera and Vovides 2006). Surprisingly, perhaps, the population of *D. merolae* from Agua Prieta, Chiapas (13) is connected to the other three subgraphs, despite its geographic location. Although situated at the extreme southeastern limit of the *Dioon* species distribution in Mexico, it displays genetic connections with *D. caputoi* and *D. edule* populations in Querétaro and Veracruz, and with *D. vovidesii* populations in Sonora, which are approximately 900 and 2,200 km away, respectively.

The Tomasellii clade is divided into the subgraphs for the North *tomasellii* (B) and the South *tomasellii* (C) populations. In these cases, only physically close populations are genetically connected, which is consistent with previous correlations between the low vagility of pollinators and bottleneck events in *D. sonorense* and *D. vovidesii* (González-Astorga et al. 2008). The only outlier in these two subgraphs is the genetic connection of *D. stevensonii* populations in Michoacán (21) to populations of *D. vovidesii* in Sonora (26), more than 1,300 km to the northwest.

Interestingly, the population graph shows genetic flow among *D. mejiae* in Olancho, Honduras (1), the *D. edule* from northern Querétaro (19), and *D. holmgrenii* populations in southern Oaxaca (16). Such flow is consistent with the results of pairwise *F*st analysis (Figure 2.4), which revealed low genetic structure ($Fst_{[1-19]} = 0.34$, $Fst_{[1-16]} = 0.16$, $Fst_{[16-19]} = 0.19$). It is important to note, however, that $Fst_{[1-19]}$ is not significantly lower than the mean pairwise *F*st, indicating a general pattern of genetic similarity among all *Dioon* populations in the region. It is also significant that populations of *D. merolae* from Oaxaca appear to have more genetic flow with other populations in Oaxaca than with sister populations in Chiapas (Mean $Fst_{[\text{Oaxaca vs Chiapas}]} = 0.39$, Mean $Fst_{[\text{among populations of Oaxaca and Chiapas}]} = 0.21$). These results suggest the presence of barriers to dispersal of seeds and pollen, perhaps the Sierra Madre del Sur or simply the 150 km distance between the populations of Oaxaca and Chiapas. This would be similar to the pattern observed among the *D. edule* populations from northern Querétaro, which have genetic flow between them but not with the *D. edule* population in Veracruz, which is approximately 300 km away.

The long-distance genetic connections among some populations revealed by our analyses could be explained by a combination of recent diversification of the genus and human-induced dispersal. Although there is as yet no direct evidence, genetic or otherwise, of human-induced seed or plant exchanges in most of the regions mentioned above, ethnographic and archaeological evidence does suggest that the unexpected long-distance (approx. 700 km) connection between *D. mejiae* populations from Olancho in Honduras and populations from the Oaxaca and Northeast subgraph (Figure 2.3c[A]) could be due to human movements of either the seeds or the plants, or both. The connection between the *Dioon* population from Querétaro and Olancho is indeed intriguing given the known ethnobotanical similarities between both regions for these species (Bonta et al. 2006, 2019). We have previously speculated that it is possible that a small number of families migrated to Honduras in the fifteenth century and could have transported a limited but diverse suite of cultural beliefs and practices involving cycads (see also Bonta et al. 2019). They preserved these practices as part of an Indigenous group that became known as "Nahoa de Honduras" (Bonta 2009). Cycads in some places are conceived as maize ancestors, an idea that probably originated in the Huasteca region in Mexico and was spread southward by highly mobile

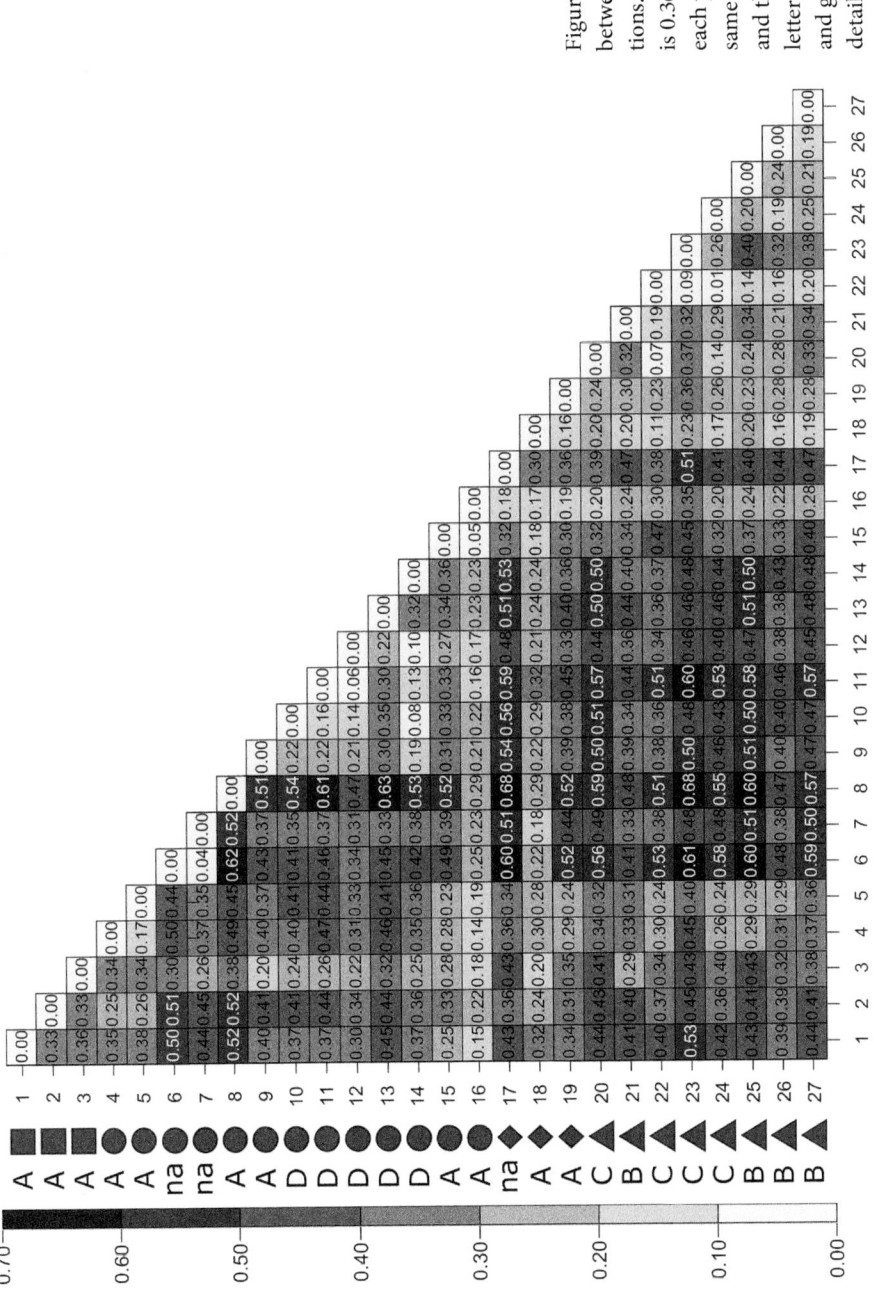

Figure 2.4. Pairwise *F*st between *Dioon* populations. Mean *F*st value is 0.36. The numbers of each population are the same as in Figure 2.3, and the shapes and the letters represent the clade and graph memberships detailed in Figure 2.3c.

Nahuatl-speaking people as far as northeastern Honduras (Bonta et al. 2019; see also the chapters by Bonta, this volume).

These cultural exchanges could very well have involved the movement of seeds and even whole plants, which would result in the population genetic connections observed in Figures 2.3 and 2.4. In various communities throughout Mexico, seeds and female cones are stocked in homes for several years and exchanged among members of local communities (Mark Bonta, personal communication; see also Vite-Reyes et al. 2010), and there is previous evidence of these practices having an impact on the demography of *Dioon* (Tristán 2012). Genetic connections among areas that are geographically isolated could be at least partially the result of human-mediated seed dispersal related to these cultural practices. Even today, seeds can be found in town markets such as Olanchito in Honduras and in the Huasteca region of San Luis Potosí, Querétaro, and Veracruz.

Other interesting evidence also points to cultural connections between human groups that continue to use cycads in Mexico and Honduras, such as the convergence in detoxification methods for seed starch (Figure 2.5), necessary for the consumption of potentially toxic cycads. Among communities in the Olancho, Yoro, and Colón Departments of Honduras, and others as far afield as Querétaro and the Huasteca region, cycad seeds are detoxified via a similar process: nixtamalization using the ash of various trees (Bonta et al. 2006; Chemín-Bassler 2000). Communities in these regions report that tree ash nixtamalization produces an ideal flour that better suits local culinary preferences and produces the best tamales, in spite of the fact that this method is distinct from the most common detoxification technique—nixtamalization with lime—used throughout Mexico and Honduras (Bonta et al. 2019).

As Englehardt et al. (this volume; see also Bonta et al. 2019) point out, there is also ample archaeobotanical evidence that suggests a deep history of cycad use, primarily as a foodstuff, in, for example, the Huasteca and adjacent regions. MacNeish's excavations revealed *Dioon* remains at various Archaic cave sites and rock shelters in both Tamaulipas and Puebla, with some cycad remains dating to as early as 4000–4500 BCE (MacNeish 1992; Smith 2005). Large *D. edule* populations still exist throughout this region into San Luis Potosí and Querétaro, some of which are quite close to MacNeish's excavated cave sites. It is tantalizing to consider that such sites may have served a similar function to the more recent cycad harvesting camps that we and others (see, e.g., Bonta et al. 2019) have documented in northeastern Mexico and in Honduras—and that were also mentioned briefly by MacNeish himself. Such a pattern of cycad exploitation in these contexts is

Figure 2.5. Preparation of cycad seeds for human consumption: *a*, removing the sarcotesta; *b*, nixtamalization of seeds using oak ash to remove toxins; *c*, seeds are washed and ground to form flour and masa; *d*, *Dioon* tamales covered with oak leaves (all photographs by Angélica Cibrián-Jaramillo).

consistent with data on cycad use among Indigenous communities in, for example, Japan or Australia (Beck 1992; Hayward and Kuwahara 2012; Kira and Miyoshi 2000; Smith 1982). Although our results cannot currently speak directly to these possibilities, they do suggest that future population genetic studies, combined with ethnographic investigations and archaeo-historical data, hold great potential for aiding the recognition, interpretation, and conservation of cycad cultures at the community level in many areas of the world. Such studies may allow us to achieve a more robust understanding of both cycad population genetics and structure, as well as the historical and modern relationships between these plants and humans.

Discussion and Conclusions

In this chapter, our principal aim was to illustrate how population genetics data can be profitably compared across disciplines to construct a deeper understanding of the potential impact of humans on the evolutionary history of cycads, as well as the reciprocal effects that cycads may have had on human culture. We suggest that previously understudied cycad-human interactions may have profoundly affected the observed genetic makeup and population structure of cycads, in both the context of ancient Mesoamerica and elsewhere.

In the first section of this chapter, we provided a brief introduction to the field of population genetics, outlining terms and methods commonly employed in this discipline. We then engaged in a very brief review of the population genetics data for cycads worldwide. Based on current scholarship, we suggest that it is probably too early to infer wider global trends or general genetic patterns for the Cycadales order as a whole. Rather, we hold that the majority of similarities in cycad population genetic patterns are more likely the result of recent diversification events common to many genera worldwide. Conversely, observed diversity may be attributed to differing habitat conditions, distinct local environments, and the unique biogeographic histories of discrete genera and species, among other factors. Of course, as we suggest here, global population genetic patterns may also be due in part to human actions (e.g., human-induced dispersal), the dynamics of which would also vary across specific regions. Yet these questions are rarely considered in the academic literature and thus continue to elude understanding.

To address these lacunae, and in light of these underreported aspects, we undertook a more detailed examination of the genetic patterns and basic biology of cycad species in Mexico, Honduras, and the Caribbean. Further, we generated *de novo* genetic data from SSRs for almost all *Dioon* species across the entirety of its distribution range in this region, with the goal of exploring genetic patterns in the more recent past, thereby facilitating the examination of the potential role of human-induced dispersal in the recent evolutionary history of cycads. Results of our analyses indicate relatively moderate genetic variation and widespread gene flow within the four primary phylogenetic clusters (clades) identified previously (Gutiérrez-Ortega et al. 2018). We should note that in our analyses, some species, such as *D. caputoi*, and some populations of *D. merolae*, did not group within the clusters we had expected. This incongruent data, however, may be due to incomplete sampling in our study, as mentioned above.

The most exciting and potentially significant conclusion suggested by our data is the genetic correspondence among discrete *Dioon* populations spread across extremely large geographic distances. Although this finding would benefit from confirmation by future research, we posit that such continuity may be directly related to human demographic movements. This suggestion is further bolstered by the wealth of evidence that strongly suggests the widespread sharing of cultural practices (such as the nixtamalization of cycad seeds; see Figure 2.5) and beliefs related to cycads, as well as parallels in linguistic terms for cycads among various Mesoamerican cultural groups, among others (see, e.g., Bonta et al. 2019; Carbajal-Esquivel et al. 2012; Englehardt et al. 2020; Vite-Reyes et al. 2010; see also the chapters by Bonta and Carrasco, this volume). A fuller interdisciplinary consideration of potential correlations between, for example, archaeological data that speak to human migrations in the recent past and these new genetic data may prove a fruitful avenue for future investigation in these and other contexts, as Englehardt et al., Pagán-Jiménez, and other contributors to this volume also suggest.

Some avenues of integrative research that we—in conjunction with many of the contributors to this collection—are currently pursuing, or consider to have particular potential, include a comparison of patterns of genetic divergence among cycads with processes of linguistic diversification. For example, given the genetic relatedness between *Dioon* populations in Honduras and Querétaro, San Luis Potosí, and the Huasteca, it is tempting to consider that perhaps these correlate with the diversification of the Nahuatl language (see, e.g., Campbell 1997; Kaufman 2001; Kaufman and Justeson 2009; Olko and Sullivan 2013). Genetic data provide information on the temporal contexts of divergence events, which may then be compared with glottochronological estimates of Nahuatl linguistic history and/or the spread of Nahuatl terms associated with cycads (e.g., teosinte; see Bonta, chapter 7, this volume). Strong correspondences between these data sets might suggest that the terms themselves were being shared and incorporated in cultural taxonomies (see Carrasco, this volume) as cycads were introduced to a region, which may be reflected genetically and suggest human-induced dispersal.

Likewise, comparing divergence in cycads against genetic data that speak to similar sub-speciation events in maize, or the introduction of maize to new areas following its domestication, may also yield potentially revelatory data. Since maize is a synthetic, artificially selected plant, its spread is almost entirely attributable to human action. If patterns of cycad genetic diversification parallel those of the emergence of new maize varieties in discrete contexts, it would suggest a human hand in the spread of both plants, albeit at different time scales, supporting the hypothesis that humans carried cycad seeds with them during archaeologically traceable migrations. Such demographic movements may also correspond to linguistic shifts (Kaufman 1976), both of which may be correlated temporally and spatially with the genetic data. This possibility is particularly intriguing given the conceptual links between cycads and maize evident in a number of Mesoamerican cultures (Bonta et al. 2019; Bonta, chapter 8, this volume).

Finally, and relatedly, it may be possible to compare genetic patterns in cycads to migration patterns in ancient Mesoamerica as revealed by human DNA. Although such studies are underway at, for example, the home institution of many of the authors of this chapter, they are in their infancy at this point. As more data on both cycad population genetics and from human DNA studies become available, it will become possible to further evaluate this suggestion. We would hope that human DNA data may also be comparatively evaluated against archaeological and linguistic evidence, but it may be that such information is really useful only at much deeper time scales. As we have learned through our collaborative efforts over the past seven years, the primary difficulty to advancing such conjunctive, interdisciplinary approaches lies in conceptualizing how our discrete data sets fit together—and this is an aspect on which we continue to labor. Yet once this groundwork has been laid, we can truly begin to design effective integrative research projects that answer the larger underlying questions about the relationships between cycad and human histories in Mexico.

In the case of ancient Mesoamerica, however, it certainly does seem likely that humans played a significant role in the recent evolutionary history of cycads, and that, likewise, cycads were incorporated into regional cultural practices at an early date. In this sense, human and cycad histories are intertwined in these contexts, at least in the recent past—and possibly even at deeper timescales. Indeed, the patterns of human-cycad interaction evident in Mesoamerica may find parallels in other parts of the world. Given the widespread use and significance of cycads in Indigenous cultures worldwide, from Australia, parts of Africa, the Pacific islands, to the Americas, it is tempting to hypothesize that cycad and human histories were intertwined in exceptionally early temporal contexts. To our minds, assuming that cycads were incorporated into human culture via multiple independent events seems a less parsimonious explanation. This speculative hypothesis may be tested via future interdisciplinary analyses that consider population genetics data for both humans and cycads and attempt to integrate such data with additional ethnographic and archaeological evidence to detect potential correlative patterns among the data sets.

On the basis of population genetic data from cycad species in Mexico and the Caribbean, we also conclude that it is perhaps too early to close the book on a single pattern in cycad population genetics. Although as with the global sample no definitive pattern emerges in the regional data, further research is necessary to more fully evaluate genetic patterns among Cycadales. As case studies accumulate, it is of course necessary to take into account the fact that the temporal scale and resolution of each study (recent past including humans or deeper past invoking climate change and biogeography) depends greatly on the type of molecular marker utilized to measure genetic variation and structure, as we outline above. This is a crucial consideration since temporal resolution directly affects the degree to which population genetics data may be correlated with other data sets. Moreover, each molecular marker offers its own advantages and disadvantages, yet the use of different markers between and among studies limits our ability to comparatively evaluate results, and thus to generalize genetic structure among various cycad species or genera. That said, the advantages offered by the specific SSR-based population genetics studies we realized on Mexican *Dioon* populations permit a more nuanced understanding of divergence and/or speciation events on much more recent time scales.

A common, although perhaps understandable, misconception regarding population genetics, and genetic data in general, is that such evidence is useful only at extremely large time scales, on the order of hundreds of thousands or millions of years—particularly for long-lived species, like cycads, with a deep and complex evolutionary history—and therefore not entirely applicable to investigations focused on more recent events. Nonetheless, as our study demonstrates, certain methods allow for finer resolutions at recent time depth and are thus capable of producing data that are ripe for comparative and integrative analyses. In the end, however, despite the somewhat idiosyncratic nature of individual population genetics studies, as many chapters in this volume, and indeed, the collection as a whole, argue, there is an urgent common need for further research. We concur wholeheartedly with our fellow contributors that conjunctive approaches hold great potential for producing new data that may be compared across disciplines—archaeology, biology, botany, and population genetics, to name but a few. Such integrative analyses are absolutely fundamental to crafting a more complete and nuanced understanding of the evolutionary history of this unique plant, as well as conceptualizing and discerning the importance of cycads in human culture and history and the mutual interrelationships between humans and cycads, in ancient Mesoamerica and beyond.

3

Zamia in the Insular Caribbean

New Insights into the Historical Ecology of an Ancient Wild Food Plant

JAIME R. PAGÁN-JIMÉNEZ

Recently, several wild plant species locally known as *marunguey, guáyiga, coontie,* or simply "zamia" in the Greater Antilles have recaptured the attention of the public in Puerto Rico and nearby islands. The fascinating history of the *Zamia* L. genus includes the fact that these plants—dangerously toxic if not processed correctly, although in some cases also highly nutritious—were prepared and consumed by Indigenous Caribbeans both before and after the Spanish conquest in the 1490s. A number of stories on the role of zamia as a foodstuff are still known and locally shared in Puerto Rico, the Dominican Republic, Cuba, and the Bahamas, although its historical relevance in Caribbean foodways has only recently begun to be recognized, primarily via paleoethnobotanical research (Del Valle 2017). *Zamia* is the cycad genus with the widest distribution in the Neotropics, and its species are found from Florida to Bolivia (Norstog and Nicholls 1997; see also Calonje et al. this volume). In the Greater Antilles, it is represented by the *Zamia pumila* complex (Pagán-Jiménez and Lazcano-Lara 2013), a distinct and monophyletic group (Caputo et al. 2004; Norstog and Nicholls 1997) ranging from one to nine species, depending on the preferred taxonomic approach (Calonje et al. 2022; González-Géigel 2003; Meerow et al. 2012; Stevenson 1987). There is no record of the natural distribution of this genus throughout the Lesser Antilles (Stevenson et al. 2003).

Grounded in the early Spanish accounts produced in Hispaniola (today's Haiti and Dominican Republic) and on historic sources of this and other islands, several ethnobotanical works have explored the economic importance that zamia likely had for the ancient Indigenous peoples of the Greater Antilles (de Boyrie Moya et al. 1957; Sturtevant 1969; Veloz Maggiolo 1973,

1992). The oldest mention of these plants was registered at the dawn of the Spanish colonization of the New World. It was early in the sixteenth century when Fray Bartolomé de las Casas first described how the inhabitants of the Amerindian cacicazgo of Higüey in the eastern part of Hispaniola made bread of guáyiga with the dough prepared from zamia root:

> The bread is made thus, convenient to know, that on rough stones such as *rallo* [a stone grater], you grate them as you would grate a turnip or a carrot on a grater from Castille [in Spain]. This results in a white dough that is then formed into globes or round buns, as big as a *bola* [or ball], which are put out in sunlight until they attain the color of bran; if left out in sunlight one and two and three days at which time they will be swollen with maggots [larvae] as if it were rotten meat, and turn as black as soot, or a washed out black more brownish; once they reach this condition, black and boiling with maggots as fat as pine nuts, shape them into flat cakes, that are like dough already as to the whiteness and toughness, like our wheat, and in a hot clay pot that is already on rocks with fire beneath it, you place the cakes to cook on one side and then the other, where simmering with the heat, the maggots fry and die, and thus are cooked. (de las Casas 1876: 261–262, translation by the author)

According to de las Casas (1876), guáyiga plants (Figure 3.1) "somewhat resemble the dwarf fan-palms of Andalucia, although they are narrower, and smoother and more delicate." The cycad described by de las Casas in eastern Hispaniola is *Zamia pumila,* the species with the widest distribution in the Greater Antilles (Lazcano-Lara 2015). The Indigenous inhabitants of this region simply collected the zamia they needed to make guáyiga from the grated dough. De las Casas also noted that in Higüey, guáyiga was even more important than maize (*Zea mays*), sweet potato (*Ipomoea batatas*), or manioc (*Manihot esculenta* Crantz) and its derived cassava bread (Veloz Maggiolo 1992).

Other interesting descriptions of the use of zamia were registered almost 270 years after the Spanish invasion by Fray Iñigo Abbad y Lasierra (1866) in southern Puerto Rico. As in the case of eastern Hispaniola, "bread" (loaves) prepared in Puerto Rico by creoles was also made with the dough from the processed underground stems of two species: *Z. pumila* and *Z. portoricensis*:

> of the root, which is like a sweet potato, [they] make bread in this way: they grate the roots [the tuberous stem] until they are well shredded, then they pile them up until they rot, breed maggots, and dry up; then

it looks like a bunch of dark red mud, being dried, they grind it to powder, which is made into buns that is relied upon for lack of maize, plantains or manioc in the time of hurricanes [. . .] This relief is very damaging to them [. . .] the years when this kind of bread is used many are killed by this accident [*referring to poisoning from the marunguey*]. (Abbad y Lasierra 1866:252, translation and comment by the author)

The historic and ethnographic records on the processing of underground stems of zamia in Hispaniola, Puerto Rico, and other islands of the Greater Antilles suggest that zamia was a regular component of late Amerindian and new Creole diets. However, processing these underground stems to render them edible is complex (Pagán-Jiménez 2007; Veloz Maggiolo 1992). *Zamia* and other genera of the Zamiaceae and Cycadaceae families produce a potent glycoside known as cycasin, which can result in severe neurological problems and even death after long-term consumption. The traditional techniques documented in the Caribbean to remove this phytotoxin are noticeably different from those used in continental American environments to eliminate this and other poisonous toxins (e.g., cyanide glycosides) from tubers, seeds, and trunks of plants (such as manioc), many of which are still used as foodstuffs in the region (Bonta et al. 2019; Dickau et al. 2007; Forno 1967). The evidence detailed above suggests that the culinary tradition

Figure 3.1. *Zamia* morphology: *a*, *Zamia pumila* growing in a volcanic substrate between the municipalities of Ponce and Peñuelas, south-central Puerto Rico; *b*, bifurcated underground stem and male cones of *Zamia erosa* naturally growing in limestone hills at the municipality of Manatí, northern Puerto Rico; *c*, underground stem of a *Zamia portoricensis* specimen from the municipality of Guánica, southwestern Puerto Rico (images by the author).

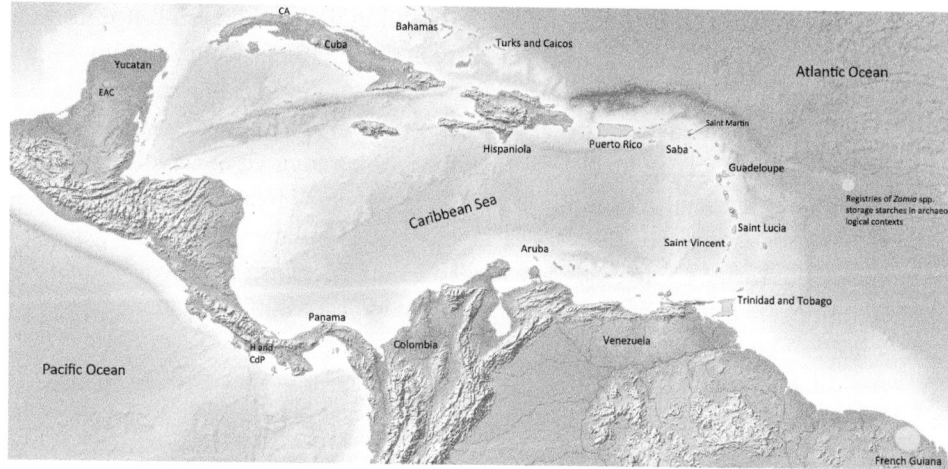

Figure 3.2. Territories of the Greater Caribbean where ancient *Zamia* spp. starches have been recovered from archaeological materials. Zamia starch grains have also been recovered from archaeological contexts in sites in northern Cuba (Canímar Abajo; Chinique de Armas et al. 2015), the Yucatán Peninsula (Escalera al Cielo; Simms 2014) and Panamá (Hornitos and Casita de Piedra; Dickau et al. 2007) (base map from Natural Earth; free vector and raster map data @naturalearthdata.com; geographic and paleoethnobotanical data by the author).

involving zamia's underground stems in the Caribbean is both ancient and unique (Pagán-Jiménez et al. 2005, 2015; Veloz Maggiolo 1992).

Recent paleoethnobotanical research in Puerto Rico (including Vieques Island) and at other sites throughout the Caribbean and northeastern South America (Figure 3.2) suggests that the precolonial inhabitants of these territories consistently accessed, processed, and produced meals from the underground stems of zamia (Mickleburgh and Pagán-Jiménez 2012; Pagán-Jiménez 2007, 2011a, 2013; Pagán-Jiménez et al. 2005, 2015, 2019; Pagán-Jiménez and Mickleburgh 2015). Documented evidence has confirmed the wide regional distribution of this culinary tradition in precolonial contexts, even in places outside the natural range of the genus. Ancient starch grains stored in the underground stems of zamia have been identified in a diverse array of food processing and cooking tools (lithics, grinding stones, ceramic pots, and griddles), as well as in human dental calculus. This evidence ranges in date from ca. 5800 BC to AD 1450.

This chapter first briefly examines current methodological approaches to starch grains that permit inferences regarding the ancient uses of zamia's underground stems in the Caribbean. A paleoethnobotanical survey then illustrates the most salient outcomes documented to date. Data obtained serve to delineate new research avenues to investigate the historical uses of zamia, as well as its natural—and perhaps cultural—distribution range in the Caribbean and beyond. Selected ethnohistoric and ethnographic information about this genus in the Caribbean is also presented to highlight the utility of this kind of data, which, in concert with paleoethnobotanical and other approaches, may elucidate research problems focused on zamia's historic phytogeography and more relevant contemporary issues of the region, such as food security. This work is expected to stimulate new multiproxy research toward a nuanced understanding of the deep historical ecology of both zamia and the ancient and modern human populations that made and still make use of this wild food plant in the Americas.

Identifying Caribbean *Zamia* through Starch Grains in the Archaeological Record

Starch grains have proven essential for the archaeobotanical identification of plants used in ancient human subsistence (Piperno 2006a). Starch is an insoluble carbohydrate formed during photosynthesis after the polymerization of several glucose residues (Bello and Paredes 1999). During this process, transitory and storage starches develop as semicrystalline structures made up of two polysaccharides (amylose and amylopectin) in certain plant organs such as leaves, stems, seeds, and roots. Transitory starches made in the chloroplasts and chromoplasts of leaves are quickly degraded by plants (in less than 24 hours) to sustain key organic processes, such as metabolism and energy production (Pfister and Zeeman 2016). In contrast, storage starches are made in the amyloplasts of seeds, rhizomes, and tubers. They are produced for similar purposes and are generally synthesized by plants only in response to environmental stresses or to produce the energy needed to germinate. The morphology, size, chemical composition, and basic structure of starches are characteristic of each species, and their particular shapes likely depend on the amount of amylose they contain (Buléon et al. 1998). Starch grains grow from the accumulation of layers of amylose and amylopectin around a nuclear point called the hilum, the "center" of the starch (Gott et al. 2006; see Figure 3.3). These layers may be evident in many starches, although their visibility varies according to the source plant, the source organ, and their morphology.

Paleoethnobotanical studies grounded in these microbotanical residues have focused mainly on the analysis of storage starches, since the organs where these are found (seeds, rhizomes, and tubers) are those used primarily as human food sources. Transitory starches found on the leaves of many plants have also been considered in this type of study to confirm their

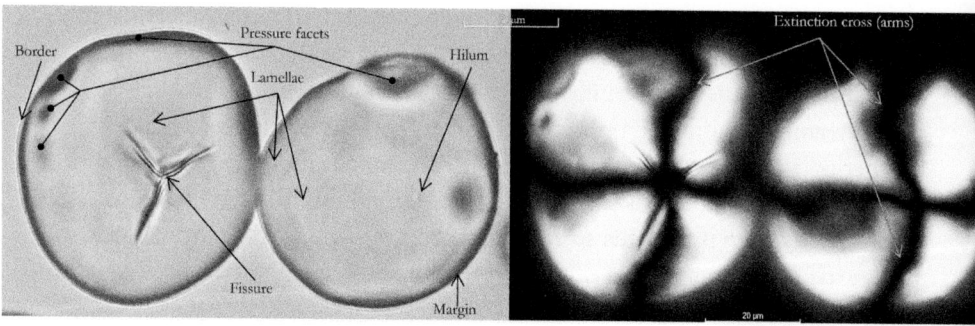

Figure 3.3. Modern oval starch grains from the underground stem of a marunguey (*Zamia pumila* L.) specimen from Puerto Rico. Morphometric features used in paleoethnobotanical starch grain analysis are indicated by arrows. Both images were taken with a polarized optical microscope in bright (*left*) and dark field (*right*). The maximum length of the starch to the left is 57 μm (micrographs by the author).

documented differences from storage starches. Morphometrically speaking, transitory starches lack features sufficient to allow reliable taxonomic ascription.

Studies have established that plant starches may be trapped and preserved for millennia in the fissures and pores of lithic, coral, ceramic, and shell tools used in plant processing and cooking, as well as in human dental calculus and coprolites (Loy et al. 1992; Mickleburgh and Pagán-Jiménez 2012; Pagán-Jiménez 2007; Piperno et al. 2009; Torrence and Barton 2006). Although it is difficult—if not impossible—to produce case by case explanations of how starch grains survive in archaeological contexts, it has been shown that a number of them may survive many physical, chemical, and biological degrading pathways (Babot 2003, 2006; Crowther 2012; Hardy et al. 2009; Pagán-Jiménez et al. 2017; Zarrillo et al. 2008). For instance, experimental studies of modern starches submitted to two of the most aggressive environments for their preservation—enzymatic digestion (e.g., α-amylase) and heat—have shown that an appreciable amount of starches may survive with good structural integrity, probably due to a combination of both the heterogeneity of the degrading agents (e.g., the nature and differential behavior of organic molecules and of the thermal transference respectively) and the highly variable environment of the matrices of the archaeological objects in which starchy organs were processed (Crowther 2012; Hardy et al. 2009; Henry et al. 2009; Pagán-Jiménez et al. 2017).

Differentiating between potential sources of starch grains is key to paleoethnobotanical analysis, since ancient cultures had access to a variety of plant resources. To do so, researchers usually rely on comprehensive reference collections of starches derived from known economic and wild plants in a given region; larger reference collections yield a higher probability of producing taxonomic matches between ancient and modern starches. Modern starches must first be characterized via morphometric criteria in order to determine diagnostic traits within taxonomic levels of interest (family, genera, and species). Although not all plants produce morphometrically diagnostic starch grains, starch morphometry generally permits reliable taxonomic correlations between starches and source plants, usually to the family or genus levels. In fewer cases, starches possess unique morphometric traits that allow for characterization at the species and even to race/variety levels (Pagán-Jiménez 2007, 2015; Piperno 2009; Reichert 1913).

The reference collection employed for analyzing archaeological materials in which zamia starch grains have been found in the Caribbean islands and northeastern South America includes more than 140 specimens comprising at least 68 genera and 90 species, mostly from the Neotropics, but also from the Old World tropics (Pagán-Jiménez 2007, 2015). Underground and seed starches from three zamia species naturally distributed in several islands of the Greater Antilles (*Z. pumila, Z. erosa,* and *Z. portoricensis*) form part of this reference collection. Consultation of additional published literature on diagnostic criteria for modern and ancient starch grains, as well as the results of experimental analyses, add further detail to identifications of starch grains in archaeological materials (Babot 2003; Henry et al. 2009; Mickleburgh and Pagán-Jiménez 2012; Pagán-Jiménez et al. 2017).

Morphometric Characteristics of Modern Zamia Starch Grains

Storage starch grains of the underground stems of three Caribbean species of zamia (*Z. pumila, Z. erosa,* and *Z. portoricensis*), as well as of other species from Cuba (Roberto Rodríguez, personal communication, 2007), are thus far the only known starches in the insular Caribbean that alternate between oval, circular, truncated, to polygonal shapes ranging in size from 1 to 95 μm (Table 3.1). This size range is among the largest yet documented in key reference collections of Neotropical plants (Dickau 2005; Pagán-Jiménez 2005; Perry 2001; Piperno and Holst 1998; Zarrillo 2012). Zamia starch grains also display other distinctive features, such as marked symmetric rings (or lamellae), radial and asymmetrical fissures, and marked distal pressure facets ranging from one to eight (Figure 3.4). The Maltese cross, or extinction cross projected by many of these starches is quite variable, though consistent within each species.[1] For example, storage starches of *Z. pumila* produce crosses mostly with marked wavy arms (Figure 3.4a1); *Z. erosa* reflects crosses with lightly curved wavy arms (Figure 3.4b1), and *Z. portoricensis* produces crosses mainly with straight, sometimes lightly curved arms (Figure 3.4c1).

Table 3.1. Main morphometric features of storage starch grains from the underground stem of three Caribbean *Zamia* species

Taxa	Size range (μm)	Mean (μm) (±SD)	No. of meas.	General shapes (bi-dimensional)	Lamellae	Fissures	Extinction cross
Zamia pumila	6–95	30 (±16)	110	Mostly truncated, circular/oval. Few polyhedral.	Both symmetric and asymmetric circles. Very few show concentric rings. Undetected in 46%.	Mostly radial fissures, stretched "m" and transverse line. A few project "Y" shapes, bifurcated lines, and "+" shapes.	Mostly cross shapes with wavy arms. Few cross shapes with straight and/or curved arms.
Zamia erosa	1–83	18 (±13.5)	103	Mostly circular, truncated and oval. Few ovate and polyhedral.	Mostly symmetric circles. Very few show dashed circles and concentric rings. Undetected in 52.3%.	Mostly transverse line, radial fissure and circular hollow. A few show "Y" shapes and "+" shapes while very few project bifurcated lines, triangular and "T" shapes or a stretched "m."	Mostly cross shapes with curved /lightly wavy arms. Few cross shapes with straight and fully wavy arms.
Zamia portoricensis	5–50	20 (±9.9)	108	Mostly circular/oval and truncated (regular, elongated, and bell-shaped). Few polyhedral including a quadrangular (elbow-like) shape.	Mostly symmetric circles. Very few show concentric rings. Undetected in 69.8%.	Fissures are scarce in this species. When registered, they are mostly radial fissures, and bifurcated and/or transverse lines. Circular hollows and triangular fissures can occur.	Mostly cross shapes with curved/lightly wavy arms. Few cross shapes with straight or fully wavy arms.

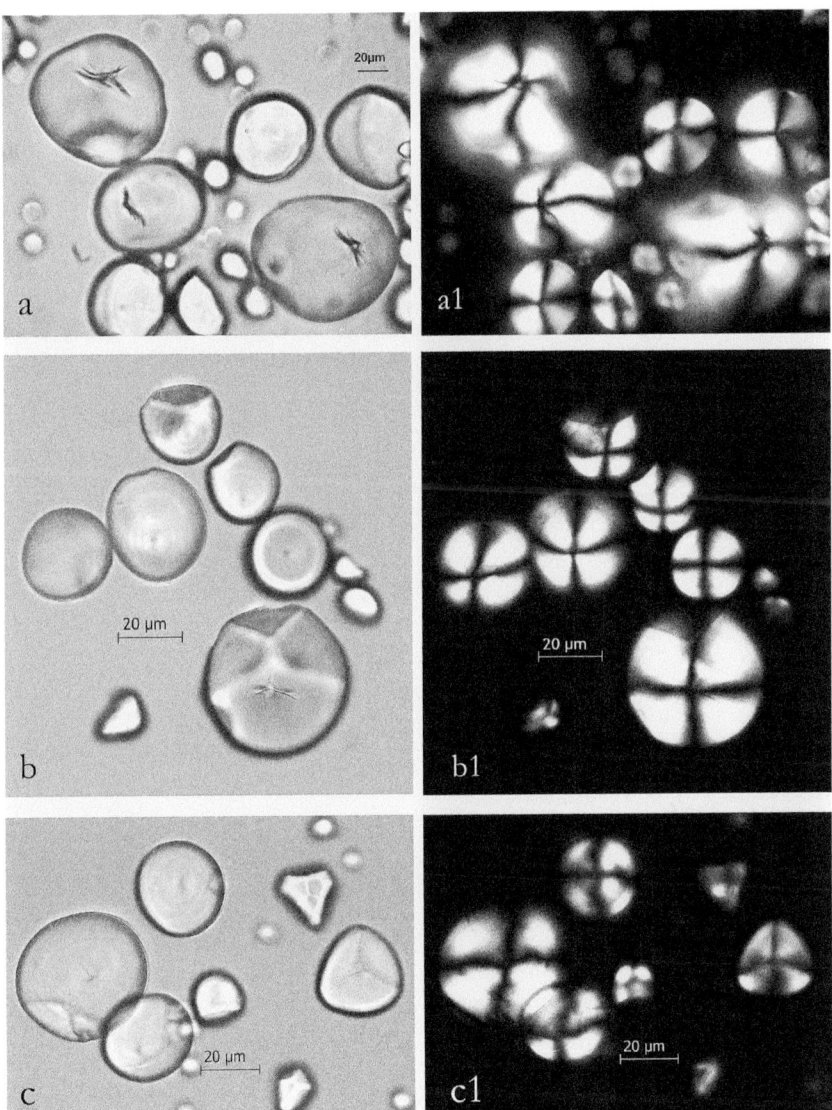

Figure 3.4. Selected modern starch grains of *Zamia pumila* (*a* and *a1*), *Z. erosa* (*b* and *b1*), and *Z. portoricensis* (*c* and *c1*) showing some of their main morphometric features. The images to the right display the same starches shown on the left, but against a dark field and with polarized light showcasing the typical extinction crosses of the three species (images by the author).

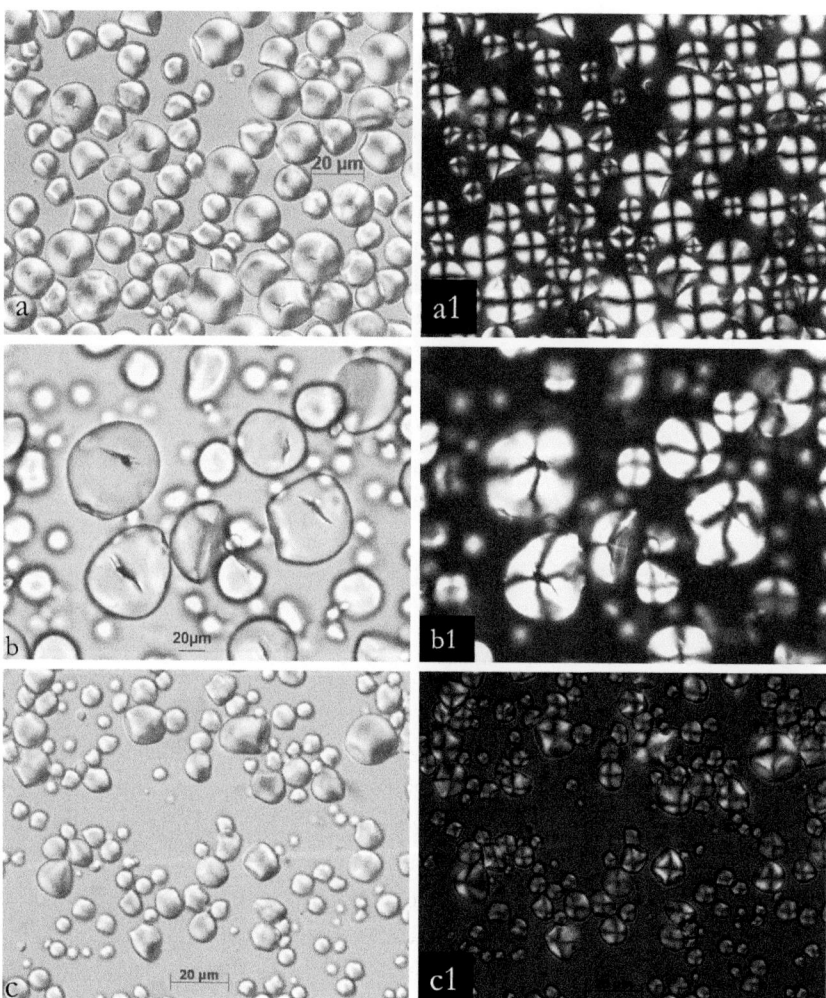

Figure 3.5. Modern starch grains from manioc (*a* and *a1*), *Zamia pumila* (*b* and *b1*), and sweet potato (*c* and *c1*). To the right, the same starches with cross-polarized light showcasing typical extinction crosses of the referred species (images by the author).

The mean size and standard deviation of zamia storage starches may sometimes overlap typical size ranges and some general shapes of storage starches from other important plants such as manioc and sweet potato (Figure 3.5; see also Pagán-Jiménez 2015); however, these morphometric features are not the only ones used to differentiate between native starches from different plant species. Many similar traits might be shared by zamia and other starchy plants, but it is the combination of key traits (Table 3.1) and the size ranges in which these occur that reliably allow distinguishing zamia storage starches.

Table 3.1 shows that zamia starches from underground stems could reach maximum sizes of up to 50, 83, and 95 μm, respectively. My own experience with these and other starches provides some relevant observations that may impact ancient starch recovery and identification. For example, basic alteration assays applied to starches of other plants such as maize and manioc have revealed that their bigger, "mature" starches tend to survive at notably higher rates and with greater structural integrity than their smaller, "immature" counterparts when subjected to different biological and mechanical degradation processes. This was the case when native maize and manioc starches were exposed to human saliva under controlled environmental conditions for seven days. In both cases, many of the smaller starch grains were completely digested during the first 72 hours by active bacteria and salivary α-amylase, while after 168 hours the surviving starches were mainly the bigger ones. The same general trend was also observed when cooked manioc flatbread was submitted to identical experimental conditions. In this case, starches of all sizes previously altered by heat (due to cooking) in a partially dry environment were much more fragile than native starches, most of which were digested rather quickly during the first four hours. However, bigger and heat-altered manioc starches (although still retaining diagnostic features) survived in notably higher numbers than smaller starches at the end of the assay. Of course, there are many other variables that could produce results distinct from these, but the key inference suggested by these qualitative observations is that bigger starches within a species seem to have greater survivability and preservational longevity. This is borne out by the evidence. Indeed, archaeological starch grains recovered at many Caribbean and other Neotropical sites clearly show that the bigger or "mature" starches from different plant sources are those that consistently survive ancient degradation processes, as well as the taphonomic processes associated with buried contexts (see, e.g., Berman and Pearsall 2008; Mickleburgh and Pagán-Jiménez 2012; Pagán-Jiménez et al. 2005, 2015; Pearsall et al. 2004; Perry 2005; Piperno et al. 2000, 2009; Piperno and Holst 1998; Zarrillo et al. 2008).

Evidence of Ancient *Zamia* Use in the Insular Caribbean and Northeastern South America

Most of the recovered ancient starch grains reliably ascribed to the genus *Zamia* in the Caribbean coincide with the pattern mentioned above; larger, mature storage starch grains have survived and thus facilitate identification as *Zamia*. Identifications of zamia starches have been based on the diagnostic features described above. Specific morphometric features used for this process are shape, size, presence and location of the hilum, presence and appearance of fissures, presence and type of pressure facets, presence and appearance of lamellae, and the appearance and differential projection of the extinction cross (Table 3.1). If recovered starches do not display known distinguishing features of modern *Zamia* specimens, identifications to the genus or species level are tentative. In such cases, "cf." is placed before the genus or species name to describe the recovered starches.

Differentiation among *Zamia* starches relies on both their currently known geographic distribution at regional (inter-island) and local (intra-island) levels, and specific morphometric data registered in recovered starches. At present, a number of wild populations of zamia are known in Cuba, the Dominican Republic, Jamaica, Puerto Rico, the Bahamas, and the Cayman Islands. However, as mentioned above there is no record of zamia's natural distribution throughout the Lesser Antilles, including Trinidad and Tobago and Aruba, and French Guiana in northeastern South America at any point in the past or present. When ancient zamia starches are recovered in areas outside their known natural distribution range, I consider two possible scenarios:

(a) Zamia was originally present and then eradicated in these regions; or
(b) Zamia starchy derivatives (or even whole plants) were acquired as exchange items from regions where this genus is still present.

Therefore, most ancient zamia starch identifications for Lesser Antillean or northeastern South American sites have not reached a species level resolution due to lack of evidence of its past and/or present natural distribution in these areas. These starches, however, fall within the accepted size range of studied modern species from the Greater Antilles, and also display distal pressure facets, marked lamellae, and/or extinction crosses typical of these modern species.

Storage starches tentatively or securely ascribed to *Zamia* spp. have been recovered in archaeological tools or human remains from 29 sites of the Caribbean and northeastern South America (French Guiana) (Figures 3.2 and 3.6, Table 3.2). Overall, these findings comprise around 7,800 years of continuous exploitation and use of zamia in a region characterized by high biodiversity and cultural heterogeneity.[2] Although zamia starches have been identified in many object types associated with different processing stages of the underground stem to render it edible, these conclusions were reached only after data from discrete chrono-cultural and geographical contexts had been unified, because of the rather opportunistic nature of many previous

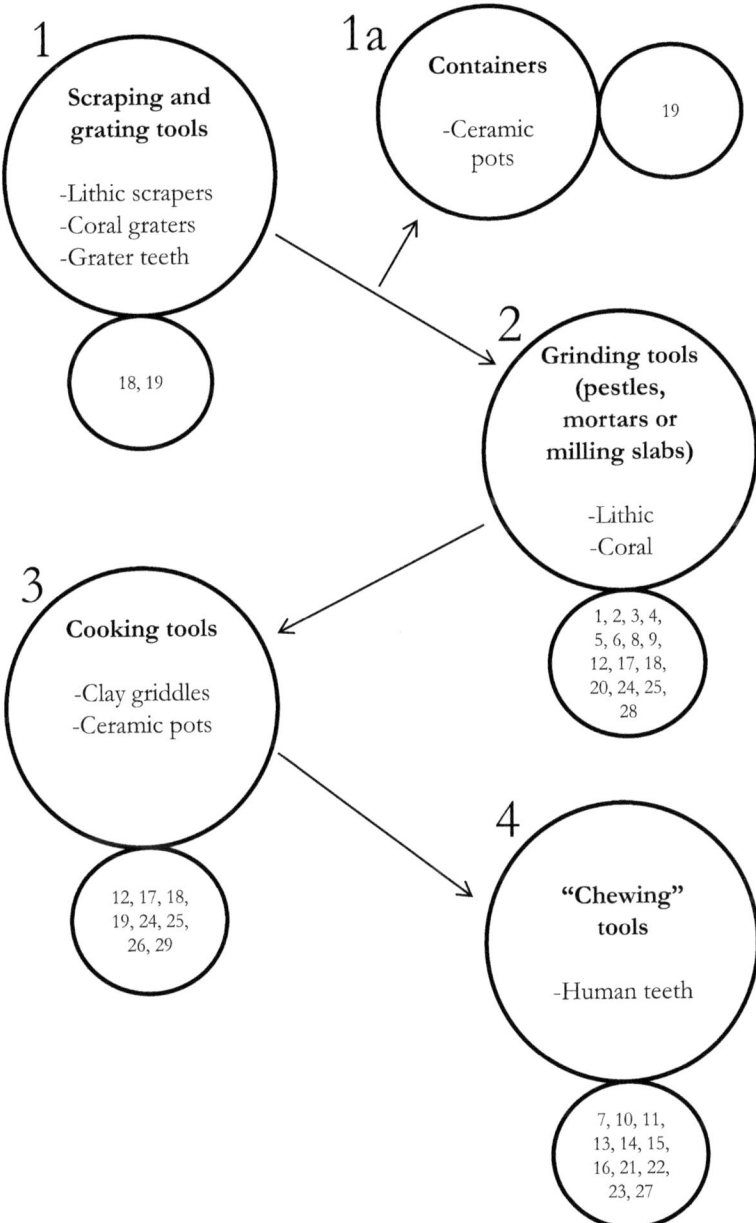

Figure 3.6. Presumed processing sequence of zamia underground stems and associated tools on which tentatively and securely identified zamia starches have been identified. The small numbers inside circles refer to the sites shown in Table 3.2 (image by the author).

Table 3.2. Sites where tentatively and securely identified zamia starches have been identified

Site name	Country	Chronological placement	Reference
1. Saint John	Trinidad and Tobago	5840–4450 BC	Pagán-Jiménez et al. 2015
2. Eva 2	French Guiana	4150–1970 BC	Pagán-Jiménez et al. 2015
3. Maruca	Puerto Rico	2900–1100 BC	Pagán-Jiménez et al. 2005; unpublished lab notes
4. Cueva Ventana	Puerto Rico	2430–1460 BC	Pagán-Jiménez et al. 2019
5. Puerto Ferro	Vieques (P. Rico)	2380–290 BC	Pagán-Jiménez et al. 2005; unpublished lab notes
6. Plum Piece	Saba	1380–1020 BC	Pagán-Jiménez et al. 2019
7. Canashito	Aruba	350 BC–AD 150	Mickleburgh and Pagán-Jiménez 2012
8. Punta Candelero	Puerto Rico	320 BC–AD 220	Pagán-Jiménez 2007
9. La Hueca	Vieques (P. Rico)	160 BC–AD 540	Pagán-Jiménez 2007
10. Maisabel	Puerto Rico	AD 250–1170	Mickleburgh and Pagán-Jiménez 2012; Pagán-Jiménez and Mickleburgh 2015
11. Argyle 2	Saint Vincent	After AD 400	Pagán-Jiménez and Mickleburgh 2015
12. Arecibo 39	Puerto Rico	AD 400–700	Pagán-Jiménez 2008
13. Manzanilla	Trinidad and Tobago	AD 400–1400	Mickleburgh and Pagán-Jiménez 2012, Pagán-Jiménez and Mickleburgh 2015
14. Anse à la Gourde	Guadeloupe	AD 500–1350	Mickleburgh and Pagán-Jiménez 2012, Pagán-Jiménez and Mickleburgh 2015
15. El Cabo	Dominican Republic	AD 600–1504	Mickleburgh and Pagán-Jiménez 2012
16. Punta Macao	Dominican Republic	AD 600–1600	Mickleburgh and Pagán-Jiménez 2012
17. King's Helmet	Puerto Rico	AD 650–780	Pagán-Jiménez 2011b
18. Punta Candelero	Puerto Rico	AD 650–1020	Pagán-Jiménez 2009
19. Playa Grande	Dominican Republic	AD 730–1680	Pagán-Jiménez 2012
20. Cueva de los Muertos	Puerto Rico	AD 850–1300	Pagán-Jiménez and Oliver 2008
21. Juan Dolio	Dominican Republic	AD 900–1500	Mickleburgh and Pagán-Jiménez 2012

Site name	Country	Chronological placement	Reference
22. Tanki Flip	Aruba	AD 950–1400	Mickleburgh and Pagán-Jiménez 2012
23. Lavoutte	Saint Lucia	AD 1000–1500	Mickleburgh and Pagán-Jiménez 2012
24. Ceiba 11	Puerto Rico	AD 1030–1270	Pagán-Jiménez 2011c
25. Edilio Cruz	Dominican Republic	AD 1160–1420	Pagán-Jiménez 2011 in Ulloa Hung 2013
26. Laguna de Limones	Cuba	AD 1200–1600	Rodríguez-Suárez and Pagán-Jiménez 2008
27. Chorro de Maíta	Cuba	AD 1250–1600	Mickleburgh and Pagán-Jiménez 2012, Pagán-Jiménez and Mickleburgh 2015
28. Utuado 27	Puerto Rico	AD 1280–1430	Pagán-Jiménez and Oliver 2008
29. Ceiba 33	Puerto Rico	AD 1410–1470	Pagán-Jiménez 2011c

Note: The site numbers are referenced inside the smaller circles in Figure 3.6.

studies, in which typically small archaeological assemblages were analyzed at each of these sites. So far, it has been impossible to study archaeological assemblages from individual sites that represent the total diversity of tools and objects associated with the presumed stages of processing, cooking, and consumption of zamia. This is a crucial task that can facilitate a better understanding of potential variability in culinary practices associated with zamia at both local and regional levels.

The oldest tentative and secure zamia storage starches have been registered at the Saint John and Eva 2 archaeological sites in Trinidad and Tobago and French Guiana, respectively (Figure 3.6 and Table 3.2). The recovery contexts were grinding tools consisting of milling/pounding slabs and conical, multifaceted pestles that also yielded starches of other important domestic and wild plants such as maize, sweet potato, chili pepper, and wild yam (Pagán-Jiménez et al. 2015). The overall size of zamia starches ranges from 27 to 42.3 mm with a mean size of 35.8 mm (±4.77 mm), and their shapes are mostly oval or truncated in centric view (Figure 3.7a–c). Some of these starches possess asymmetrical or deep transversal fissures at the hilum area. Lamellae are formed by concentric, mostly undulated rings, while the extinction cross shows both straight and wavy arms, sometimes combined in a single starch. In Saint John, for example, one of the starches recovered in a multifaceted pestle and tentatively ascribed to the *Zamia* genus is circular in centric view and lightly oval after rotation; however, its size, lamellae, and straight extinction cross are consistent with those observed in starches produced mainly by *Z. portoricensis* and *Z. erosa*. Another zamia starch grain recovered in a milling slab of Eva 2 (Figure 3.7b–b1) is diagnostic at least to the genus level. It has a truncated shape that is common in modern zamia and also projects a wavy extinction cross, a characteristic typical of modern starch assemblages of the three species mentioned above, but most common in *Z. pumila*. A large, circular zamia starch with a similar size and extinction cross was recovered from a milling slab at Eva 2. It was registered together with smaller, unidentified starches (Figure 3.7c) that could not be assigned to *Zamia* spp. This pattern is commonly seen in large modern zamia starches, which are often surrounded by many smaller "immature" grains.

At these particular sites, zamia identifications could not be assigned to species level because of the lack of historic and modern botanical reference data on this genus from the southern Caribbean and northeastern South America. However, these findings suggest that this genus was accessed by unknown means by the ancient inhabitants of these sites. Two processing possibilities emerge when grinding tools are seen from their presumed functional utility: (a) underground stems might have been pounded to make a paste or dough, or (b) dough previously produced by unknown means was ground while still fresh or previously dried. Unfortunately, no other plant processing tools from these sites were analyzed to better understand the steps by which zamia derivatives were processed. Saint John is, together with Banwari Trace (both in Trinidad and Tobago), the earliest archaeological site so far known in the Caribbean region. Advanced cooking technologies such as the use of ceramic tools were not registered there; however, hundreds of unique clusters of fired-altered rocks interpreted as cooking pits (van den Bel et al. 2018) were identified in contexts associated with studied tools at Eva 2. It was only in later occupation phases of this site (between 2200 and 1970 cal. BC) when some Late Archaic "crude" ceramic vessels were found. The absence of ceramic cooking technologies in Saint John, and the presence of dozens of cooking pits in Eva 2, suggest that zamia food derivatives previously processed by means of pounding and grinding at these sites were ultimately cooked directly over a fire or on perishable grills.

Similar starch traces have been noted on grinding tools recovered from later preceramic sites in Puerto Rico, Saba, and Cuba (Figure 3.6 and Table 3.2; see also Chinique de Armas et al. 2015), some of which could be identified at the species level (Figure 3.7d–h). For example, at the site of Puerto Ferro (Vieques, Puerto Rico), irregular oval *Z. pumila* starches exhibiting key diagnostic features of the species (size, lamellae, and extinction cross) were retrieved from two edge grinders (Figure 3.7d–e), while another

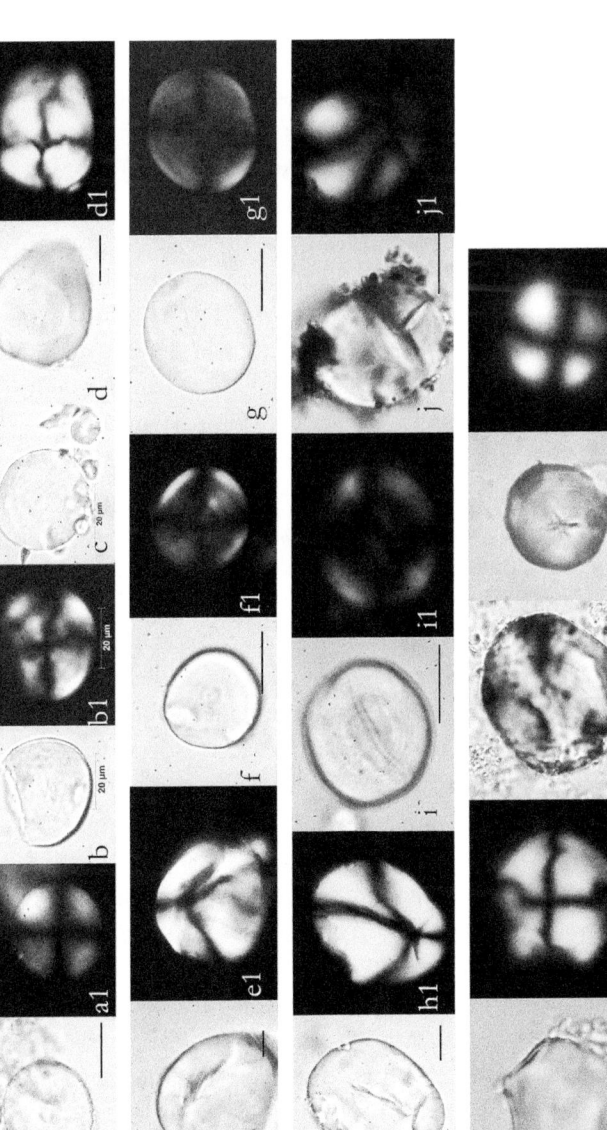

Figure 3.7. Ancient zamia starches from Caribbean and northeastern South American sites: *a* and *a1*, oval *Zamia* spp. starch (Saint John, Trinidad and Tobago); *b* and *b1*, truncated *Zamia* spp. starch (Eva 2, French Guiana); *c*, oval *Zamia* spp. starch with smaller grains probably of the same plant source (Eva 2); *d* and *d1*, irregular oval *Z. pumila* starch with marked lamellae (Puerto Ferro, Vieques); *e* and *e1*, irregular oval *Z. pumila* starch with pronounced fissure (Puerto Ferro); *f* and *f1*, oval and partially depressed *Zamia* cf. *portoricensis* starch (Puerto Ferro); *g* and *g1*, oval *Zamia* spp. starch with marked lamellae (Maruca, Puerto Rico); *h* and *h1*, irregular oval *Z. pumila* starch with marked fissure and perpendicular striations (Maruca); *i* and *i1*, oval *Z. pumila* starch with marked lamellae (Punta Candelero, late Saladoid context, Puerto Rico); *j* and *j1*, irregular oval *Zamia* spp. starch (Playa Grande, Dominican Republic); *k* and *k1*, irregular quadrangular *Zamia* spp. starch diagnostic to the genus level (Manzanilla, Trinidad and Tobago); *l*, irregular oval *Zamia* spp. starch with central depression and marked lamellae (Argyle 2, Saint Vincent); *m* and *m1*, oval *Zamia erosa* starch with stellate fissure and marked lamellae (Vega Nelo Vargas, Puerto Rico). Scale bars represent 20 μm in all images (images by the author).

circular starch (oval after rotation), morphometrically consistent with those commonly produced in *Z. portoricensis,* was recovered in a conical pestle (Figure 3.7f–f1). These three artifacts are associated with both early and late occupations of the site, respectively.

Likewise, Maruca, another site in Puerto Rico, yielded zamia starches securely identified to the species level. These were retrieved from an edge grinder and a pestle with pecked ends, both ascribed to an early occupation of the site (ca. 2900 BC). At least one big oval starch here was securely assigned to *Z. pumila* while another one was identified at the genus level (Figure 3.7g–h). These starches fall within expected size ranges and also exhibit lamellae and extinction crosses characteristic only of zamia. At both sites, starches of other economic plants, such as manioc, maize, bean, yam, and canavalia, among others, were also identified (Pagán-Jiménez et al. 2005). In this case, and that of the site of Cueva Ventana, the presence of zamia starches is logical, since these sites are located within the natural distribution range of both *Z. pumila* and probably *Z. portoricensis* in southern Puerto Rico, and *Z. erosa* in northern Puerto Rico respectively (Pagán-Jiménez and Lazcano-Lara 2013). However, for Puerto Ferro and Plum Piece (Saba) there are no current registries of any zamia species. In the northern, more humid limestone region of Puerto Rico only *Z. erosa* is found. The other two species are distributed across the limestone, serpentine, and volcanic substrates of the drier south and southwestern part of Puerto Rico. Interestingly, Puerto Ferro is situated in southern Vieques, which is also characterized by similar geological and drier environmental conditions.

Human dental calculus of an individual from the late Archaic site of Canashito in the Lesser Antilles yielded a starch tentatively ascribed to zamia. As in the previous cases, no ceramics were associated with this find. Other starches identified in dental calculus attest to the access, processing, cooking, and consumption of key economic plants like maize and probably sweet potato (Mickleburgh and Pagán-Jiménez 2012). Many of these other starches, including that of zamia, show clear signatures of alteration by pressure (probably due to grinding) such as radial striations and edge cracking, while two of them also show light twisting and a faint central fold consistent with heat damaging in a partially wet environment. As in the case of Saint John (Trinidad and Tobago), Eva 2 (French Guiana), and Plum Piece (Saba), this zamia starch could not be identified at a higher taxonomic level of resolution. There are no historic or modern botanical references attesting to the presence of this genus in the southern Caribbean.

Beyond these results, further evidence for the processing and consumption of zamia and many other starchy plants such as sweet potato, maize, bean, and manioc has been found at many so-called agroceramic age sites throughout the region. At such sites, varied plant processing tools (scrapers, graters, "chewing" tools) are found in greater concentrations than at preceramic sites studied to date. Lithic and shell scrapers, as well as coral and lithic grater teeth that possibly attest to the initial stages of peeling and grating of tubers, have yielded tentative and secure zamia storage starches at late ceramic age sites in the Bahamas (Ciofalo et al. 2018), Puerto Rico, and the Dominican Republic (Figure 3.7i–j). Interestingly, some of these sites (e.g., Punta Candelero, Playa Grande) also revealed zamia starches in other tools presumably used in later processing steps, such as the grinding/pounding of zamia dough (edge grinder/milling slab), the reception/storage of dough or "squeezed" substances (ceramic container), and the cooking of starchy derivatives (cooking pots).

Paleoethnobotanical data have also been recovered from many other sites of the Lesser and Greater Antilles. The artifactual contexts of these findings suggest that methods for processing, cooking, and consuming of zamia derivatives were relatively consistent through space and time, even in territories where wild *Zamia* populations have not been registered, such as those sites in the southern Caribbean. Starches recovered from the dental calculus of three individuals from Manzanilla in Trinidad and Tobago (Figure 3.7k) and one from Tanki Flip in Aruba confirm that foodstuffs made with underground stems of zamia were eaten in this part of the Caribbean islands from at least AD 400 to 1400. This scenario strengthens starch grain data from zamia obtained at much earlier archaeological sites in this and nearby continental areas (e.g., Saint John and Eva 2), thus suggesting that ancient human agency is associated with the presence of this genus, or at least starchy derivatives made from it, in the region. Other areas in the mid-Lesser Antilles where wild *Zamia* populations have never been registered have also revealed the ancient consumption of zamia derivatives. Dental calculus retrieved from human burials postdating AD 400 in sites like Argyle 2 in Saint Vincent (Figure 3.7l), Lavoutte in Saint Lucia, and Anse à la Gourde in Guadeloupe have produced starches tentatively or securely identified to *Zamia* genus level (Mickleburgh and Pagán-Jiménez 2012; Pagán-Jiménez and Mickleburgh 2015).

The remaining evidence for the processing, cooking, and consumption of zamia in late agroceramic sites comes from Puerto Rico, the Dominican Republic, Cuba, and the Turks and Caicos islands (see Ciofalo et al. 2019). Natural populations of several *Zamia* species have been documented on these islands, which would have allowed an easier procurement of these plants. Some of these islands are those in which early European chronicles detailed the presence and varied uses of this genus among Amerindian and creole populations. Apart from scraping, grating, and "container" tools, other kitchenware linked to processing (grinding/pounding) and cooking of zamia and other starchy plant derivatives have yielded storage starches of this genus and several species (Figure 3.7m). Among these lithic or coral grinding tools are pestles, milling slabs, and mortars, while the cooking tool assemblage consists of clay griddles and ceramic cooking pots. Ancient human dental calculus from some of these Greater Antillean agroceramic sites has also yielded zamia starches indicative of consumption.

Emerging Research Problems on Zamia's Historical Ecology in the Greater Caribbean

The results synthesized above suggest that zamia played an important role as a wild food source in the Caribbean for more than 7,800 years, as it appears to have done in Mesoamerica (see Englehardt et al., this volume). Paleoethnobotanical data on zamia have provided new insights into the temporal depth of the use of these plants by many regional sociocultural traditions. Nevertheless, several sites outside the known natural range of *Zamia*, such as those in the Lesser Antilles and northeastern South America, present complex research questions that have yet to be conclusively addressed: How did inhabitants of those areas access zamia? Did *Zamia* species exist in the past within territories far from their currently known natural distribution range?

As mentioned previously, potential explanations of the presence of archaeological zamia starch grains outside the genus' natural range include (a) that it was originally part of the paleoflora of these territories and subsequently eradicated; and (b) that its starchy derivatives (or even whole plants) were acquired via exchange with regions where this genus existed (see Mickleburgh and Pagán-Jiménez 2012; Pagán-Jiménez 2007). Regarding the latter possibility, recent research has established that ancient Caribbean island peoples participated in intense pan-Caribbean mobility and exchange networks that included northern and middle South America, as well as Central America. Indeed, the earliest peopling of Caribbean islands seems to have been impacted by the previous acquisition and use of new domesticated plants in coastal continental settings (Pagán-Jiménez et al. 2019). Such domesticates were portable resources that appear to have been consciously translocated to the islands, following specific cultural and adaptive behaviors, even when they initially could have provided low return rates. Similarly, continental raw materials and finished products (e.g., jadeite, other semiprecious stones) reached the Caribbean islands by means of well-established pan-Caribbean exchange networks (Rodríguez-Ramos 2010; Rodríguez-Ramos and Pagán-Jiménez 2006). Moreover, zamia's underground starches have been recently identified in a Late Classic period (AD 600–1020) Maya site in Yucatán (Simms 2014), which could reflect the movement of insular Caribbean culinary practices through broader pan-Caribbean reticulated interaction networks. Thus, the translocation of *Zamia* species in the form of whole plants, as partially processed products, or even as a culinary-ideological asset into and across the Caribbean is eminently plausible at any point in the region's history, regardless of potential dietary significance (or lack thereof).

In contexts outside of the known distribution range of *Zamia*, however, other cultural factors could have motivated the inter-island movement of these plants or their derivatives. Considering the antiquity of zamia starch grains at sites like Saint John and Eva 2 in the Caribbean, as well as the early presence of starches, seeds, and other cycad remains at archaeological sites outside the Greater Caribbean (e.g., Hornitos and Casita de Piedra, Panama, dated to ca. 5600 BC [Dickau et al. 2007], Coxcatlán, La Perra, and other cave sites in Mexico, dated to ca. 4700 BC [Bonta et al. 2019]), it is possible that these plants formed part of ancient, preagricultural diets in the Americas (see also Englehardt et al., this volume). Thus, although zamia could have been a staple wild plant due to its natural abundance in some regions within the Caribbean at the time of European contact, it is also very likely that, in addition to economic significance, this plant and its derivatives possessed deeper symbolic values, as part of ancestral foodways and histories (see Carrasco, this volume). In this sense, zamia plants and/or foods could have been exchanged as prestige items or to reinforce group culinary identities. Pan-Caribbean and inter-island movement of other material items associated with personal, religious, or political prestige are common in the regional archaeological record (see, e.g., Oliver 2009; Ostapkowicz et al. 2012). Nonetheless, prior to the formulation of theoretical interpretations for these and other possible regional interaction dynamics, researchers must first assess zamia's availability in those regions in which this genus is not known to occur—or to have occurred—naturally.

New Directions for Research on Zamia's Historical Ecology

The identification of zamia starch grains in archaeological materials elucidates the processing and consumption of these plants and their derivatives by ancient human groups. In the Caribbean, these findings confirm the ancient use of zamia at the studied sites, but do not provide data that speak to its natural availability in those areas. Ancient zamia starches in Puerto Rico have been recovered from archaeological sites located outside the accepted distribution range for the three known species of the island. Interestingly, a recent study based on the geographical search for old toponymic names for zamia in Puerto Rico (Pagán-Jiménez and Lazcano-Lara 2013) has provided firm evidence that notably expands the traditionally accepted natural range for the genus on the island, thus showing that in the past this plant was certainly available for human use and consumption in "unexpected" areas.

Small, island-scale findings of modern *Zamia* populations growing wild in previously unknown areas of Puerto Rico provide the basis for exploring new research avenues to better understand the historical ecology of zamia in this and other Neotropical regions. Paleoecological and paleoethnobotanical studies using pollen and phytoliths in the Caribbean area, but specifically in territories where ancient zamia starch grains have been recovered, should consider microbotanical zamia remains to evaluate possible natural occurrence of this genus outside its current distribution range. *Zamia*, like other Cycadales, produces quite uniform pollen grains that are easily distinguishable from other pollen assemblages (Dehgan and Dehgan 1988; Schwendemann et al. 2009). Pollen preservation is not a major issue (Frederiksen 1978), since fossil cycad and zamia pollen has been registered in records millions of years old from various environmental contexts (Slater and Wellman 2015). Most zamia pollen grains are boat-shaped or prolate spheroidal, and elliptical with disulcate aperture when seen in polar/distal view (Figure 3.8). Their common mean length oscillates between 24 and 27 μm. Maximum mean sizes of pollen from selected *Zamia* species are 21 μm (*Z. furfuracea*), 25 μm (*Z. herrerae*), 26 μm (*Z. integrifolia*), and 28 μm (*Zamia* spp., Jamaica) (Hamada et al. 2015). Based on the above data, and that reported in specialized literature on zamia, it may be assumed that other Greater Caribbean *Zamia* species produce pollen grains with similar morphometrics (González-Géigel 2003). Hence, zamia pollen could be a reliable source of information to determine its phytogeographical history in the Caribbean region. Studies aimed at finding and interpreting these grains in ancient sediments and soils, however, should take into account that most if not all species are insect pollinated (Norstog 1987; Tang 1987). This means that low pollen counts are expected in sediment cores even in waterlogged sampling areas (ponds, lakes, lagoons) close to current wild populations of zamia.

Previous studies on phytolith production and variability in gymnosperms such as cycads have demonstrated that these plants show low silica abundance (Trembath-Reichert et al. 2015), although this fact is usually not of

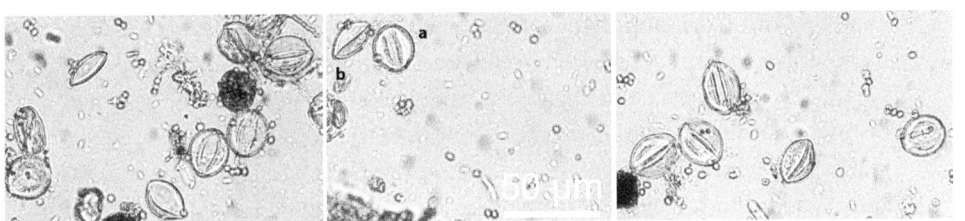

Figure 3.8. Modern pollen assemblage of *Zamia integrifolia*. Pollen *a* (*center image*) is in polar view; *b* is in equatorial view. These non-acetolyzed pollen grains were photographed by the author with an optical microscope.

taxonomic significance (Piperno 2006b). Nonetheless, systematic phytolith studies of modern Caribbean zamia need to be carried out in different plant segments and organs to verify and, if necessary, document any silicate structure that could serve as a reliable biomarker of the genus or even of different species. In short, the careful consideration of both zamia's pollen and phytoliths in paleoecological and paleoethnobotanical registries (Castilla-Beltrán et al. 2018; Lane et al. 2009) could confirm the presence or absence of *Zamia* populations in areas beyond its accepted natural distribution range. This will provide additional frameworks through which to better approach future research on its evolutionary pathways, geographic distribution, and dispersal dynamics at regional and local levels (Calonje et al. 2019; Meerow et al. 2012; Zonneveld and Lindstrom 2016).

Besides the recovery and identification of ancient zamia starch grains in ancient plant processing tools, the correlation of ethnohistoric and ethnographic data with specific archaeological objects (see Figure 3.9) may reveal potential research avenues by which to acquire additional evidence of zamia's ancient uses. One of the most promising approaches is the analysis of preserved chemical signatures in ancient artifacts or human remains. Cycads produce unique chemical compounds and metabolites that could potentially be recovered and identified in archaeological contexts (see Englehardt et al., this volume). However, the extent to which these unique compounds and metabolites survive in buried contexts has not yet been properly assessed. Analytical chemistry techniques previously applied to the study of archaeological residues have proven reliable for detecting lipids, alkaloids, amino acids, and metabolites from animals and plants in archaeological artifacts. Experimental and archaeological pottery studies, as well as human dental calculus, have shown that many of these compounds are quite stable, even after aging; their identification on and recovery from archaeological objects is thus theoretically possible (Fankhauser 1997; Hardy et al. 2012; Velsko et al. 2017; Zarrillo et al. 2018). Based on historic, ethnohistoric, and ethnographic data, Figure 3.9 synthetically suggests which Caribbean archaeological artifacts are likely associated with the different stages of ancient zamia processing and consumption, and what kind of clue or signature could be expected on them. Other proxies such as ancient plant DNA analysis could also be explored on such artifacts, and in human dental calculus, although DNA preservation biases must be taken into account when applying these techniques, especially in tropical archaeological contexts (Weyrich et al. 2017, 2015; Zarrillo et al. 2018; Ziesemer et al. 2015; see also Cibrián-Jaramillo et al., this volume).

Ethnohistoric and ethnographic data on zamia are also valuable sources of information for the elaboration of paleoecological, paleoethnobotanical, archaeological, and resource conservation interpretive frameworks (see Bonta, chapter 8, this volume; Calonje et al., this volume). The historic information in particular provides clear notions regarding the distinct sociocultural and geographic contexts into which zamia was integrated at discrete points in time. Additionally, this information could be carefully weighted to produce reliable models that may assist in interpreting other precolumbian sociocultural and environmental contexts, such as those typically targeted by paleoecology, paleoethnobotany, and precolonial archaeology in the Caribbean. Ethnohistoric and ethnographic sources also provide similar, though probably more accurate, information on the recent historical ecology of zamia. Thus, the combination of these sources can aid cross-contextual comparisons of zamia processing and preparation techniques, better situating these practices within ever more defined cultural and environmental frameworks. In sum, such conjunctive approaches between archaeology, ethnography, paleoethnobotany, and ethnohistory have the potential to improve our understanding of zamia's historical ecology, as well as the interrelationships between these wild plants and people.

Moreover, modern ethnographic (and ethnobotanical) work on the interactions between people and zamia would provide paramount local [ecological] knowledge on this genus (Bonta et al. 2019; see also Bonta, chapter 8, this volume), which would augment efforts to address current and future challenges in the Caribbean, such as food security. Considering that this region is increasingly affected by unprecedented climatic and environmental changes, revisiting the place of ancient resources in regional cultures may prove productive—especially a resource like zamia, which the data show has been used for millennia in various cultural (ancestral food as prestige/identity validation) and economic (easily procured staple starch/famine food) contexts. For example, eighteenth to mid-twentieth century ethnohistoric records in Puerto Rico suggest that zamia was probably the most important

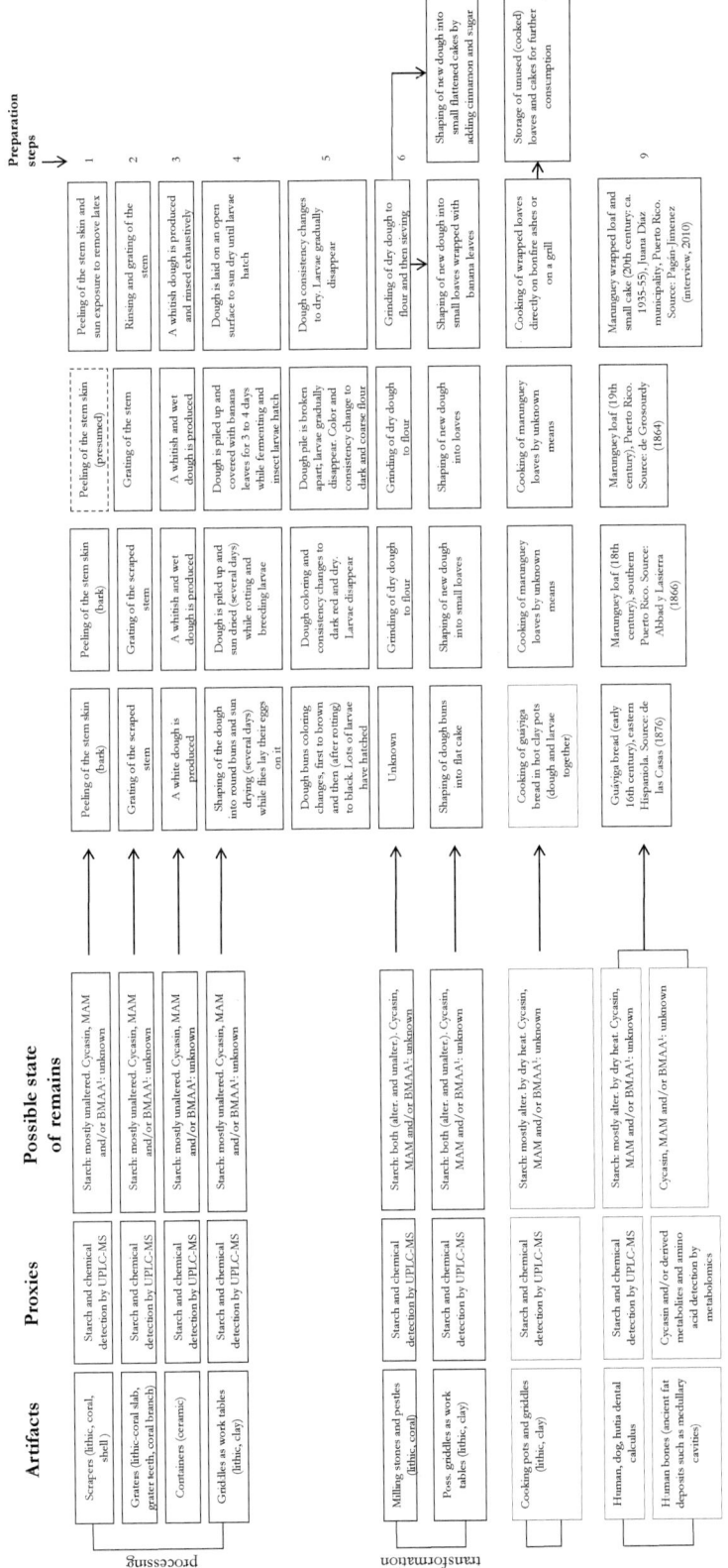

Figure 3.9. Reconstruction of early to late colonial processing stages of zamia in the Greater Antilles, potential archaeological correlates, and suggested methods to be applied to recover ancient zamia signatures in archaeological contexts.

source of carbohydrates in times of tropical storms and hurricanes, due to the loss of main cultivated food plants such as maize, plantain, or manioc. Thus, if Puerto Rican groups relied on the procurement of wild zamia during times of climatic and environmental crisis until the mid-twentieth century, this may still be a feasible option in this and other islands to confront future food shortages such as that recently experienced after the catastrophic Hurricane María in 2017 (Community Science 2019; Mares 2019).

Zamia's Histories, Present, and Future

For thousands of years, from pre-Hispanic to modern times, a wealth of evidence suggests that zamia was a significant wild food supply for various Indigenous groups in the insular Caribbean, northern Central America, and northeastern South America. Zamia remained a crucial dietary staple during precolumbian times even when other high-yield plants such as maize, manioc, and sweet potato were available, and this continued to be the case within colonial contexts framed by both large-scale monocultures (exportation) and small-scale polyculture production (local consumption). The peoples that used zamia were distributed throughout rich and diverse environments in the lowland Neotropics of the Caribbean region, territories that provided a broad spectrum of natural culinary resources. Available resources and exploitation technologies appear to have been exchanged via extended pan-regional networks, in which maritime voyaging was pivotal to support a dynamic and interconnected system (Rodríguez-Ramos 2010; Rodríguez-Ramos et al. 2013; Rodríguez-Ramos and Pagán-Jiménez 2006).

Even with the limited array of plant-related tools studied in the region to date, paleoethnobotanical data have begun to show that in some particular sites of the northern Antilles, and also in Trinidad and Tobago and French Guiana, zamia together with maize and sweet potato (among others) were notably ubiquitous food plants. Others, such as manioc, yam (wild and domestic *Dioscorea/Rajania*), and cocoyam (*Xanthosoma* spp.) seem to have played a secondary but consistent role in regional foodways (Pagán-Jiménez 2013). In Puerto Rico and the Dominican Republic, for example, starch grain analysis has shown that maize and zamia are consistently the most ubiquitous food plants evident in archaeological materials through time and space (Pagán-Jiménez 2007, 2009, 2012), and they figured prominently in the culinary identities and practices of various Indigenous groups.

During and after the European invasion of the Caribbean region, zamia continued to be used. In the Dominican Republic, traditional ways of making foodstuffs from zamia were registered by *cronistas,* and its economic importance was asserted in cacicazgos such as Higüey (de las Casas 1876). After the forced translocation of enslaved Africans into the Caribbean, incredible processes of transculturation occurred (Anderson-Córdova 1990, 2017), some of which involved the merging of ancient foodways from the Caribbean, Africa, and Spain. In this context, zamia was incorporated into reinvented culinary practices by Africans and their descendants, which resulted in the creation of new culinary traditions, typically associated with the Caribbean. In these new foodways, instead of focusing on the grated dough from the underground stems of zamia for producing finished foodstuffs (such as flatbread or loaves), starch was carefully extracted to be used as the main component of other dishes while the squeezed dough was discarded. New Caribbean foodstuffs made with zamia starch are known as *arepa de guáyiga* (a kind of bread), *chola* (a kind of wrapped loaf made with previously cooked starch), and *hojaldre* (a puff pastry) (de Boyrie Moya et al. 1957; Veloz Maggiolo 1992). The way of processing zamia storage starch to make these new foodstuffs is similar to other culinary traditions surrounding these plants among Indigenous peoples in the Bahamas, Florida, and the southeastern United States (e.g., the making of coontie bread; Swanton 1913). These postcolumbian foodstuffs are still produced in such regions.

Puerto Rico also underwent profound transformations that resulted from Spanish and US invasions, as well as from slavery and indentured agricultural labor. In the eighteenth century, zamia was procured, prepared, and consumed, as described to Abbad y Lasierra (1866), referenced in the introduction to this chapter. According to ethnohistoric documents, its importance apparently lay in the fact that it was a scarcity food easily accessed and consumed in times of food crisis, mainly after tropical storms and hurricanes hit. By the mid-twentieth century, methods and techniques for the procurement, processing, and preparation of zamia appear to have changed little since they were first described to Europeans some 270 years previously. This fact indicates that the culinary knowledge surrounding zamia was efficiently transmitted through generations, thus suggesting that, more than maintaining a purely theoretical knowledge, people likely targeted and consumed these plants on a regular basis, far beyond times of crisis. The 1950s witnessed, apparently for the first time in thousands of years of Puerto Rican history, the cessation of practices of use, management, and transmission of traditional knowledge surrounding zamia. Sparse ethnographic fieldwork in south-central Puerto Rico, one of the regions where zamia was historically used, indicates that traditional subsistence activities surrounding this genus began to die off at that time. The last carriers of this ancestral culinary tradition are people currently ranging from 80 to 100 years old. Sadly, traditional local and ecological knowledge about zamia appears to be coming to an end without being properly transmitted, in either theoretical or practical terms, to newer generations of Puerto Ricans.

The succinct histories of zamia presented here suggest some potentially significant avenues of future research. Interdisciplinary research and conservation practices involving combined efforts from the natural and social sciences would help to fill in the gaps of information in zamia's historical ecology. This will lead to a better understanding of a deep cultural history that stretches back thousands of years and, almost incredibly, managed to survive both the European and US colonization of the Caribbean, as well as industrialization and fierce capitalism. Comprehensive management plans directed toward the study, possible conservation, and sustainable management of zamia and attached cultural values must be supported by much needed ethnographic work with communities in which direct traditional ecological knowledge on this plant may reside (see Calonje et al., this volume). In Puerto Rico, this ethnographic fieldwork is critically urgent and must be undertaken before the lived experiences of the last direct carriers of local knowledge about zamia are lost forever. Such research could provide new opportunities, both for zamia and for societies that preserve knowledge of its use, for example, in facing uncertain climatic and economic futures. In the Caribbean context zamia was, and still is, an invaluable wild food supply that needs to be comprehensively studied and secured.

Notes

1. The Maltese cross or extinction cross is an optical phenomenon produced in some crystalline or semicrystalline structures when observed under cross-polarized microscopy. In paleoethnobotanical analysis, this feature is used together with other optical and morphometric elements to differentiate starches from other crystal-like particles, but also to sort out and differentiate starches produced by different taxa that display a diversity of extinction cross patterns.

2. To learn more about the dietary importance that Zamia may have had for the peoples and archaeological sites already studied, or compared with other recovered taxa at such sites, I encourage the reader to consult the original works referred to throughout this chapter, but especially those referenced in Figure 3.6 and Table 3.2.

4

Ancient Mesoamerican Agricultural Strategies

The Role of Cycads and *Phaseolus* Beans in
the Development of Intensive Field Cultivation Systems

AMBER M. VANDERWARKER

Agriculture in ancient Mesoamerica is often described in broad terms as a dedicated focus on maize, beans, and squash, often termed the Three Sisters or Trinity. Likewise, the development of agricultural systems is presumed to proceed toward greater cultivation intensity over time as populations rise and technology changes. These broad strokes, while perhaps useful in textbooks, obfuscate the incredible variability in agricultural strategies across Mesoamerica in both time and space. Beyond a focus on the Three Sisters has been an even larger emphasis on the importance of maize, which has also hampered research aimed at understanding other types of foods and production strategies (e.g., avocado domestication, development of agroforestry systems). Given this bias, coupled with the rapidly dwindling botanical diversity across the broader region, it behooves us to expand our scale of analysis—not only to fill gaps in archaeological knowledge, but also to contribute knowledge about the natural history and prehistoric uses of plants as a means to bolster conservation efforts for currently endangered species, which include a variety of cycads.

My purpose in this chapter is to present a type of cultivation strategy that ancient Mesoamerican farmers may have used in some times and places: the intercropping of maize with cycads. As with legumes (Fabaceae), species in the cycad order (Cycadales) can fix nitrogen, which significantly benefits any plants growing nearby by providing them with more access to this important element for plant growth and development (see Calonje et al., this volume). My argument in this chapter is both temporal and spatial: first, I suggest that early farmers may have intentionally planted their maize fields with transplanted cycads or alongside naturally occurring cycads, prior to the adoption and cultivation of *Phaseolus* beans in some regions of Mesoamerica; second, given that cycads fix nitrogen at much higher rates than beans, I suggest that farmers living in ecological zones in which cycads are better adapted may have maintained this practice longer, adopting beans much later or perhaps never at all. In other contexts, farmers may have intercropped with beans and cycads under different field conditions. I develop these hypotheses below and offer some future directions for ways in which they can be tested with archaeological data. It is first important to provide some details about the growth and habitat requirements of relevant cycad species, in order to understand the cycads' relationship to maize cultivation. Understanding the process of biological nitrogen fixation (BNF) is critical to the argument and is reviewed. Following this, I discuss our current knowledge of the timing of domestication events specific to maize and beans. The broad range of agricultural strategies employed throughout ancient Mesoamerica are then presented, with an emphasis on how cycads and beans could have been incorporated into these variable cultivation strategies. Embedded throughout the larger discussion are the broader economic, cultural, and ritual contexts in which people have used cycads in both the ancient past and present day. Ultimately, as was the case with manioc (*Manihot esculenta*), we must actually design our research projects to include the possibility of finding cycads in archaeological contexts if we truly want to expand our understanding of ancient Mesoamerican agriculture.

Cycads, Beans, and Biological Nitrogen Fixation

The cycads are botanically categorized as a taxonomic order (Cycadales) encompassing a primitive group of gymnosperms that have been around since the Permian period (Goel and Khuraijam 2015; Vovides and Iglesias 1994; Whitelock 2002). This family of slow-growing dioecious plants includes more than 300 species worldwide, growing native in the Old World, South America, Mexico, the Caribbean, and various Central American countries (Bonta et al. 2019; Goel and Khuraijam 2015; Pérez-Farrera et al. 2006; Vovides, González, Pérez-Farrera, et al. 2004). Of all geographic locales, Mexico ranks second in cycad species diversity and hosts three distinct genera (*Ceratozamia, Dioon,* and *Zamia*) from the family Zamiaceae that encompass at least 70 species (Bonta et al. 2019; Goel and Khuraijam 2015; Pérez-Farrera et al. 2006; Vovides and Iglesias 1994), although botanists continue to identify new species (e.g., Salas-Morales et al. 2019).

Cycads can be found in a broad range of habitats, but most well-adapted cycads in Mexico occur in areas of steep, rocky terrain with long hot summers and low levels of seasonal rainfall (Bonta et al. 2019; Goel and Khuraijam 2015; Whitelock 2002). Generally, they are native to various ecological zones, including cloud forests, open areas (grasslands, savannas), forests (evergreen tropical, seasonally dry tropical), and some mangroves (Vovides, González, Pérez-Farrera, et al. 2004). As a group, members of the *Ceratozamia* genus are found primarily in densely shaded forests, dominated primarily by oaks, but can also be present in rainforests, cloud forests, and pine forests; within these woodlands, *Ceratozamia* species inhabit steep and rocky slopes composed of limestone or clay substrate (Whitelock 2002). *Dioon* cycads are adapted primarily to dry, rocky, steep slopes in both exposed and shaded locales, and many can be found in true desert environments; species of this genus are also occasionally found in open, stunted oak forest (Whitelock 2002). *Zamia* species are by far the most broadly adapted of the Mesoamerican cycad genera and inhabit rainforests, savannas, coastal sand dunes, tidal swamps, and deserts (Whitelock 2002). Cycads are sometimes confused with palms, and some mature species develop large cones that resemble maize ears; some cycad kernels also resemble maize kernels (Bonta et al. 2019; Bonta and Osborne 2007; see also Carrasco, this volume). Today these plants are highly sought after as ornamentals, and their illegal poaching has led to a significant population decline that has reached endangered status for some species (Mapes and Basurto 2016; Pérez-Farrera et al. 2006; Whitelock 2002).

Most seeds and stems of cycads contain neurotoxic and carcinogenic chemicals, requiring extra processing to remove their toxicity, or consumption may result in illness or death (Bonta et al. 2019; Goel and Khuraijam 2015; Whitelock 2002). Two primary toxic compounds include cycasin and macrozamin, which are known to cause both liver damage and cancer; when consumed in small amounts over the long term, these compounds may also lead to various neurological disorders (Whitelock 2002:47–48). Cross-cultural methods of toxin removal from cycad seeds and stems include drying in the sun (often for several days), soaking in water, fermentation, or some combination of all three techniques. At some point during the detoxification process, the cycad seeds and stems are sometimes ground into a fine powder and then added as an ingredient to other foods or beverages (Whitelock 2002:48), which we know is true of some Mexican cycads (Mapes and Basurto 2016).

Modern uses of cycads include recreational, nutritional, medicinal, ceremonial, and agricultural, in addition to their ornamental utility (Bonta et al. 2019; Goel and Khuraijam 2015; Santi et al. 2013). The large seeds of *Dioon spinulosum*, located throughout southeastern Mexico, are sometimes modified into toys or whistles for children (Bonta et al. 2019:20; Whitelock 2002:171). Indeed, one potential toy was encountered in MacNeish's cave excavations in Tamaulipas, attesting to the antiquity of this practice (MacNeish 1958, cited in Bonta et al. 2019:20). Found in the mountainous cloud forests of Chiapas and Oaxaca, *Ceratozamia mexicana* is used for its seeds, which are roasted as a snack or boiled, ground, and added to maize for tortillas (Mapes and Basurto 2016). Whitelock (2002:47) notes the use of stems of *Dioon sonorense* for *sotol* production in Sonora, Mexico. Along Mexico's Gulf coast, people gather the seeds from the cones of *Dioon edule* and grind them as an ingredient used in tamales, atoles, and tortillas after they are boiled with ash and lime (Mapes and Basurto 2016; note that this soaking process represents a critical stage in detoxification, as described above). Given the analogous appearance of cycad cones and maize ears, I find it interesting that the seeds of this cycad are processed similarly to the way maize is traditionally nixtamalized (see also Bonta et al. 2019:13); this preparation association could indicate that the human-mediated relationship between these two plants may extend far into the past, a topic I discuss below.

In terms of human health, cycads are also generally noted to improve kidney function and the symptoms of rheumatism, diarrhea, tuberculosis, neuralgia, skin lesions, and common colds, in addition to relieving hypertension (Goel and Khuraijam 2015:352; Whitelock 2002:51); however, it is unclear whether the Mexican cycads are among those that yield such results. Some Neotropical cycad species have narcotic and/or hallucinatory effects; the plant parts used for this purpose are referred to as *peyote* (Bonta et al. 2019:20). Bonta and colleagues (2019:20) observed at least two modern human groups in the broader region that ingest various cycad parts for narcotic purposes. These researchers also note that cycads have a rich tradition of use in celebrations and ceremonies, often those related to renewal and life-crisis events. For example, the Chiapanec people take annual pilgrimages to various sacred sites where large, dense stands of *Dioon merolae* grow; visitors harvest the leaves sustainably and carry large bundles back to their communities, the arrival of which begins a series of public celebrations (Bonta et al. 2019:21; Pérez-Farrera and Vovides 2006).

Throughout Mexico and Central America, local beliefs in which ancient people cultivated cycads are common, and today people incorporate more than 26 species into their land management practices (Bonta et al. 2019:14). Some groups even intercrop their maize with cycads, with the goal of balancing cycad cone production with maize production (Bonta et al. 2019:14–15).

Bonta and colleagues (2019:27) note that "cycads are often said to 'pull down' moisture from the clouds to the *milpa*"; this statement demonstrates local recognition that intercropping cycads with maize improves maize growth, even if the catalyzing process of biological nitrogen fixation is not known (see below). We also know that at least one cycad species (*Dioon edule*) is a C_3 plant that switches to a CAM photosynthetic pathway when it undergoes water stress (Vovides, Etherington, Dresser, et al. 2002); such growth flexibility under variable water conditions makes it an excellent plant for intercropping with others, even in times of drought, and especially given its nitrogen-fixing capabilities. The Teenek people of San Luis Potosí consider *Ceratozamia latifolia* to be the "spirit of the *milpa*"; referred to as *konlif*, the plant grows both inside and along the perimeter of maize fields, "protecting and strengthening maize" (Bonta et al. 2019:25). A final point regarding the relationship between maize and cycad production regards ethnographic reports that cycad seeds help deter crop pests; farmers concoct a poison for rodents from the raw, toxic seeds (Bonta et al. 2019:16).

In addition to the cropping complementarity between maize and cycads, there is a rich cultural linkage between maize and cycads throughout the broader region in which cycads are "conceptualized as types of maize, friends of maize, and ancestors of maize" (Bonta et al. 2019:8; Bonta and Osborne 2007:143; see also the chapters by Bonta, this volume). Indeed, extensive ethnographic interviews revealed that there were cultural and linguistic associations between maize and 20 species of cycads throughout Mexico and Central America (Bonta et al. 2019:23). In addition to the general resemblance between cycads and maize in terms of both kernels and cones/ears, maize and cycads can be food substitutes for one another under different circumstances. Among some groups, cycad seeds and stems are consumed when maize is in short supply or vice versa; other groups consume cycads under general food shortage conditions, and still others prize cycads as a special ingredient of great value (Bonta et al. 2019:13; Bonta and Osborne 2007:147). It seems clear that there is a strong cultural association between maize and cycads across the broader region, and while variable in nature, this association has implications for understanding the early onset and intensification of agriculture.

One of the unique biological traits found in cycads, and of particular relevance here, is their symbiotic relationship with various cyanobacteria that results in biological nitrogen fixation (BNF). Sufficient nitrogen is absolutely necessary for plant development and is a limiting factor in plant growth (Islam and Adjesiwor 2018; Santi et al. 2013). Plants cannot access atmospheric nitrogen (dinitrogen gas), and thus the inert form of nitrogen (N_2) must be converted into a form usable by plants (e.g., ammonia, NH_3), which can then be absorbed through plant root systems (Bano and Iqbal 2016; Meeks 1998; Santi et al. 2013). BNF is one way in which atmospheric N is converted to a usable form. The BNF process occurs through a symbiosis between the plant host and bacteria, the latter which develop as nodules on the plant's roots; these bacteria encode nitrogenase, an enzyme essential to the conversion process (Islam and Adjesiwor 2018; Santi et al. 2013). When the symbiotic bacteria die, they release nitrogen-containing compounds accessed by the host plant for growth and development (Bano and Iqbal 2016). Because the process of BNF releases usable N into the soil, both the host plant and nearby nonhost plants will benefit from improved growth and yield resulting from greater nitrogen uptake (Bano and Iqbal 2016; Islam and Adjesiwor 2018; Meeks 1998; Santi et al. 2013). The bacteria responsible for the nitrogen conversion benefits through enlargement (relative to nonsymbiotic counterparts), reduced carbon uptake, and absorption of other key nutrients (Santi et al. 2013). Ultimately, BNF increases nitrogen use efficiency (NUE), recovering losses of nitrogen from the soil, leaving it more productive and fertile (Bano and Iqbal 2016; Islam and Adjesiwor 2018).

BNF occurs differently in legumes and cycads in terms of both type of symbiotic bacteria and rate of conversion. In legumes, the primary bacteria catalyzing BNF are in the *Rhizobia* genus, which form nodules directly on the legume's roots (Bano and Iqbal 2016; Santi et al. 2013). In cycads, cyanobacteria (or blue-green algae) are responsible for initiating BNF (Meeks 1998; Santi et al. 2013); indeed, cycads represent the only group of gymnosperms that can form symbiosis with cyanobacteria for BNF (Santi et al. 2013). Cyanobacteria are a diverse group of prokaryotes that are ubiquitous in terrestrial and aquatic environments (Santi et al. 2013); the genera that most commonly form symbiotic relationships with the Mexican cycads include *Nostoc* and *Anabaena* (Halliday and Pate 1976; Meeks 1998; Santi et al. 2013). The cyanobacteria inhabit the coralloid roots of cycad plants, resulting in their enlargement into coralloid masses (Bonta et al. 2019; Halliday and Pate 1976; Meeks 1998). For cycads, BNF conversion is less complex and conversion rates much higher than in legumes, which results in more usable N made available to the cycad hosts and other plants growing nearby (Bano and Iqbal 2016; Halliday and Pate 1976; Meeks 1998).

Not only do cycads fix nitrogen at significantly higher rates than legumes, but they are also long-lived perennials, which means they continue to fix nitrogen throughout their lives, continually renewing the soil in which they grow (Halliday and Pate 1976:356; Islam and Adjjesiwor 2018). It is notable that the coralloid roots of cycads grow toward the ground surface, as opposed to deep into the soil, which means the converted N is released directly into the soil's upper layer for faster absorption by both the host plant (cycad) and any plants within its vicinity (Bano and Iqbal 2016; Islam and Adjesiwor 2018). This continual conversion of N_2 into NH_3 by cycads has important implications for ensuring and/or increasing yields of plants intercropped with them. As mentioned above, studies have shown that maize yields can increase significantly when maize is cultivated in proximity to nitrogen-fixing plants, including cycad plants specifically (Bano and Iqbal 2016:597; Islam and Adjesiwor 2018:103; Santi et al. 2013:754). The amount of nitrogen transferred to neighboring plants varies between 0% and 73%, determined by proximity of the BNF plant to the crop in question (Islam and Adjesiwor 2018:102). The closer the partner crop is planted to the BNF plant, the more N is transferred to it, leading to higher yields. When the BNF crop is an annual (legume) and not a perennial (cycad), then successional cropping will also confer a benefit for any crop that is planted subsequent to harvest of the BNF plant (Islam and Adjesiwor 2018). In the case of the cycads, however, juvenile plants will fix considerably less nitrogen (Halliday and Pate 1976); thus, partner crops will benefit most when planted in close proximity to mature cycads.

Finally, it is important to consider the effects of fire on this process, as early cultivation practices included forest clearance throughout many regions of Mesoamerica (see below). After initial field clearance, farmers likely engaged in regular controlled burns as a means of field preparation and maintenance. Burning leads to mixed results in terms of cycad growth and regeneration. Cycads form replacement sets of coralloid roots at irregular intervals, but most notably after burning (Bonta et al. 2019:3; Grove et al. 1980; Halliday and Pate 1976: 349); ethnographic data from Honduras demonstrate that local inhabitants are aware of this process, and ancient peoples likely were as well (Bonta et al. 2019:15). When cycads are subject to fire, root production and growth are stimulated; these new roots, however, will need to mature over the long term to become a host to cyanobacteria (Bonta et al. 2019; Grove et al. 1980; Halliday and Pate 1976). Bonta and colleagues (2019:15) note that cycads "are often the only plants to survive in burned-over maize fields after all other vegetation has been eliminated." Thus, early forest clearance in Mesoamerica likely set the stage for increased cycad growth and subsequent population spread that would have had clear long-term benefits for maize cultivation. As farmers realized the effects of planting their maize seeds in cycad-rich areas, they likely would have spared any cycads in their fields during controlled burns and weeding activities. This scenario is further explored below in both regional and temporal contexts.

Domestication of Maize and Beans

Given my focus on maize and N-fixing plants, I restrict my discussion of domestication events to maize and beans, as these are the two domesticates most relevant here. We now know from recent genetic studies that maize was domesticated only once (7150 BC). The progenitor was most likely from a population of wild teosinte (*Zea mays* ssp. *parviglumis*) growing naturally in the lower Balsas River valley of Guerrero, Mexico (Bennetzen et al. 2001; Matsuoka et al. 2002). Research conducted in the Neotropical lowlands demonstrates the presence of maize in the form of starch grains, phytoliths, and pollen as early as 6750 BC in the central Balsas River valley (Piperno et al. 2009; Ranere et al. 2009), 5150 BC along the Gulf coast of Mexico (Pohl et al. 2007; Pope et al. 2001), and 4550 BC in Chiapas, Mexico (Neff et al. 2006). These data often co-occur with evidence of burning consistent with field clearance and preparation (e.g., Kennett et al. 2010; Neff et al. 2006). In Panama, we have paleoecological evidence of slash-and-burn farming as early as 5850 BC in the central region (Piperno et al. 2007) and by 5050 BC in the western part of the country (Dickau et al. 2007).

There are also abundant data from northern South America that point to a relatively fast dispersal out of Mexico and through Central America. Some of the earliest maize starch grains in South America come from northwest Colombia and date between ~6000 and ~5600 BC (Pagán Jiménez et al. 2016; Santos Vecino et al. 2014). In Ecuador, directly dated starch grains from food residues in early ceramics affirm the presence of maize and other crops in domestic contexts around 4050–3050 BC (Zarrillo et al. 2008; *sensu* Pearsall 2002; Pearsall et al. 2003, 2004; Piperno 2003); more recently dated maize phytoliths from this region push maize's entry into Ecuador back to ~6000 BC, similar to its entry into Colombia (Piperno 2011; Stothert and Sánchez 2011). Once maize dispersed into northern South America, it was quickly transported to several Caribbean islands (see Pagán-Jiménez, this volume). Microbotanical research by Pagán-Jiménez and colleagues on the islands of Trinidad and Tobago and in French Guiana has identified the earliest presence of maize sometime around 5700 BC, only 300 years after its first identification in coastal South America (Pagán-Jiménez, this volume; Pagán-Jimenéz et al. 2015). Collectively, these and other microbotanical studies suggest an early domestication of maize, followed by its rapid dispersal throughout the lowlands of Mesoamerica, Central America, northern South America, and the Caribbean after 6750 BC (e.g., Bryant 2007; Holst et al. 2007; Pearsall 2002; Perry et al. 2007; Piperno et al. 2000, 2002; Piperno and Stothert 2003; Pohl et al. 2007; Pope et al. 2001; Rosenswig et al. 2013; Sluyter and Dominguez 2006; Wahl et al. 2006; Zarrillo et al. 2008).

Our knowledge of bean (*Phaseolus* spp.) domestication is less advanced than for maize, and there remain gaps in what we know about the timing and location of domestication. At some point, human-directed selection culminated in a shift from perennial vines to annual bushes, although the timing of this genetic shift is unclear (Gepts and Debouck 1991; Kaplan 1981; McClung de Tapia 1992; Smartt 1988; Smith 1998). Currently, the earliest known domesticated bean (in macro form) is only 2,300 years old (Kaplan and Lynch 1999), which suggests that domesticated beans were adopted relatively late into existing systems of cultivation; most Mesoamerican farmers had been cultivating maize for thousands of years by this point. Genetic evidence has provided the bulk of our knowledge and supports two independent centers of domestication—one in Mesoamerica and the other in the South American Andes (Chacón et al. 2005; Piperno and Dillehay 2008). Within Mesoamerica, several bean types can be found, including common bean (*Phaseolus vulgaris*), tepary bean (*Phaseolus acutifolius*), runner bean (*Phaseolus coccineus*), lima bean (*Phaseolus lunatus*), and year-long bean (*Phaseolus dumosus*) (Piperno and Smith 2012). A recent genetic study (Kwak et al. 2009) proposes that common beans were initially domesticated in the moist habitats of west-central Mexico, a location close to the Balsas River valley, where maize appears to have originated (see Piperno et al. 2009). Given this overlap in geographical origins, we should not be surprised that most research concerning bean dispersal and use continues to emphasize beans in the context of the Three Sisters (e.g., Bonomo et al. 2011; Hart et al. 2002; Iriarte 2007). It is widely accepted, however, that maize, beans, and squash did not evolve in tandem as a single package (see Rosenswig 2015). Moreover, given that beans were added to maize cultivation thousands of years after people began growing maize, it is almost certain that some early farmers discovered the benefits of planting maize in ecological zones that were naturally rich in cycads.

Agricultural Strategies in Temporal Context

Archaic Period Burning and Cultivation

In addition to understanding the growth and habitat requirements of various cycad species, assessing the plausibility of this argument requires a review of Mesoamerican agricultural strategies, which varied considerably across space and time (Table 4.1). I present this admittedly brief synthesis of

Table 4.1. General ranges for major periods in Mesoamerican prehistory

Postclassic	AD 900–1521
Late Classic	AD 600–900
Early Classic	AD 250–600
Late/Terminal Formative	~400 BC–AD 200/300
Middle Formative	~1000–400 BC
Early Formative	~2000–1000 BC
Archaic	~8000–2000 BC

agriculture in a diachronic format to generalize broad changes across regions. While there is clear evidence that maize was domesticated during Archaic times, as early as 6750 BC (Piperno et al. 2009; Ranere et al. 2009), we do not observe evidence for burning and initial forest clearance for cultivation until ~3,500 years later, toward the end of the Archaic period. While there is evidence that foragers engaged in burning as part of landscape management earlier in time (e.g., Zizumbo-Villarreal and Colunga-García Marín 2010; Zizumbo-Villarreal et al. 2012), such burning does not appear to have been intended for agricultural field clearance until ~3100 BC, when evidence for both disturbance and burning is accompanied by maize remains (Kennett et al. 2010; Voorhies 2004:66; see also Rosenswig 2015). We can consider this initial shift toward field clearance as an early step toward intensification (e.g., Zizumbo-Villarreal and Colunga-García Marín 2010; Zizumbo-Villarreal et al. 2012). There is abundant evidence for Archaic period deforestation and controlled burning in other regions as well, including the Gulf Coast (Goman and Byrne 1998; Piperno and Pearsall 1998; Pohl et al. 1996; Sluyter and Dominguez 2006), the Maya lowlands (Lesure 2008), and western Mexico (Zizumbo-Villarreal and Colunga-García Marín 2010; Zizumbo-Villarreal et al. 2012). In some regions, however, evidence for burning does not appear until after the transition to the Formative period, when it is clear maize cultivation was already in full swing (e.g., Basin of Mexico [Lesure 2008]).

Controlled burns of forest patches to clear space for cultivation would have resulted in increased soil nutrients from the resulting ash, which would have benefited maize yields (Boserup 1965), especially early in a region's agricultural history when there was more fuel to burn (e.g., primary and secondary forests). If controlled burns were initiated in areas with naturally occurring cycads, then fires would have catalyzed coralloid root growth, leading to long-term increases in BNF in these Late Archaic maize fields. Over time, farmers would have noted the differences in maize yields between crops planted in proximity to cycads versus those planted without, or at a distance from, cycads. As noted above, there is ethnographic evidence that modern farmers understand this result even if they do not know the exact mechanism that causes it. One can imagine the long-term outcomes of Late Archaic farmers sparing cycads (during burning and clearing), culminating in a practice of maize/cycad intercropping; such a strategy would involve planting maize in proximity to naturally occurring cycads and/or transplanting mature cycads in systematic patterns (e.g., rows) around which maize could be cultivated. Given the long-lived perennial nature of cycad plants, it is possible we may see the legacy of systematic cycad transplantation in the modern landscape, in much the same way that we observe drained fields or terraces originally constructed by ancient people.

Late Archaic agricultural strategies thus encompassed practices of burning, multi/intercropping, rainfall irrigation, and likely some field rotation; in regions where cycads were central to maize cultivation, BNF would have permitted longer-term cropping before requiring a fallow period to recoup nutrient loss. Indeed, Thomas Killion (1992:6–7) notes that if there is abundant, prime land available for cultivation (which would have been the case in various regions of Mesoamerica given relatively low Late Archaic population levels), then settlements are likely to be dispersed with cultivation focused on infields nearby the residence; such a pattern would be conducive to cycad intercropping, as mature cycads are long-lived perennials that are difficult to transplant.

Unfortunately, evidence of ancient Mesoamerican cycad use is rare in archaeological contexts, as macro-plant parts are unlikely to preserve unless carbonized, and unlikely to be identifiable if preserved through carbonization (given the high starch content which leads to bubbling and distortion). Reported cycad macro-plant remains come from caves and rockshelters and have preserved primarily through desiccation. Several caves in Tamaulipas, Mexico that were originally excavated by Richard MacNeish have yielded abundant macro-remains of *Dioon edule* (Hanselka 2011; MacNeish 1958; see also Englehardt et al., this volume). In total, nearly 3,000 specimens of *D. edule* were identified in the Tamaulipas caves, including leaf bases, bracts, and seeds, with seeds composing the vast majority of the sample. In the desiccated plant assemblage from El Gigante rockshelter in Honduras, I recently identified seeds of *Zamia* sp. While analysis is ongoing and publications are still forthcoming, it is notable that the *Zamia* seeds were encountered at multiple stratigraphic levels, indicating their use through multiple periods of occupation (see Kennett et al. 2017).

Developments in archaeological starch grain analysis over the past two decades have permitted recent identifications of cycad residues on tools and human teeth that were not previously detectable. Research by Atchison and

Fullagar (1998) in Australia paved the way by demonstrating not only that cycads produce starches but also that cycad starches preserve on ancient archaeological tools. Ruth Dickau's (2010; Dickau et al. 2007) microbotanical work in western Panama has revealed the presence of *Zamia* starches (possibly *Zamia skinneri*) on preceramic tools by 5000 BC. It is clear that Archaic Panamanians were processing *Zamia*, but whether it was used as food, medicine, or for some other purpose is unknown.

Zamia starch grains have also been identified on three different ground stone tools at the Terminal Classic site Escalera al Cielo, Yucatán (Simms 2014). Recent research in the Caribbean by Pagán-Jiménez and colleagues (Mickleburgh and Pagán-Jiménez 2012; Pagán-Jiménez et al. 2005; see also Pagán-Jiménez, this volume) reveals evidence that ancient people living in the Dominican Republic, Cuba, and Trinidad were consuming cycads from 5800 BC onward. While this latter finding is not particularly relevant to the case at hand, it nevertheless demonstrates that *Zamia* starches can and do preserve in human dental calculus, a circumstance which provides direct evidence of consumption.

Formative Period Intensification

The transition into the Formative period around 1800 BC can be characterized as a time of continued intensification (see Table 4.1). At some point during this approximately 2,000-year period, many Mesoamerican groups shifted to a diet dominated primarily by maize, introduced *Phaseolus* beans to their cropping systems, and invested more time in field preparation and maintenance. The shift to a Formative period reliance on maize is documented in many regions, although the timing varies according to the setting. Along the Gulf Coast, relative maize abundance did not increase significantly until the Late Formative period (VanDerwarker 2005, 2006). This transition occurred earlier on the Pacific Coast, during the Middle Formative (Rosenswig et al. 2015). Unfortunately, there has been no systematic assessment of the archaeobotanical literature in Mesoamerica to assess when beans entered cropping systems in different regions, and thus our knowledge of the spread and adoption of *Phaseolus* beans lags far behind what we know of maize. As mentioned, the earliest known domesticated bean is only 2,300 years old (~300 BC), which places it firmly in the Middle–Late Formative transition. If this specimen truly represents one of the earliest domesticated *Phaseolus* beans, then its adoption during the Middle/Late Formative may well correlate with increases in maize abundance in some regions (e.g., the Gulf Coast [VanDerwarker 2006] and Honduras [Morell-Hart et al. 2014]). One wonders if the addition of beans into a multicropping or successional cropping strategy was in part responsible for increases in overall maize abundance, via the introduction of BNF resulting in greater maize yields. Perhaps in regions in which maize abundance increased significantly *in advance of* bean adoption (and which lacked other intensive strategies), such increased yields may indicate that farmers employed a strategy of intercropping maize with cycads.

Formative farmers also employed landscape modification by creating and maintaining ridge-and-furrow systems, in which crops were planted on the ridges, with excess rainwater funneled into the furrows. This intensive strategy represents a simple form of water management and would have required periodic maintenance to re-mound the soil ridges and excavate the furrows. Unfortunately, ancient ridge-and-furrow systems are much less visible on the modern landscape than terraces or irrigation canals, making it difficult to assess the geographic extent of their usage during the Formative period. Fortunately, ancient volcanic eruptions and subsequent ash deposition onto Formative agricultural fields have enabled the preservation of ridging in some locales. In the Sierra de los Tuxtlas of southern Veracruz, the eruption of Cerro Mono Blanco during AD 150–250 blanketed a broad swath of the region in ash, preserving at least two ridge-and-furrow segments at the sites of Bezuapan (Pool 1997) and La Joya (Arnold 2000). Interestingly, no ridges were encountered under ashfall from the previous two eruptions (1250–900 BC and ~150 BC) at either site, suggesting that the ridge-and-furrow strategy was initiated sometime during the first half of the Terminal Formative period along the Gulf Coast. Perhaps the most well-known preserved ridge-and-furrow systems in Mesoamerica are those identified in and around the farmstead of Cerén in El Salvador (Sheets 1982; Zier 1992). The practice of field ridging appears to have been first adopted here during the Late/Terminal Formative period (400 BC–AD 300), possibly coeval with its adoption along the Gulf. Field ridges have also been documented at the site of Kaminaljuyu in Guatemala under Late Formative period house structures (Bebrich and Wynn 1973; Reynolds and Cardenas 1973).

With respect to the Gulf, it is intriguing that the first ridge-and-furrow systems appear around the same time that *Phaseolus* beans were introduced and maize abundance increased. It is plausible that Gulf Coast farmers initially planted their maize around native or transplanted cycads, then shifted to intercropping with beans once beans entered the region. Given that *Phaseolus* beans do not require detoxification prior to consumption, they may have become a substitute for cycad kernels given their high nutritional value and low relative processing time. Beans do not, however, come close to approaching cycads in terms of the amount of nitrogen they fix. Indeed, farmers may have introduced field ridging to offset the reduced BNF efficiency associated with a shift from cycad to bean intercropping. Water is the single biggest determinant of nitrogen in the soil (Handley and Raven 1992; Rimski-Korsakov et al. 2009), and by managing water to ensure optimal plant uptake, farmers by extension optimize nitrogen uptake as well.

Diversification of Farming Strategies during the Classic Period and Beyond

Classic period agriculture in Mesoamerica is perhaps best understood within the conceptual framework of landesque capital, a culmination of the long-term investment into permanent landscape change (Feinman 2006). These enduring changes to the landscape vary regionally; as most changes involved water management, the tangible form of landscape alteration would have depended on water availability. For example, the Classic Gulf Lowlands are riddled with relict drained fields in frequently inundated areas that are still visible today; constructing a drained field system is a strategy to both deal with excess water and redirect it throughout the system (Sluyter 1994). Drained fields are sometimes referred to as raised fields as well, as it is difficult to assess whether the planting rows were raised or the field furrows excavated (or both) without soil profiles exposed through archaeological excavation. Regardless of its label, visible ancient drained field systems are widespread throughout Mesoamerica and have been documented along the Gulf lowlands (Siemens 1983; Sluyter 1994; Stoner 2017), in the Maya lowlands (McAnany 1992; Sluyter 1994), Maya highlands (Sluyter 1994), Basin of Mexico (Feinman 2006; Killion 1992; Siemens 1983; Sluyter 1994), and Central America (Siemens 1983; Zier 1992). As cycads are less likely to be found in flat, low lying, inundated areas, I doubt that cycad intercropping would have been a significant agricultural strategy utilized in drained field systems.

In contrast to flat lowlands in which drained systems are found, hilly or mountainous locales are known for their terraced fields, the construction of which allowed farmers to increase the availability of arable land. Systems of terraces can be found in both lowland and highland settings and were constructed and used primarily during Classic and Postclassic times (Evans 1992; Feinman 2006; McAnany 1992; Sluyter 1994). Drier environmental zones with greater topographical relief, whether open or forested, would have served as ideal natural habitats for cycads. One can envision a scenario in which hillsides were converted to terraced fields during which cycads were preserved and managed in place alongside other crops of economic import. Indeed, in drier environments, BNF can potentially offset the lower nitrogen levels in the soil that result from lower water content, as N is positively correlated with water (Handley and Raven 1992; Rimski-Korsakov et al. 2009). One can imagine the inverse for fields that were heavily irrigated via extensive hydraulic systems. Because increasing the regularity of water availability for any crops will result in more available nitrogen for plant uptake, BNF is less necessary to plant growth in permanently irrigated versus nonirrigated fields. In situations in which farming systems combined terraced hillsides with irrigated valleys, it is plausible that farmers intercropped with cycads on the terraces and with beans in the valleys.

Future Directions: Testing the Cycad Hypothesis

I have raised more questions in this chapter than I have answered. But from these questions, it is possible to create testable hypotheses to assess cycad use in Mesoamerican agriculture through time and across space. First, a comprehensive review of the Mesoamerican archaeobotanical literature is needed to create a spatial and temporal map of cycad and bean specimens identified in archaeological assemblages. This step is absolutely critical to ideas about timing of use, transition from cycad to bean intercropping, and the complementarity of the cycad/terrace and bean/valley cultivation hypothesis. Future archaeobotanical analysis should require a microbotanical component, with an explicit attempt to identify cycad starches and pollen from soils and tools (see Pagán-Jiménez, this volume). Given that cycad starches can also be recovered from dental calculus, every effort should be made to extract and identify starches from human teeth; this method is the only one that will unequivocally demonstrate cycad consumption by humans. In humid tropical and subtropical settings, it is rare for human bones to preserve and be recovered archaeologically, but teeth are more frequently encountered under poor conditions of preservation.

Understanding the role cycads played in agricultural practices across Mesoamerica requires botanical surveys targeting archaeological sites, specifically relict fields and terraces to assess their presence and growth in these locales. If cycad intercropping was integral to terrace agriculture in some places, then we would expect to see distributions of mature descendent populations growing on or around some terrace systems. Pairing a botanical survey with a systematic plan to sample soil from relict terraces and fields for microbotanical analysis is another essential step in testing the ideas presented in this paper. Given the increasingly endangered status of many Mesoamerican cycads, it is imperative to conduct these botanical surveys sooner rather than later.

For decades, a focus on maize has eclipsed broader examinations of cultivation systems, which has resulted in an underdeveloped understanding of agricultural practices across time and space. Thanks to the work of many scholars, we have a firm foundation for understanding the domestication, adoption, and intensification of maize. We now need to expand our lens to consider how other resources and technologies were engineered and deployed in relation to maize agriculture. By magnifying our focus, we not only enrich our ancient narratives, but also contribute to the examination of sustainable systems of agriculture that may have relevance to farmers in our modern era of unpredictable climate change.

5

The Archaeology of Cycad Use in Ancient Mesoamerica

Old Data, New Methods

JOSHUA D. ENGLEHARDT, EDDER D. BUSTOS-DÍAZ,
EMANUEL BOJÓRQUEZ QUINTAL, LUIS ROJAS ABARCA,
LUIS R. VELÁZQUEZ MALDONADO,
ESTEBAN SÁNCHEZ RODRÍGUEZ,
AND FRANCISCO BARONA-GÓMEZ

Researchers interested in elucidating cycad use among ancient human groups in the recent archaeological past—within the last ten to twelve thousand years—face significant obstacles, in Mesoamerica and elsewhere. As the chapters in this volume attest, cycads played—and continue to play—significant roles in Indigenous cultures, agroecological practices, foodways, subsistence strategies, and epistemologies, both in the Mesoamerican region and in other contexts (Beck 1992; Bonta et al. 2019; Bradley 2005; Hayward and Kuwahara 2012; Khuraijam and Singh 2012; Kira and Miyoshi 2000; Patiño 1989; Radha and Singh 2008; Smith 1982; Thieret 1958; Veloz Maggiolo 1992). Nonetheless, cycad remains and archaeological materials potentially associated with cycad use are often unrecognized or overlooked in archaeological assemblages. Far more frequently, researchers focus on the origins and development of better-known domesticates, such as maize (*Zea mays*; see, e.g., Benz 1999; Blake 2015; Matsuoka et al. 2002; Piperno et al. 2009; Pohl et al. 2007) and other "traditional" Mesoamerican cultigens (e.g., beans [*Phaseolus* spp.], squash [*Lagenaria siceraria* and *Cucurbita* spp.], etc.).[1] Thus, we face a severe lack of evidence and hard data.

Relatedly, although evidence from previous archaeological investigations may exist, buried deep in reports or archives, finding it requires the

Herculean task of comprehensively reviewing the entirety of the archaeobotanical literature for a given region. Alternately—an even more daunting proposition—one could pore through hundreds of thousands of excavated materials in dispersed collections to possibly identify cycad remains missed (or misidentified) by the excavators. And this possibility assumes the original investigators included an archaeobotanical component and moreover knew how to identify cycad remains, and that such materials were retained and not simply tossed aside with the backfill. For a region as large as Mesoamerica, it is difficult to overstate the monumentality of such a review. As Amber VanDerwarker (this volume) notes, the scale of this kind of work is overwhelming—let alone the following necessity of collating the data to discern spatial and temporal patterns.

Finally, even if one accepts the potential significance of cycads in preagricultural subsistence strategies, we lack—in Mesoamerican contexts—conventional methods by which to detect cycad remains in archaeological materials. In contrast to, for example, cacao (see Hurst 2006), there is no established method for the determination of cycad traces on archaeological objects. Although, as we discuss below, several promising potentialities exist (e.g., starch grain analysis), these have not been applied commonly in the context of possible Mesoamerican cycad use (but see Simms 2014 for a notable exception). Nor have functional-typological studies of material cultural assemblages considered the potential roles of tools and other objects in the harvest or processing of cycads in this region. Other more innovative methodological possibilities require further technical and conceptual advances to become feasible, as well as baseline comparative data against which to interpret results, since we lack a great deal of basic evidence regarding the fundamental biology, genetics, and chemistry of this ancient plant lineage.

We do not make these observations to elicit sympathy for the relatively poor status of cycads in Mesoamerican archaeology—or scholarship in general—nor to reprimand colleagues or straw men, in archaeology or other disciplines, for failing to consider cycads in their research designs. A focus on the origins of food crop agriculture is understandable and, indeed, a principal goal of archaeological investigation, since domestication and the development of agriculture is one of the most significant cultural and evolutionary transitions of the Holocene (Smith 2005:9438). Nonetheless, theoretical research over the last 30 years has suggested that the horticultural management of wild foods was critical in the transition to agriculture in many contexts (see, e.g., Ford 1985; Freeman et al. 2015; Gremillion et al. 2014; Terrell et al. 2003; Winterhalder and Kennett 2006). In this sense, and echoing VanDerwarker (this volume), we merely seek to highlight what we consider an understudied subject that presents potentially fruitful avenues and may shed light on a central subject of archaeological investigation.

To that end, in this chapter, we undertake a critical review of the limited archaeological evidence that speaks to the presence of cycad remains in the material record of ancient Mesoamerica. We then present a series of methodological proposals that may facilitate the identification of cycad remains in archaeological research, emphasizing the need not only for methodological advances, but also for the acquisition of detailed biological, genetic, and chemical data for extant cycad species. We conclude with a brief discussion of how these methods may assist in elucidating the place of cycads in the cultures and agroecological systems of ancient Mesoamerica, as well as the potential role of cycads in processes of domestication and the transition to agricultural lifeways.

Archaeological Evidence of Cycad Use in Ancient Mesoamerica

As the above comments make clear, the material record associated with cycad use in the Mesoamerican past—both before and after domestication events and the development of agriculture—is exceedingly thin. As Bonta et al. (2019; see also Hanselka 2011, 2017) reported, the vast majority of available evidence derives from excavations by Richard S. MacNeish at a series of cave sites and rockshelters in the states of Tamaulipas and Puebla (Figure 5.1). The projects behind these excavations were specifically archaeobotanical in nature and aimed to elucidate the origins of Mesoamerican agriculture. Unsurprisingly, these sites also yielded data whose analyses have produced "the most significant advances in understanding the transition to food production" in Mesoamerica (Smith 2005:9438). Although analyses (and reanalyses) of recovered materials (see, e.g., Hanselka 2017; MacNeish 1964, 1992; Mangelsdorf et al. 1967; Smith 1997, 2005; see also chapters 9–11, and 15 of MacNeish [ed.] 1967) did indeed produce such advances, it is noteworthy that cycad remains were also recognized and documented by MacNeish in his research at both extremes of the Sierra Madre Oriental. These materials have received far less attention in the scholarly literature, despite suggesting some potentially surprising conclusions.

Tamaulipas Cave Sites

In the late 1940s and early 1950s, MacNeish undertook archaeological fieldwork in Tamaulipas, focusing primarily on cave sites in the Infiernillo Canyon of the Sierra de Tamaulipas and the Ocampo region, 120 km

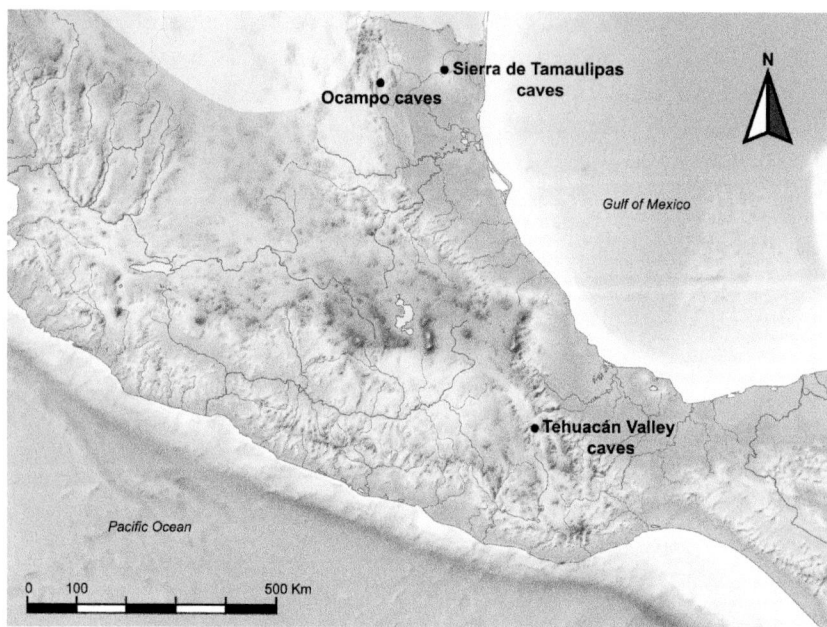

Figure 5.1. Dry cave sites in Mesoamerica that have yielded evidence of cycad remains in archaeological contexts (map by Joshua D. Englehardt, based on data reported by MacNeish [1958:20, fig. 5], cf. Hanselka [2011:4, fig. 1.2, 246, fig. 8.1]).

west-southwest in the Sierra Madre Oriental (Blake 2015:79; Hanselka 2011; MacNeish 1958:20, fig. 5). Excavations at the La Perra (Tmc 174), Nogales (Tmc 82), and Diablo (Tmc 81) caves in the Sierra de Tamaulipas and the Ojo de Agua (Tmc 274), Valenzuela's (Tmc 248), and Romero's (Tmc 247) caves in Ocampo (Kelley 1954a, 1954b, 1954c; MacNeish 1954a, 1954b, 1954c, 1958) produced a critical data set spanning a 9,000-year sequence that revealed a rich inventory of cultivated foods (Table 5.1). These include early specimens of key Mesoamerican domesticates, such as maize remains dating to as early as 4405 BP (Mangelsdorf et al. 1967:33–34; adjusted via AMS to 3930 ± 50 BP by Smith 1997) and cucurbits at as early as 6440 BP (Mangelsdorf, MacNeish, and Willey 1964:430; AMS dated to 5540 ± 60 BP by Smith 1997).

MacNeish's work also uncovered a considerable amount of plant remains associated with wild or managed (undomesticated; see discussion of the term "wild" in the Introduction) food plants, which by volume greatly surpassed the domesticated crops central to later Mesoamerican foodways. Of particular note, excavations recovered a large number of *Dioon edule* cf. *angustifolium* remains (Kelley 1954a; MacNeish 1954a, 1954b, 1954c, 1958; Figure 5.2). These included 14 bases, 31 bracts, and 1,621 seeds in the La Perra/Flacco-Guerra phase, 618 seeds, 100 bracts, and 56 bases in the Laguna/La

Table 5.1. Chronology of Tamaulipas cave sites and associated domesticate remains

Ocampo chronology		Sierra de Tamaulipas chronology		
Phase	Age Range (cal. BP)	Phase	Age Range (approx. BP)[a]	Cultigens
San Antonio	500–200	Los Angeles	750–200	
San Lorenzo	1100–500	La Salta	1450–950	Common bean (1285 BP)
Palmillas	1900–1100	Eslabones	1750–1450	
La Florida	2400–2000	Laguna	2600–1750	Butternut squash (2750 BP)
Mesa de Guaje	3600–3000	Almagre	3950–3450	
Guerra	4400–3600			Maize (4405 BP)
Flacco	5200–4400	La Perra	4950–3950	Cushaw squash (5035 BP)
Ocampo	6000–5200	Nogales	6950–4950	Pepo squash (6310 BP)
				Bottle gourd (6440 BP)
Infiernillo	9000–7600	Lerma	9950–8950	
		Diablo	13,950–11,950	

Source: Data from MacNeish 1958; Mangelsdorf, MacNeish, and Willey 1964; Mangelsdorf et al. 1967; Smith 1997 (cf. Hanselka 2011:6, table 1.2, 104, table 5.1).
[a] Insecure, uncalibrated dates.

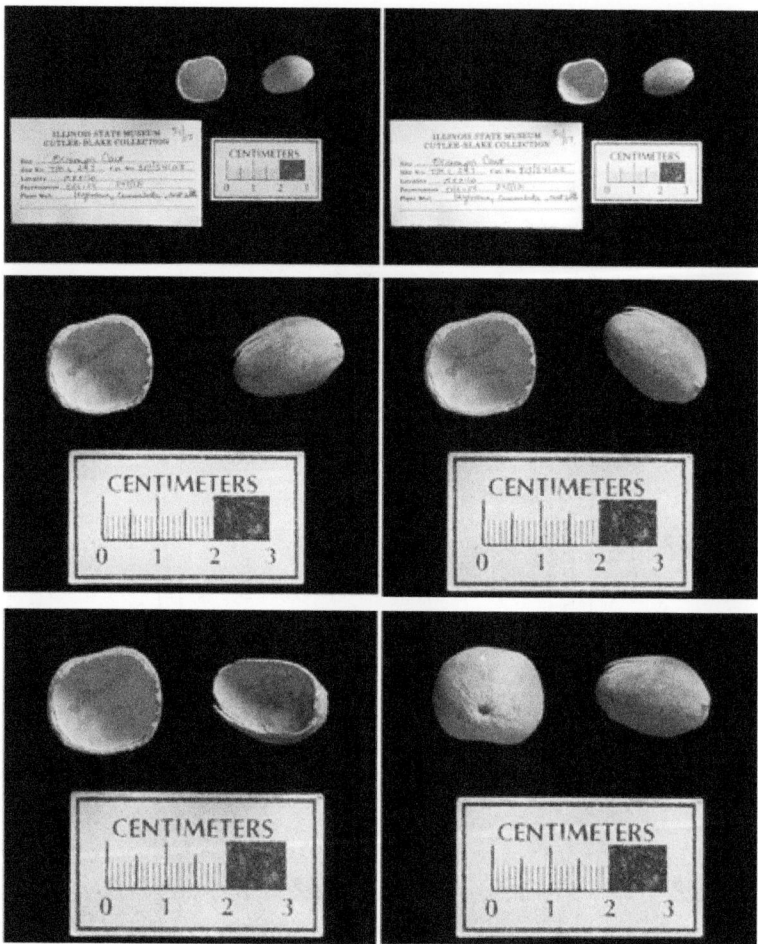

Figure 5.2. *Dioon edule* seed fragments recovered from MacNeish's excavations at the Ocampo cave sites (photographs by Doug Carr, courtesy of the Illinois State Museum).

Florida phase, and 689 seeds in the Los Angeles/San Antonio phase (MacNeish 1958:144–149, fig. 47[5]; see also Hanselka 2011:229, 236, 2017), as well as a curious object fashioned from a *Dioon* seed that MacNeish (1958:104, 95, fig. 33[2]) called a "nut on a stick," whose function remains unknown.[2] MacNeish (1958:144, 149) reported that *Dioon* was the dominant plant remain from the La Perra through Laguna phases in the Sierra de Tamaulipas, although he mentioned this almost in passing, as if it is a relatively unimportant distraction that obscured the concomitant appearance of maize.

Table 5.2. Key elements in the diet of the hunter-gatherers in the Sierra de Tamaulipas cave sites, showing significance of cycads in early subsistence strategies

Phase	Radiocarbon dates (approx.)	% of total volume of wild plants	Total of all food (L)	*Dioon* (L) and % of total food	Domest. plants (L)	Maize (L) and % of total food
La Perra	4500–4000 BP	76	19.77	8.10 (41%)	1.84	0.85 (4%)
Laguna	2000 BP	51	23.72	3.09 (13%)	9.28	5.91 (28%)
Los Angeles	500–200 BP	42	15.57	3.44 (22%)	6.17	5.42 (35%)

Source: Data from MacNeish 1958:147–148, table 20.

It would appear, however, that MacNeish may have missed the forest for the trees. In fact, as early as 4500 BP, *Dioon* made up 41% of the edible plant remains encountered in the Sierra de Tamaulipas caves, whereas maize composed only 4% of total foodstuffs (Table 5.2; see also Bonta et al. 2019:table 2)—and recall that cucurbits had likely been domesticated by this time (see Table 5.1; Smith 1997). At the Tamaulipas cave sites, wild foods by volume greatly surpassed the domesticated crops. It is clear that at these cave sites, cycad seeds played a significant dietary role, particularly in the earliest occupational phases. Moreover, as data from the more recent Laguna and Los Angeles phases demonstrate, cycads continued to be exploited in significant quantities even *well after* the domestication of maize—and other food crops—at these sites. Despite the potential significance of these findings, the study of the materials recovered from MacNeish's excavations in Tamaulipas (Mangelsdorf, MacNeish, and Galinet 1964; Mangelsdorf, MacNeish, and Willey 1964; Mangelsdorf et al. 1967) and more recent reconsiderations of this material (e.g., Smith 1997) seem to have glossed over these data, focusing instead on the evidence related to domesticates. While this may be understandable (and forgiven), it is unfortunate that, as Hanselka (2011:7) notes, "the original excavators did not retain the majority of wild plant materials encountered [in the Tamaulipas caves]." It is therefore, as Hanselka continues, no longer possible to comprehensively reconstruct the preagricultural diet of the groups that occupied these sites—or the role that cycads may have played in that diet.

Tehuacán Valley Rockshelters

Following his work in Tamaulipas, MacNeish undertook excavations at the Santa Marta cave in Chiapas (Blake 2015:79–80) before beginning his landmark interdisciplinary investigation of agricultural origins and cultural development in the Tehuacán Valley in 1960. At the rockshelters and caves of

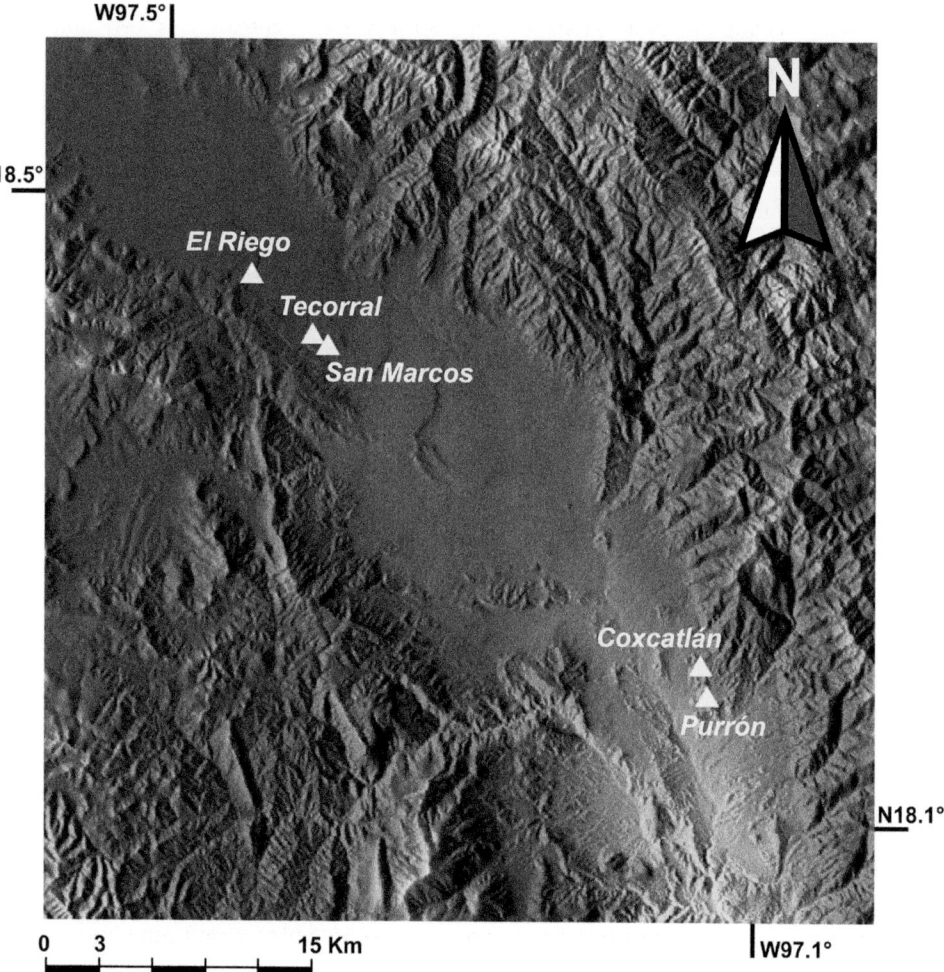

Figure 5.3. Approximate locations of cave sites excavated by MacNeish in the Tehuacán Valley, Puebla (map by Joshua D. Englehardt, based on data reported in Mangelsdorf, MacNeish, and Galinat 1964 and MacNeish [ed.] 1967).

Coxcatlán, El Riego, Purrón, San Marcos, and Tecorral (Figure 5.3), MacNeish excavated an enormous amount of material, including even earlier specimens of maize and cucurbits (dated to 3450 BC and 5960 BC, respectively; Smith 2005:9444, table 3; cf. Mangelsdorf, MacNeish, and Galinat 1964:540, fig. 2) than he had encountered in Tamaulipas. MacNeish's research produced yet another millennial sequence for the valley's occupation, and provided still more data on the development of traditional Mesoamerican domesticates in the Archaic period. Together with data from the

Tamaulipas excavations, as well as evidence from the Guilá Naquitz cave in Oaxaca (Flannery 1986; Piperno and Flannery 2001), researchers were able to create a clearer vision of the origins of agriculture in Mesoamerica.

Once again, however, MacNeish encountered cycad remains in the Tehuacán excavations. Although not found in the quantities reported in MacNeish's investigations in Tamaulipas, cycad remains were present in two of the Tehuacán caves (Smith 1967:table 26). Smith (1967:235) mentions that broken *D. edule* seed coats, although not abundant (approximately 31 seeds), "showed that the plant had been fairly persistently used over some thousands of years, primarily by the occupants of Coxcatlan Cave" (see also Hanselka 2011:237). Curiously, however, Smith goes on to state that, although *D. edule* provides a starch food when cooked, "it could never have been important in the diet of the Coxcatlan Cave people." Exactly how he reached these seemingly contradictory conclusions, especially in light of the Tamaulipas data, remains unknown. The majority of the analyses of the excavated materials (MacNeish [ed.] 1967; Mangelsdorf 1974; Mangelsdorf, MacNeish, and Galinat 1964; Mangelsdorf, MacNeish, and Willey 1964), as well as more recent reanalyses (e.g., Smith 2005), have focused on the history of domesticates revealed in the Tehuacán excavations.

Discussion

The material record of cycad use in ancient Mesoamerica is thus, as VanDerwarker (this volume) asserts, quite limited. We reiterate that this is eminently understandable. Cycad remains are not easily preserved in most contexts, and their detection does not generally form part of most archaeological research designs. Archaeologists often do not know how to identify them, and distinct research questions drive those studies in which cycad remains are most likely to be encountered.

Nonetheless, the archaeological evidence, thin as it may be, clearly demonstrates the alimentary importance of cycads as early as 6000 BP, suggesting that cycads were used persistently for several thousand years by Archaic populations. Indeed, at the Tamaulipas caves, cycad seeds played a significant dietary role and were the dominant plant remains, particularly in the earliest occupational phases. The critical point is that by the end of the Archaic period (to 4000 BP), at a time before or synchronous with the development of modern maize (and other traditional Mesoamerican domesticated food crops), cycads formed a major component of Mesoamerican dietary strategies and were consumed in quantities surpassing or similar to other wild plant resources, such as maguey and various palms, forming a major component of the diet in these regions. Moreover, as data from more recent occupational phases demonstrate, cycads continued to be exploited in significant quantities even *after* the domestication of maize.

When considered in light of the ethnographic data that contextualizes middle-range archaeological research, such conclusions are unsurprising. Ethnographic research has revealed a variety of uses for cycads, from utilitarian to ritual-religious, in modern Mexican communities (see, e.g., Bonta et al. 2019; Carbajal-Esquivel et al. 2012; Pérez-Farrera and Vovides 2006; Sifuentes de Ortiz 1983; Vite-Reyes et al. 2010; see also the chapters by Bonta in this volume). As a botanical resource that occurs in diverse ecosystems, by far the most common use of cycads reported in contemporary ethnographic studies is as a foodstuff, be it as a staple starch or as a famine food. This evidence aligns with data from other parts of the world (Beck 1992; Bradley 2005; Hayward and Kuwahara 2012; Khuraijam and Singh 2012; Kira and Miyoshi 2000; Radha and Singh 2008; Smith 1982; Thieret 1958). Thus, we should *expect* to find cycads in archaeological, particularly preagricultural, contexts.

Of particular note, ethnographic accounts show that the processing of cycad seeds involves nixtamalization with ash or lime, to remove toxins derived from cycad starch. Common preparations of cycads include as tortillas and tamales—foods "traditionally" associated with maize—as well as bread (Bonta et al. 2019; Carbajal-Esquivel et al. 2012; Vite-Reyes et al. 2010). These facts, combined with the archaeologically documented temporal depth of cycad use in Mesoamerica, suggest the possibility that key components of the maize complex have origins that predate the domestication of that plant, and were developed originally for the processing of cycads. Further, given the formal similarities between cycad cones and maize cobs (Figure 5.4), as well as the symbolic significance of cycads in Mesoamerican thought (see Carrasco, this volume), it may be that cycads served as a type of mental template for the domestication of maize itself (see Bonta 2010b; Bonta et al. 2019; Englehardt et al. 2020).[3]

None of this is meant to suggest that cycads were the paramount foodstuff of preagricultural foodways, or that archaeologists should drop all their current projects and rush to investigate cycads. However, in light of these intriguing preliminary data, combined with recent theoretical advances that place increased emphasis on the management of wild foods in "nonlinear" transitions to agriculture (Freeman et al. 2015), we would certainly suggest that cycads deserve a closer look in archaeological research, echoing VanDerwarker's (this volume) call. The difficulty then becomes a methodological one: how can we identify cycad remains in archaeological assemblages? It is this question to which we turn below.

Figure 5.4. Comparison of cycad cones and maize cobs. *a,* Evolution of maize (image by Joshua D. Englehardt after original photograph in Mangelsdorf 1974:182, fig. 15.25); *b,* desiccated *Zamia* strobili, Universidad Autónoma del Estado de Hidalgo cycad strobili collections (photograph by Joshua D. Englehardt); *c, Zamia furfuracea,* Veracruz (photograph by Angélica Cibrián-Jaramillo).

METHODOLOGICAL PROPOSALS FOR THE IDENTIFICATION OF CYCAD USE IN ANCIENT MESOAMERICA

Since its inception, one of the central goals of archaeology has been to elucidate diachronic changes in human cultures and the reciprocal relationships between culture and environment. Recent technological advances and fundamental changes in the focus of modern archaeological research have facilitated the integration of earth sciences such as paleoecology and geochemistry with archaeology (Brüssow 2020; Mayle and Iriarte 2014; Pastor et al. 2016). Such approaches emphasize the synthesis of cultural and environmental information into a cohesive interpretation of human cultural ecology and its impact on soils, landscapes, and agroecological systems. Although cycads have not been treated extensively in the archaeological literature, techniques for the detection of cycad remains in archaeological contexts already exist, and others can be developed or adapted as necessary. What remains is to determine what such methods can uncover when they are applied in archaeological research.

Some methods that may be used to detect cycads in archaeological remains are relatively well established, and they simply have not been applied consistently in Mesoamerican contexts—or have been applied to detect other types of botanical samples. For example, VanDerwarker (this volume) details the potential value of starch grain analysis (see also Pagán-Jiménez, this volume, for examples of the practical application of this method). This method has been employed with success in Caribbean and Central American contexts to detect *Zamia* starches in archaeological materials (e.g., Dickau et al. 2007; Pagán-Jiménez et al. 2005). *Zamia* starches have also been detected in stone tools at a Late Classic period Maya site (Simms 2014).[4] Likewise, Mickleburgh and Pagán-Jiménez (2012; see also Pagán-Jiménez, this volume) have demonstrated the exciting possibility of identifying cycad starches in human dental calculus. These methods are reviewed at length in other chapters in this volume, and elsewhere, so we will not engage an extended discussion of them here, although we do concur with VanDerwarker that more consistent application of microbotanical analyses is sure to reveal valuable new evidence on human cycad use. Rather, in the desire to complement the previous studies and suggestions of our colleagues, we offer below some preliminary considerations of other methodological techniques that may be incorporated into investigations aimed at studying the role of cycads in ancient Mesoamerica.

Ethnoarchaeology and Experimental Archaeology

While cycad seeds, cones, and other remains are relatively easily identifiable to the expert eye, microscopic traces that may be encountered on archaeological objects are considerably more difficult to detect, since we do not know exactly which artifacts, if any, were commonly used to process cycads in the past. Starch grain analyses, like other techniques discussed below and elsewhere, are very powerful methodological tools, yet to apply them we must first be able to identify the artifacts on which such traces may be found (see Pagan-Jiménez, this volume, Figure 3.6 and Table 3.2 for a useful starting point based on Caribbean data). In this regard, ethnoarchaeological research and experimental archaeology can be invaluable sources of information. Ethnoarchaeological investigations on cycad use have been carried out in Australia (e.g., Beck 1992; Bradley 2005; Smith 1982), where these plants were used in myriad ways and occupied an important place in the cultures of which they formed a part. These studies have been very useful for the identification of stone artifacts used in the process of seed detoxification—a crucial step in the preparation of cycad-based food—as well as providing valuable information regarding the detoxification process itself (see the chapters in David et al. [eds.] 2006). Likewise, ethnographic and ethnobotanical research in many parts of the world (e.g., India, Japan, the Caribbean; see Hayward and Kuwahara 2012; Khuraijam and Singh 2012;

Kira and Miyoshi 2000; Patiño 1989; Radha and Singh 2008; Veloz Maggiolo 1992; see also Pagán-Jiménez, this volume) provides clues to the types of objects employed in cycad processing that may be encountered in the archaeological record. We believe that similar investigations in Mesoamerican contexts will assist in the identification of artifacts that may have been used during cycad processing, for food preparation, or otherwise. The work of Bonta and colleagues (Bonta et al. 2019; Bonta, chapter 8, this volume; see also Diego-Vargas 2017; Vite-Reyes 2012; Vite-Reyes et al. 2010) represents an excellent start, but a greater number of more focused studies are needed.

Identifying the objects used in cycad processing is crucial, but knowing how cycad tissues were processed is just as important, since such data can provide invaluable clues as to which types of microscopic remains may be preserved on artifacts. To understand the chemical transformations that cycad tissues go through during the detoxification process, we consider experimental archaeology a potentially fruitful avenue of research. Such an approach allows for the reconstruction of cycad processing methods in a controlled manner, such that the chemical reactions that occur during it—and their outcomes—can be observed and measured, and their effects quantified. This is a crucial first step in establishing a reference database for starches and phytoliths produced via cycad processing (*supra* n4), which is necessary for the correct identification of cycad remains via micro- and macrobotanical analyses. Baseline data for cycad starch grains, phytoliths, tissues, and even yeasts could be established by applying techniques such as optic or scanning electron microscopy to experimentally produced materials or evidence (Aouizerat et al. 2019; López-Montalvo et al. 2017; Wang et al. 2016), particularly to support the bioarchaeological identification of microorganisms associated with cycads (Brüssow 2020).

To that end, and following the studies of Pagán-Jiménez and colleagues on the effects of processing on starch morphometry in the Caribbean (Pagán-Jiménez et al. 2017), we have invited members of a Teenek community from San Luís Potosí—one of the few Indigenous groups in Mexico to still prepare and consume cycads on a regular basis—to process cycads in controlled laboratory conditions. We believe that the outcomes of this experimental work will assist in establishing baseline data that may be applied and refined in future studies, as well as preliminary functional typologies that may further elucidate the types of artifacts on which cycad remains are most likely to be found. Likewise, such data will permit the comparative evaluation of ancient cycad starch grains identified on archaeological objects against modern reference collections. Evaluating variability in morphological and morphometric characteristics may provide clues to processing and preparation methods and the tools employed therein (Reinhart 2020; see also Pagán-Jiménez, this volume, Figures 3.6 and 3.9, and Table 3.2)

Soil Chemistry and Isotopic Nitrogen Measurements

VanDerwarker (this volume) mentions the possible use of cycads as nitrogen fixers in early agricultural practices, highlighting the potential for cycad intercropping with other cultigens prior to the domestication of other nitrogen-fixing food crops, such as beans and legumes. Cycads are well known for their ability to grow in adverse conditions, due in part to their ability to form symbiotic relationships with nitrogen-fixing microorganisms, namely cyanobacteria (Bustos-Díaz et al. 2019; Lindblad 2008). Thanks to this adaptive characteristic, cycads are often able to thrive in depleted soils that are not conducive to the growth of other plant species. More importantly, cycads can positively alter the level of bioavailable nitrogen in the soil, which can be advantageous for surrounding crops (Watanabe et al. 1977). To evaluate the possible role of cycads in ancient horticulture and agriculture—and, potentially, domestication processes—and to measure the potential effects of cycads on soils in which more commonly known cultigens were planted, edaphological studies and analyses of soil chemistry may prove useful.

The mineralogical analysis of soil samples is well established in archaeological science. Many previous methods, however, such as those reliant on inductively coupled plasma (ICP) technologies, require very large samples and are destructive, laborious, and expensive (Pastor et al. 2016; Rawal et al. 2019). Considering the study of the role of cycads in nitrogen fixation, it is preferable for the soil and plant samples to be analyzed in situ, or, alternately, for collected samples to be homogenized to obtain a reliable laboratory analysis. In both cases, Portable X-ray fluorescence (pXRF) spectrometry has proven an effective technique for rapid physicochemical assessment of archaeological, soil, and plant samples (Rawal et al. 2019; Tadeu Costa Junior et al. 2019). This archaeometric method of elemental analysis is both accurate and cost-effective and has been deployed in a variety of in situ and ex situ contexts (Brent et al. 2017; Cooper et al. 2020; Hunt and Speakman 2015; Ravansari et al. 2020; van der Ent et al. 2019). We thus suggest that pXRF may prove useful in the archaeological investigation of cycads, provided that reliable comparative data are available.

Plant-driven changes in the physicochemical composition of soils have been thoroughly studied (see, e.g., Brinkman et al. 2010; Ehrenfeld et al. 2005; Kardol et al. 2013; Neina 2019), and while most of the effects that a plant like the cycad could have on the soil can—and should—be studied, they are not generally considered in archaeological research designs. Further, although soil samples are often collected in archaeological field research, at the moment there is no reliable method of which we are aware to measure the potential effects of cycads on soil samples or cores from archaeological contexts (although pXRF offers potential). We nonetheless reiterate, however, that the lack of such methods is likely due to the fact that researchers often do not look for such data, rather than to the total absence of evidence, especially given the extensive ethnographic and (limited) archaeological records of cycad use in Mesoamerica dating to the Archaic period.

It is also true that a major technical challenge in this respect originates from the fact that cycad-cyanobacteria symbioses occur in highly specialized, transitory organs called coralloid roots (see Introduction, this volume, Figure 0.4). Although not all cycads have such root structures, all cycads have the potential to produce this tiny organ. We therefore postulate that a better understanding of coralloid roots may assist in overcoming this methodological hurdle, since the basis of this symbiosis suggests the existence of broader microbial diversity beyond cyanobacteria, which together would imply evolutionary signatures associated with metabolic traits (Gutiérrez-García et al. 2019; Suárez-Moo et al. 2019) that have greater potential to leave traces in the fossil, archaeological, or edaphological records.

To address this lacuna, we propose here two methods that may be used to detect cycad root structure symbioses in edaphological soil cores. The first is the measurement of variable levels of isotopic nitrogen ($\delta 15N$) in biological remains, such as bones, teeth, plant remains, and even ancient soil samples. Because distinct nitrogen-fixing vectors will produce differing amounts of nitrogen isotopes in organic remains, such variability, when measured in plant or soil remains, can be a good indicator of the type of nitrogen fixing or uptake system that resulted in a given amount of a specific nitrogen isotope encountered in a particular sample (Szpak 2014). This idea is supported by the recent proposal of the use of cycad foliage as an archive of the isotopic composition of atmospheric nitrogen (Kipp et al. 2019). Such data, in turn, should provide clues that may be used to detect the presence of nitrogen-fixing cycads in ancient soils and untangle their potential role in early agricultural practices.

A second possibility is the detection of siderophores, microorganic iron-chelating compounds that are secreted by microorganisms such as bacteria and fungi in plant root structures. Siderophores, typically seen as specific for iron (but also able to mobilize other metals, e.g. Cu, Zn, etc.), are excellent indicators of the presence of microorganisms in soil samples from modern and archaeological contexts (Chakra et al. 2019). Recent studies have demonstrated the ways in which bacterial and fungal communities in the rhizosphere correlate with soil physicochemical properties (see, e.g., Lee et al. 2019; Nelson et al. 1988). Since specific rhizospheric interactions between fungal and microbial consortia—characteristic of specific plants—produce distinct chemical signatures in siderophores, measuring such chemical compositions in archaeological soil samples—via mass spectrometry, for example—should produce baseline data indexical of specific microbial consortia, derived from particular plant root structures, including coralloid roots, within discrete edaphological contexts. Such data, in turn, may reveal the presence of cycads in ancient soils and, when considered in conjunction with other data sets (e.g., starches, phytoliths, siderophores derived from other plants, etc.), potentially illuminate their place in ancient agricultural strategies.

The primary difficulty with each of the methods proposed above is the lack of baseline data against which measurements can be comparatively analyzed. It is also possible that variable levels of nitrogen isotopes and/or siderophores that indicate specific, distinct metals may be due to differing edaphological characteristics, or possibly to unknown microbial interactions taking place within the coralloid roots. Regarding siderophores, it is unclear whether they can be reliably detected in ancient samples, since their endurance in soil matrices is unknown: their abundance is highest in the rhizosphere (Akers 1983; Buyer and Sikora 1990; Nelson et al. 1988), but it is unlikely that rhizospheric soil would be conserved in archaeological contexts for any length of time. To evaluate these potentialities, and to produce initial baseline measures, the authors are currently collaborating on a multidisciplinary research project that has as one of its goals the characterization of organic consortia within ancient soil remains. As part of this work, we have isolated and cultured siderophore-producing bacteria from different cycad tissues, including coralloid roots, and also from cones and the guts of insects that fed and/or pollinate cycads. In addition, we have recently carried out analyses via scanning electron microscopy–energy dispersive X-ray spectroscopy (SEM-EDS) to map the location of specific elements within cross sections of coralloid root structures and to link elements such as Fe to particular tissues where cyanobacteria are located (Figure 5.5). We are exploring other methods (see, e.g., Kopittke et al. 2020; van der Ent et al. 2018) that may be used for the visualization of elements in cycad tissues. The next step is to culture siderophores in ancient soils to establish preliminary criteria by which particular organic consortia may be correlated with specific cycads. At the very least, we expect to produce a tentative measure of presence/absence of cycads in archaeological soil cores. It is hoped that

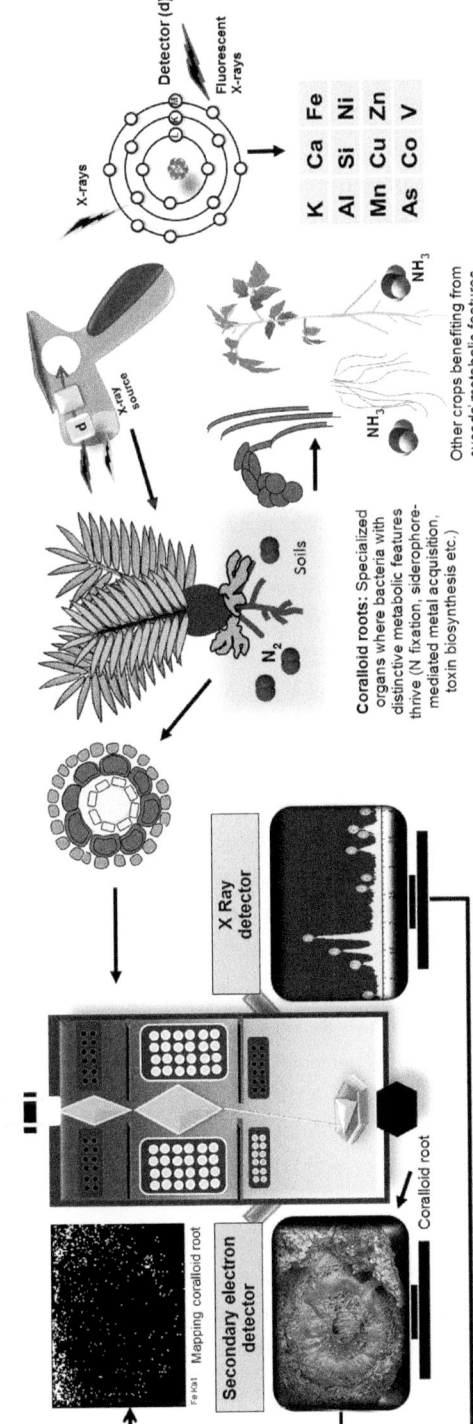

Figure 5.5. SEM-EDS and pXRF: multi-elemental analytic methods to *visualize* and *quantify* elements in soils and cycad tissues. *Left panel*: the images obtained under SEM correspond to secondary electrons emitted after interaction with the sample. The X-rays generated in a sample subjected to electronic bombardment (SEM-EDS) permit determination of element presence and concentration. Using SEM-EDS samples are scanned with an incident electron beam to produce the characteristic fluorescent X-rays. *Right panel*: pXRF is widely used for environmental assessment of soils and plants, both in situ and ex situ. Elements are detected based on characteristic fluorescent X-rays, which are generated by passing the specimen through a focused beam of high-energy X-rays (XRF) or electrons (SEM-based EDS) (image by the authors).

such methods and data will inform future investigations and methodological refinements, particularly in conjunction with paleogenomic evidence.

Paleogenomics and Archaeogenetic Analyses

The final methodological suggestion that we wish to discuss here involves paleogenomics and archaeogenetics (see, e.g., Amorim 1999; Pauling and Zuckerkandl 1963). Although generally applied to investigations of human evolution (see, e.g., Renfrew and Boyle 2000), as Cibrián-Jaramillo et al. (this volume; see also Zeder et al. 2006) note, paleogenetic techniques have recently begun to be applied to questions surrounding the evolutionary history of many plant species—including cycads—in a variety of contexts (e.g., Calonje et al. 2019; González et al. 2008; Gutiérrez-Ortega et al. 2018; Kyoda and Setoguchi 2010; Meerow et al. 2012; Prado et al. 2016; Zonneveld and Lindstrom 2016). These exciting new data lead us to believe that genomic evidence, including proteins and metabolites encoded by genes (i.e., biomarkers), can play a crucial role in the identification of cycad remains in archaeological samples. Key gene functions, if related to distinctive phenotypes or physiological traits (such as nitrogen fixation) can provide a means by which to identify the proverbial needle in the haystack, since the evolutionary history of cycads and their interacting and/or symbiotic microbes are distinct from other plants with which they coexist.

Many plants, especially those with agricultural and cultural uses (whether domesticated or not), were spread by humans throughout history. Because of this, the scattering of their populations does not always parallel the natural dispersion patterns of such plants (see Cibrián-Jaramillo et al., this volume; Pagán-Jiménez, this volume). The patterns of migration for these plants might, in fact, closely mimic human movements, since it is highly probable that ancient human groups carried, intentionally or unintentionally, seeds of the plants with which they had close relationships. The genetic study of modern cycad populations using reliable genomic markers (e.g., chloroplast DNA) may suggest indicators of such a migration, such as founder effects and admixture. Temporal patterns revealed through the analysis of population splits and genetic divergence events in cycads, in turn, may be compared against archaeologically identifiable demographic movements (also visible via the study of human DNA), linguistic markers (e.g., the spread of shared terms for cycads [see Bonta, chapter 7, this volume]), or the transmission of cultural practices associated with the cycad preparation complex (e.g., ash nixtamalization [see Cibrián-Jaramillo et al., this volume, Figure 2.5]). Correlations among spatial and temporal patterns in these discrete datasets could suggest human involvement in the dispersal and development of specific cycad genera or species in the ancient Mesoamerican past.

Kyoda and Setoguchi (2010) employed a similar methodology to describe the genetic diversification and spread of *Cycas revoluta* Thunb. throughout Japan's Ryukyu archipelago (see also Chang et al. 2019), and their results may suggest a human role in these processes (Englehardt and Carrasco 2022). Their research focused on modern cycad populations, using genetic markers such as chloroplast and mitochondrial DNA. These markers are useful because of their high number of copies in biological samples, and the low variability of their sequences (low mutation rate). To be of analytic use, the very same characteristics are also required of potential markers in archaeobotanical remains, along with short length (80–300 base pairs [bp]) and prior knowledge of their existence in ancient populations. Schlumbaum et al. (2008) suggest examples of appropriate genetic markers potentially found in archaeobotanical materials, such as *rbcL* and *trn* introns and intergenic spacers—noncoding intervening sequences of DNA within a gene (Kinniburgh et al. 1978)—which can be used in the analysis of cycad remains, especially when these are derived from seeds and pollen, which are natural reservoirs of DNA. Likewise, specific cyanobacteria, considered in concert with or independent of plant genomics, can be exploited for this purpose.

The largest obstacle to such an approach in Mesoamerican contexts, however, is the lack of a reference genome for endemic cycad species in the region. Further, ancient DNA is often scarce, due to its natural decay and highly fragmented status, owing largely to the generally unfavorable conditions in which cycads populations occur in Mesoamerica. In addition, samples of ancestral genetic material obtained from the tropics and subtropics are hard to sequence and interpret, even when a reference genome is available to correctly assemble sequences (*de novo* assembly techniques are unreliable due to the aforementioned limitations). Promising recent studies by Calonje et al. (2019), González et al. (2008), Gutiérrez-Ortega et al. (2018), and Prado et al. (2016) on the phylogeographic genetic relationships among Mexican cycad species, however, are beginning to fill in these lacunae (see also Cibrián-Jaramillo et al., this volume). Nevertheless, the identification of the specific genes that mediate interactions between plants and their associated microbial consortia is only just beginning (see, e.g., Gutiérrez-García et al. 2019). A final problem is that of scale: the vast temporal depth of genetic data may not provide sufficient resolution at smaller, archaeological scales to be useful. Nonetheless, these difficulties do not preclude the use of specific genomic markers that can be more easily assembled and studied,

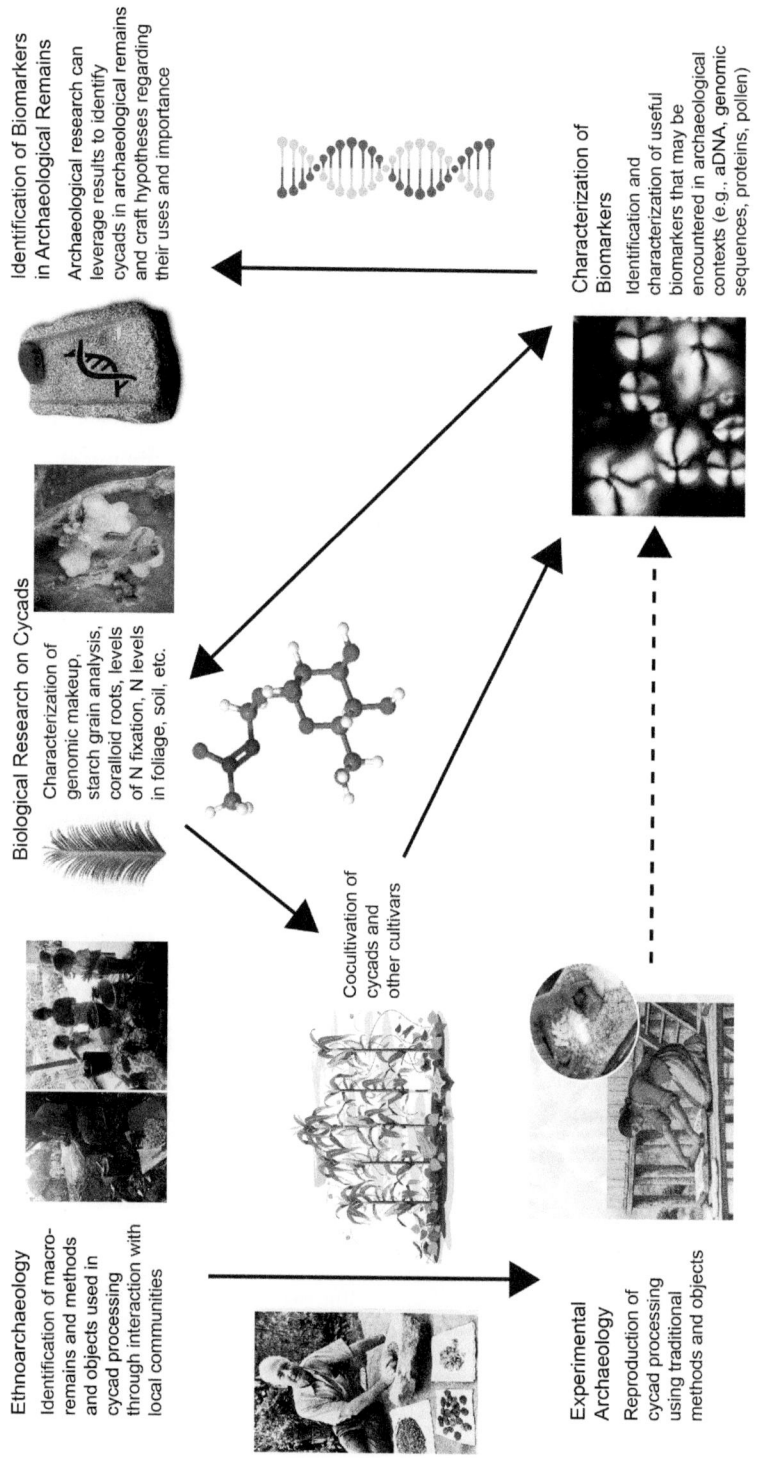

Figure 5.6. Integrated methodological proposal for archaeobotanical, biological, paleogenomic, ethnoarchaeological, and experimental archaeology research on cycad use (image by the authors).

assuming that they can be recovered. In any case, the mere identification of cycads in archaeological remains through this or any of the aforementioned techniques (as integrated in Figure 5.6) is crucial to develop investigations and test archaeological hypotheses surrounding cycad use in ancient Mesoamerican cultures.

An alternative to DNA as a biomarker for the archaeobotanical analysis of organic remains is use of the physiological products coded by genes. Proteins and metabolites, including a broad range of molecular entities derived from metabolism (e.g., carbohydrates, peptides, lipids, etc.), are chemically more stable than DNA and thus have been successfully exploited as biomarkers to investigate archaeological remains (see, e.g., Cappellini et al. 2010; O'Donoghue et al. 1996). Cycads, as such a unique group and order of plants, surely contain an array of such potential biomarkers; for example, proteins, peptides, or lipids involved in symbiotic nitrogen fixation, ancestral biochemical adaptations, or specialized metabolites. The identification of such biomarkers is facilitated by the increased sensitivity of modern spectroscopic techniques, such as mass spectrometry, which has become the norm for searching in situ, in field conditions, for specific biomarkers in complex and unprocessed samples (see Cappellini et al. 2010 and references therein)—even those that contain a matrix of target chemicals, plus environmental contaminants such as soil and decomposed organic material from other organisms. Although powerful, these techniques carry a caveat in regard to their use in research on archaeological materials: they can reliably detect only known markers. Therefore, a crucial first step is the identification and characterization of putative biomarkers via the study of contemporary cycad populations (see Figure 5.6).

For example, we recently have employed direct analysis real-time mass spectrometry (DART-MS)[5] to identify the nonprotein amino acid β-N-methylamino-L-alanine (BMAA) in cycad roots, as well as in the wing and stomach of butterflies of the genus *Eumaeus*, whose larvae are known to feed on cycads (Contreras-Medina et al. 2003) (Figure 5.7). Initial results are promising and may offer an alternative and more precise technique by which to detect this neurotoxin (cf. Bishop and Murch 2020). BMAA is produced by various taxa of cyanobacteria and other microorganisms (Kerrin et al. 2017), and it has even been suggested, by cycads themselves (Marler et al. 2010). We believe that we can use BMAA as a biomarker to identify the presence of cycads in archaeological remains. Via DART-MS, we may even be able to monitor diachronic changes in the levels of BMAA or other biomarkers (Yew 2019), for example, during the processing or preparation of cycads by nixtamalization. Despite the potential benefits of DART-MS spectrometry

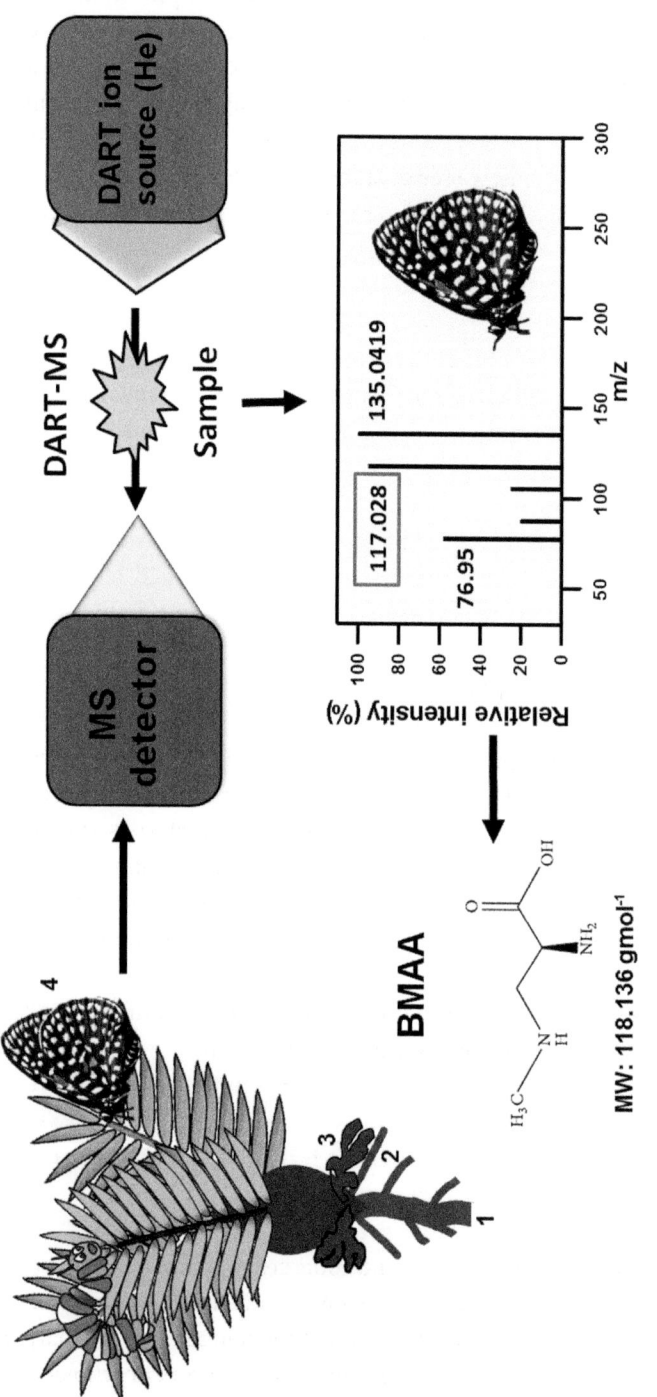

Figure 5.7. Direct analysis real-time mass spectrometry (DART-MS) in cycad research. Cycads form three types of roots, which can be sampled: (1) a primary tap root similar to the root system of most terrestrial plants; (2) lateral roots; and (3) coralloid roots. Other associated biological elements can also be sampled, e.g., herbivores such as butterflies (4). BMAA stands for beta-methylamino L-amino acid, a neurotoxic non-proteinogenic amino acid produced by various taxa of cyanobacteria. Different cycad organs and associated insects may accumulate BMAA. Mass spectrometry can help to identify and quantify BMAA and other biomarkers in intact samples, in a non-destructive form, providing a means for archaeobotanical investigation (image by the authors).

(Feider et al. 2019; Manfredi et al. 2016), it is not often employed in experimental archaeology. We hold that investigations such as these can increase the resolution of the approaches suggested above and, in conjunction with archaeological or ethnobotanical contextual data, may assist in the identification of cycad remains in, for example, archaeological soil samples. Thus, and despite some limitations, we see great potential for the incorporation of paleogenomic and edaphological analyses in the archaeological investigation of ancient cycad use, in Mesoamerica and elsewhere.

Final Thoughts and Future Directions

In this chapter, we have reviewed the admittedly limited evidence for cycad use in ancient Mesoamerica and offered some observations on potential analytic methods that may assist in the detection of these botanical remains in the archaeological record. As made clear in the first part of this chapter—and in other chapters within this volume—scant archaeological data notwithstanding, cumulative evidence points to the use and significance of cycads in ancient Mesoamerica, a conclusion that parallels the important place of cycads observed and documented in other areas of the world. The data thus suggest the utilitarian and symbolic use of this botanical resource since at least the Pleistocene–Holocene transition through the present, even if there is precious little scholarly, particularly archaeological, attention devoted to cycads. And as the second part of this chapter demonstrates, methods and techniques to detect cycad remains in archaeological materials exist—the question is, as VanDerwarker (this volume) points out, refining and applying them. Adding to that suggestion, we would also highlight the necessity of producing baseline data to facilitate comparative analyses in Mesoamerican contexts.

Cycad remains are not often recognized—and rarely sought—by archaeologists. The majority of archaeological investigation that is most likely to uncover cycad remains is focused, more often than not, on a different set of research questions. However, recent research at the El Gigante rockshelter in Honduras (Kennett et al. 2017) has revealed *Zamia* seeds in the archaeobotanical assemblage (Amber VanDerwarker, personal communication, 2019). This project has further produced exciting new data that add substantially to discourse surrounding the use of wild food crops in the preagricultural diet of this region—and possibly even ancient cycad DNA (Amber VanDerwarker, personal communication, 2020) and other data that can be analyzed via the methodologies proposed here. In this sense, we concur wholeheartedly with the suggestion by VanDerwarker—an author of that study and a contributor to this volume—that "we must actually design our research projects to include the possibility of finding cycads in archaeological contexts if we truly want to expand our understanding of ancient Mesoamerican agriculture" (VanDerwarker, this volume).

In that sense, a secondary aim of this chapter—and indeed, of this collection as a whole—is to increase awareness of the potential research value of this subject. It is our hope that doing so, in conjunction with methodological and conceptual advances, will ensure that archaeologists do not overlook the important place of wild plant materials in research on preagricultural diets and the origins of domesticates, as was the case with the original excavators at, for example, the Ocampo caves (Hanselka 2011:7). We echo VanDerwarker's (this volume) call that future archaeobotanical analyses incorporate a microbotanical component, with an explicit attempt to extract and identify cycad starches and pollen from soils and tools, as well as from dental calculus. Again, the methods exist; we simply need to apply them.

We should also note that our aim in this chapter is not meant to be prescriptive; rather, we actively seek to practice what we preach. The authors, in conjunction with many of the contributors to this volume, have been collaborating for several years on interdisciplinary research aimed at developing conceptual and methodological frameworks that will assist in forming a more robust understanding of the place of cycads in ancient Mesoamerican foodways, epistemologies, and cultural practices. At present, we are employing several of the techniques outlined above in a new research project, to both produce needed comparative data and refine the proposed methodologies. We are also poring through the Peabody Museum's online archives of MacNeish's excavations at the Tamaulipas caves and the Tehuacán Valley (https://peabody.pastperfectonline.com/bycreator?keyword=MacNeish%2C+Richard+S) in the hope of recovering information on the cycad remains encountered in these projects.

Of course, it is clear that key questions remain, and much work remains to be done if we are to fully comprehend the historical uses and roles of cycads in ancient Mesoamerica. On the one hand, it is imperative to continue investigating the evolutionary biology and history of cycads with modern, postgenomic tools (see Rai et al. 2003 and Wu and Chaw 2015 for examples of such "-omics" approaches to cycad research), taking into account not only the genome of this group of plants (currently being deciphered by the Plant Genomics Consortium of the New York Botanical Garden's International Plant Science Center), but also their proteomes and metabolomes, by themselves, and during biotic and abiotic interactions with microorganisms, insects, other plants, and/or environmental stresses. Such investigations could provide data that may be translated into relevant and informative biomarkers, which can then be used to develop nondestructive screening methods for large collections of archaeological objects, which could even be deployed in situ. Likewise, ethnoarchaeological and experimental archaeological studies can and should be developed to produce baseline data, establish functional artifact typologies for objects potentially used in cycad processing, and develop hypotheses that can be tested with archaeological data.

Finally, it is important to reiterate the point that conceptual advances are also necessary. Our own collaborative investigations have shown that although limitations in technological approaches can be overcome (or their consequences at least diminished), a more thorough and complete conceptual framework for the ancestral uses of cycads is a necessary first step, since a focus on, for example, cultivation sites vs. ceremonial sites and/or contexts of food preparation will carry distinct implications vis à vis research design. It is also paramount to address the issues of scale that may problematize comparisons between discrete data sets (e.g., archaeological vs. genomic), to which we briefly alluded above. In our experience, an iterative feedback loop informed by multiple disciplines presents the most promise for overcoming these obstacles, and for allowing for the incorporation of technical advances from, for instance, chemistry or genomics, in the archaeological study of cycads in ancient Mesoamerica. This chapter, like this volume, is intended to serve as a preliminary step in the investigative process.

Notes

1. But see Chinique de Armas et al. (2015); Mickleburgh and Pagán-Jiménez (2012); Pagán-Jiménez et al. (2015); or Pagán-Jiménez, this volume, for excellent examples of paleobotanical research on cycad use in prehistoric Caribbean contexts.

2. Although we cannot offer a definitive interpretation regarding the function or significance of this particular artifact (perhaps it was a spindle[?]), its presence does suggest other potential uses for cycads.

3. Mark Bonta originally brought this idea to our attention, and we have continued to pore over this intriguing possibility—and how it might be evaluated—through discussions with Bonta and his collaborators, as well as—and including—many of the contributors to this volume.

4. The absence of a reference database for starches of *Dioon* and *Ceratozamia*—the most common cycad genera in Mexico—is a major impediment to the application of starch grain analysis in Mesoamerican contexts (cf. Pagan-Jiménez 2015).

5. DART-MS is a rapidly developing mass spectrometry technique with ionization at ambient atmospheric pressure, useful for the rapid determination of low molecular weight compounds in different matrices, with a minimum pretreatment of the sample (Cody et al. 2005; Feider et al. 2019; Hajslova ct al. 2011).

6

The Maize God Revisited

Iconology of Cycads, Maize, and Fertility Deities in Mesoamerican Art

MICHAEL D. CARRASCO

> The tsalaam Thipaak (*Zamia* sp.) bordering the trail through the fallow reminds the passerby of his obligation to the godpowers for the bountiful crop of maize now maturing in his milpa.
>
> Alcorn 1984:95

In this chapter I suggest that cycads and the sacred-maize-ancestor (SMA) concept (Bonta et al. 2019; also see Bonta, chapter 8, this volume) played a significant role in mythology, agroecological systems, and Formative period iconography. While maize clearly informs the imagery of many Mesoamerican fertility deities, their representations and the mythology in which they figure draw from a diverse range of taxa and phenomena, especially at their initial appearance in the Early and Middle Formative periods. This visual material provides a major avenue for understanding Indigenous botanical taxonomies that then may be compared to linguistic, ethnohistorical, and contemporary data. Working between these data sets reveals a conceptual world in which there are not just different names for specific things, but the categories themselves are different (Debaene 2013:35). Indigenous terms and classes posit taxonomies that do not align with Western ones; rather, they denote completely different categories whose semantic range often includes items that make sense within Indigenous systems because the items in a given category stand in morphological, functional, or symbolic relation with each other (Viveiros de Castro 2009:241).

In some cases the saliency of certain plants compels the use of their names and associated terminology as meta-labels for a broad category. Identifying this pattern in language and symbolism allows us to review the iconographic information from a new perspective, one that permits iconological analyses that better employ Indigenous categories. The hypothesis is that if the taxonomy uses a marked taxon as a meta-label for other members of a taxonomic class, we might expect the depiction of the key taxon to denote either itself or the class. In current art historical studies, a one-to-one iconic, isomorphic relationship between the image and the referent is often assumed, but this assumption fails to account for hybrid depictions and often leaves emic categories underexplored. Here, I specifically examine the maize-cycad relationship from a number of perspectives to reveal the complex classificatory schemes that connect them to one another and to other things, categories, and even time.

The alimentary evolution documented in numerous mythohistories, most famously expressed in *Historia de los Mexicanos por sus Pinturas* (1941:233) and the *Leyenda de los Soles* [*Legend of the Suns*] (León-Portilla 1966; Moreno de los Arcos 1967), shows that Indigenous epistemologies and narratives of world creation often dealt with this culinary evolution (Alcorn et al. 2006:603, 608; Braakhuis 2009:16; Carrasco 2015, 2020). Indeed, we find alimentary categories include foods appropriate to particular eras of world creation. In the present era maize often plays a central role, but not always, as we will see through a detailed investigation of cycads and the differences between iterations of the origin-of-corn or Corn Master[1] narratives (Chevalier and Bain 2003:173) and etiological narratives of creation. Corn Master narratives are widespread across Mesoamerica and can be found among the Nahua (Chicōmexōchitl, Sintiopiltsin [lit. ear-of-corn-god-son-honorific]), Teenek (Thipaak), Soke (Homshuk), Mije (Kondoy), and Classic period traditions, among others.

While it would take more space than is available here to fully chart the differences between each of these narratives in their specific cultural contexts, it is clear that this imagery and mythology is often more inflected by regional traditions and changing foodways across time than is generally appreciated (Alcorn et al. 2006:608; Arnold 2005; Hepp 2019; Ochoa 2010). The focus for instance on *Dioon edule* (*chamal*) among the Xi'iuy (Pame) (see Bonta, chapter 8, this volume) versus the central role played by maize in the K'iche' Maya *Popol Vuh* clearly shows regional expressions of structurally similar mythology. This and other examples demonstrate that etiological narratives were responsive to differences in regional foodways (Ochoa 2010) so that, while mythologies often preserved basic structural paradigms, the elements within them changed, as we will see with particular clarity in variations in the Corn Master narratives discussed later in this chapter.

In light of this, we might consider "maize" gods as fertility deities that personify processes and the notion of "foodness" or alimentation more than as personifications of specific taxa per se, even if a shared identity between maize and a particular deity is often the case, especially in the Maya iconographic record. On this point, Alcorn (1984:71–72) notes a conceptual difference between the epistemologies of the Mayan-speaking Teenek, living in the Huasteca, and the West. She finds that:

> While Western scientific knowledge is built upon impersonal physical laws governing natural events, Teenek knowledge, in addition to its general pragmatic acceptance of the limitations imposed by natural laws, relies heavily on a personalization of natural forces in order to gain control over them on the possibility of thereby circumventing natural laws which usually govern the activity of these forces. The forces are personified as godpowers known only through their manifestation in action, often by their ability to transform something from one state to another.

Thus, we might consider fertility deities as amalgams and hybrids of various elements and as personifications of change. Doing so complicates explanations that do not adequately consider ethnobotanical categories, the ecological, social, and symbolic processes that deities personify, or which posit a simple one-to-one or isomorphic correspondence between maize, or any other "single" referent, and specific deity representations. In this there is an attempt to build on a consistent line of commentary about Mesoamerican deities that sees them as overlapping or in states of change and transformation. Thus H. B. Nicholson (1971:408) suggests that "[t]his great legion of deities was organized around a few fundamental cult themes, although they greatly overlapped and no clear line can be drawn between them." Karl Taube (1992:1) likewise describes Yucatec religion as "strongly polytheistic, with a myriad of divinities with frequently overlapping if not competing attributes and functions," and David Stuart (2017:248) states that "[d]eity imagery is hardly susceptible to neat, well-bounded categories, but reflects instead a complex philosophy where the sacred was manifested in multifaceted ways." Using the maize-cycad complex example, I present a partial framework for understanding how deities related to one another and to the natural and cultural worlds, as well as how they transform as they move through various mythological temporalities.

Building on the pathbreaking work of Mark Bonta, María Teresa Pulido-Silva, and their colleagues (Bonta 2007, 2009, 2012, this volume; Bonta and Osborne 2007; Bonta et al. 2006, 2019; Diego-Vargas 2017; Vite-Reyes 2012), I examine these broader topics through the circumscribed example of cycads and maize, their symbolic overlap or convergence in the ethnohistorical and ethnographic record, and the possible implications of their close association for the development of the iconography and symbolism of Formative period fertility deities. I consider the deities these plants embody in light of cognate entities from later Mesoamerican traditions with an eye for the contribution that cycads might have made to this symbolic and iconographic complex, thereby adding conceptual complexity to Formative period iconography of sprouting imagery, particularly God II or the Olmec Maize God (Taube 1996). My basic premise is that early depictions of the maize god, or what should perhaps better be called the Fertility Deity Complex, present a range of conceptually related taxa and processes instead of a one-for-one depiction of personified maize (see Chevalier and Bain 2003:178–179). I further propose that this symbolic and iconographic complex might align closely with the sacred-maize-ancestor concept proposed by Bonta et al. (2019), especially in those areas in which cycads are culturally significant. I do not suggest that this entity personifies a cycad instead of maize—although that might certainly be the case in some examples—but rather that there was an early and sustained conceptual convergence that placed these plants into the same taxonomic category and genealogical lineage. Understood in this way, not only does the richness of this material increase, but also we may shed light on a central issue of how maize comes to be of ritual significance prior to its alimentary dominance. Cycads create a formal category or morphological model of a paradigmatic *centli/cintli* or "cob" that eventually would be filled by maize. While this specific point remains speculative, defining what is and historically has been included in Corn Master narratives and their earlier visual iterations broadens the interpretation of such material because it pulls into focus those seemingly idiosyncratic elements that do not entirely fit the maize paradigm, but are critical for isolating and defining Indigenous axes of meaning.

In dealing specifically with cycads I further argue that the plants themselves and the godpowers they incarnate represent a set of transformations in mythological symbolism. These symbolic connections and transformations are summarized in part in Figure 6.1. I will examine these relationships in more detail throughout this chapter; however, suffice it to say that cycads appear to serve as a bridge or mediator between pre- and post-maize foodways. They also appear to be imbricated in a web of symbolism that is especially intertwined with the creation of the present age. From an agroecological standpoint, their unique place in ecological systems and dioecious nature (see Bonta, chapter 8; Calonje et al., this volume), edibility yet toxicity, regrowth

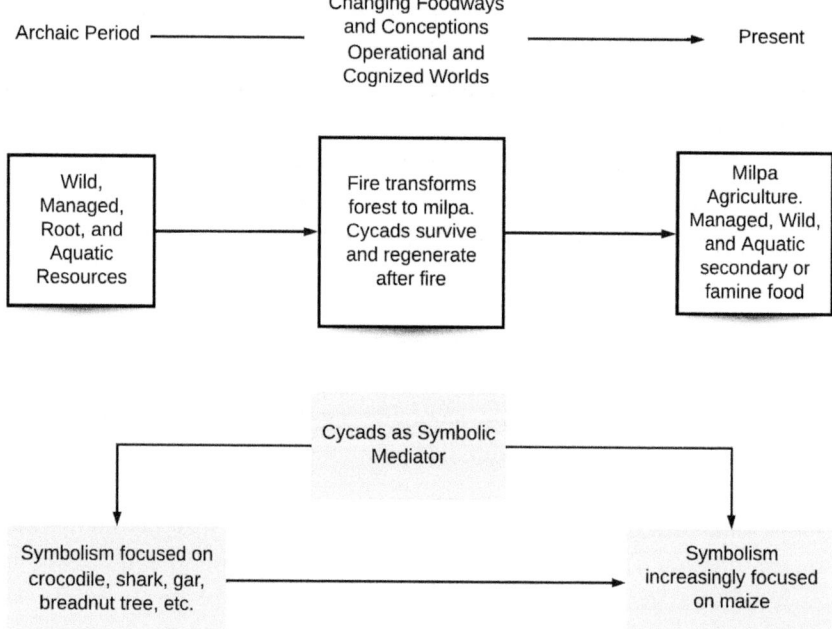

Figure 6.1. Cycads in agroecological mythological symbolic regimes over time (created by author).

after fire, and the strobilus's morphological resemblance to the maize cob, among other properties made them, as Claude Levi-Strauss (1962:132) has put it, especially *"bonnes à penser"* (good [things] to think [with]).

Cycads in Mesoamerican Foodways and Mythology

As other chapters in this volume have outlined in greater detail, cycads (Cycadales) are an ancient order of gymnosperm, one family (Zamiaceae) of which is found in Mexico. In ancient Mesoamerica and modern Mexico they have played a significant role in regional agroecology, Indigenous foodways, cultural practices, and beliefs, in ways comparable to their cultural and culinary position in the Caribbean, Africa, Australia, and Asia (Beck 1992; Bonta and Bamigboye 2018; Bradley 2005; Hayward and Kuwahara 2012; Khuraijam and Singh 2012; Kira and Miyoshi 2000; Mickleburgh and Pagán-Jiménez 2012; Pagán-Jiménez et al. 2015; Patiño 1989; Radha and Singh 2008; Smith 1951; Smith 1982; Thieret 1958; Tōyama and Ankei 2015; Veloz Maggiolo 1992). In Mesoamerica, cycads occur in diverse regions spanning from Tamaulipas to Tabasco and into Chiapas; they are found as well in the Yucatán and into Guatemala (see Calonje et al., this volume). They thrive in diverse ecosystems including xeric shrubland, cloud forests, and tropical lowlands, among others (see Cibrián-Jaramillo et al., this volume). Their use is also varied and ranges from alimentary to decorative (Alcorn 1984; Bonta 2010a; Bonta et al. 2019; Sifuentes de Ortiz 1983; Tristán 2012; Valdez 2009; Vite-Reyes et al. 2010). They figure prominently in mythologies, regional festivals (Pérez-Farrera and Vovides 2006), and foodways (Bonta 2010b; Bonta et al. 2019; Carrasco 2012, 2015; Diego-Vargas 2017). Cycads are culturally significant plants that, with few exceptions (Alcorn 1984; Bonta et al. 2006, 2019), have largely been overlooked in the study of Mesoamerican cultures and foodways. The lack of scholarly focus on Mesoamerican cycads appears to stem from their toxicity, which, while requiring processing similar to that of manioc (*M. esculenta*; Cox and Sacks 2002; Whiting 1963, 1989), seemingly has diverted attention from them, despite considerable evidence indicating their utilitarian and symbolic use from the Pleistocene–Holocene transition to the present. Archaeological data from dry cave sites in the Tehuacán Valley and Tamaulipas offer proof of the dietary importance of *D. edule* and *D. angustifolium* as early as 4500 BCE (Bonta et al. 2019; Englehardt et al., this volume; Englehardt et al. 2020). In the 1950s and 60s Richard MacNeish's excavations of these caves provided one of the most detailed records of the dietary strategies of Archaic and Early Formative period peoples in Mesoamerica yet discovered. In addition to finding evidence for the three domesticated crops central to later Mesoamerican foodways, he also uncovered a rich inventory of wild, managed, and cultivated foods, which by volume surpassed the better known domesticates: corns, beans, and squash (see Englehardt et al., this volume).

For example, at the Tamaulipas cave site beginning as early as 4500 BP, *D. angustifolium* played a significant role in the Archaic and Formative period diet. As MacNeish (1958:144) notes, *Dioon* seeds were the "dominant plant remains." Indeed, cultivated or wild foods represent 76 percent of the vegetal food by volume while domestic crops composed only 9 percent in the La Perra horizon (2495 BCE ± 280, Libby 1952). In terms of volume, *Dioon* seeds were followed by *cuauhmochitl* (*Pithecellobium confins*), a kind of leguminous tree, maguey (*Agave* sp.), and palms (*Sabal* spp.). Maize was also found but it appears to have been chewed, popped, or possibly used for the stalk's sugar content and fermented (Iltis 2000, 2006; Smalley and Blake 2003).[2] At the earlier Tehuacán Valley sites, cycad remains were not as abundant, although Smith (1967:235) notes that they remained "fairly persistently used over some thousands of years." Despite this assessment he nonetheless then concludes that cycads could not have played a significant dietary role because of their toxicity. He was apparently unaware that cycads continued to be consumed in some areas of Mesoamerica.

The critical point is that as early as the Archaic era (5500–4000 BP), at a time before the development of modern maize, the archaeological record indicates that cycads were a major part of the Mesoamerican diet alongside such plants as maguey and various palms (Brown 2010:105; Meléndez Guadarrama et al. 2012). Cycads were used in quantities surpassing or similar to these other plant resources that came to play significant symbolic roles in later Mesoamerican societies. Likewise, the linguistic and ethnohistoric records suggest that cycads of various species were culturally significant in numerous ways and continue to be essential components of ritual (Pérez-Farrera and Vovides 2006) and foodways among various groups, including the Teenek (Alcorn 1984; Bonta 2007, 2010a), Nahua (Bonta et al. 2006; Vázquez Torres 1990), and Xi'iuy (Tristán 2012), among others.

The maize-cycad complex is found with greatest intensity (see Bonta, this volume, Figure 7.1) along the Sierra Madre Oriental and in Chiapas and the Gulf of Mexico lowlands. Over this range we find a consistent association between cycads and maize preserved most clearly in cycad terms and foodways (Bonta, this volume). For example, within this area Nahuas use the term *teosinte* to name cycads (C. *fuscoviridis*), a relationship that Francisco Hernández documented in his botanical research in the 1560s (see Bonta, this volume; Hernández 1946:866). Over this area, in fact, teosinte refers exclusively to cycads as opposed to Z. *mays parviglumis*.

While the term teosinte (cf. "tiusinte," a term developed to disambiguate cycads from the grass) has traditionally been associated solely with Z. *mays parviglumis* (or *Tripsacum* gamagrasses and occasionally to *Setaria*), in some cases it designates an entirely different plant. In other cases these terms, as we will see, refer not simply to the maize plant and associated deities, but also to a broader category into which a number of affiliated things can be placed. Take as another example the Ch'ol Mayan word **ña'al*. This term is cognate to the Classic Maya word *nal* that is nearly exclusively associated with maize. However, in Ch'ol it refers to a "god of abundant plants and animals . . . it is said that it appears concretely in the form of maize, beans, chickens and pigs. The ancient idols of the Chol were made for these gods. [*dios de la abundancia de plantas y animales . . . Se dice que aparece en forma concreta en el maíz, frijol, pollos y puercos. Los ídolos antiguos de los choles fueron hechos por este dios.*]" (Aulie and Aulie 1978:62). Even in the Maya cultural context, in which maize was clearly both a dominant crop and of singular symbolic significance, the meaning of this term left sufficient semantic range to include not simply other flora but animals as well. Thus, while the choice to use terms such as "sacred ear" or "earth monster" to name either maize or cycads varies across regions, the mythologies in which these entities figure are widespread throughout the Gulf Coast and Mesoamerica. These are precisely the kinds of words that seem to reveal a category into which these two plants and others, even possibly gar fish (*Atractosteus spatula* or *Lepisosteus tristoechus*), may be placed conceptually (Carrasco 2020; Ochoa 2010). Which one was chosen likely depended on the cultural saliency of the particular plant in a given location. In this light when we consider the antiquity of the use of cycads and their symbolic associations with maize across several language families, we should also consider what role they might have played in the region's mythologies and artistic traditions, particularly the earliest ones that developed at a time when it appears maize was just becoming the dominant food. The rich mythologies concerning the creation of the present age and the origins of maize offer particularly fruitful areas to examine the place of cycads within Mesoamerican thought.

Pre-Creation Ecologies and Foodways

The Mesoamerican etiological mythologies that survive in colonial and ethnographic sources describe the origin of maize as co-occurring with the creation of the current world age and the formation of people. Maize becomes human flesh through its consumption, or in myth by a deity modeling it into human form. Thus humans are beings of maize. However, people of previous creations survived on other foods, ones that were considered wild or possibly managed, as compared with the human-dependent maize. Through this these mythologies encode what might be thought of as historically specific ecologies and foodways. While it is not always clear if these narratives represent ex post facto sequences or indeed present relic foodways or practices, the correspondence of these past foods to ones thought to have played a significant role in the Archaic and Early Formative diet is suggestive of their antiquity.

Among the most detailed descriptions of this culinary evolution are those found in the *Historia de los Mexicanos por sus Pinturas* and *Leyenda de los Soles,* which record the foods eaten by people in each of the four world-eras before the fifth, current one (see Table 6.1). From these central Mexican examples, we look east to the Huasteca to examine similar mythologies among Teenek Maya speakers, eventually moving south to include again Nahua speakers living in the regions of the Tuxtlas, as well as Soke and Mije versions of the Corn Master narrative. In this central Mexican scenario, as described in the 1530s manuscript titled *Historia de los Mexicanos por sus Pinturas*

Table 6.1. Tenochtitlan sequence of the five ages

Name of Sun	Presiding deity	Human / Food	Fate of humanity	Type of destruction
Naui Ocelotl (Four Jaguar)	Tezcatlipoca	Giants / acorns, roots, and wild foods	Eaten by jaguars	Jaguars
Naui Ehecatl (Four Wind)	Quetzalcóatl	Humans / pine nuts (*acocentli*)	Transformed into monkeys	Hurricanes
Naui Quizhuitl (Four Rain)	Tlaloc	Humans / aquatic seed (*acecentli*)	Transformed into dogs, turkeys, butterflies	Fiery rain
Naui Atl (Four Water)	Chalchiuhtlicue	Humans / wild seeds (*cincocop*,[a] probably *teocentli*)	Transformed into fish	Great flood
Naui Ollin (Four Movement)	Tonatiuh	Humans / Maize (*centli*)	To be devoured by *tzitzimeme* (celestial monsters)	Earthquakes

Source: Adapted from Berdan 2014:table 7.3.

[a] The line that refers to the food of this period in the *Historia de los Mexicanos por sus Pinturas* reads "macehuales comían en este tiempo de una semiente como maíz que se dice *cintrococopi*" [the people at this time used to eat of a maize-like grain called *cintrococopi*]. Citing Molina, Phillips (1884:647) suggests that *cintrococopi* is derived "from *cintli*, spindles (Mazorcas), full of dry and cured maize and *cocopatic*, something that burns the mouth greatly." Caso (1953:27) felt this term was a substitute for *teocentli*: "Indeed, the last plant cited, *cencocopi*, is none other than *teocentli*, a plant so similar to corn that it has come to be considered the wild ancestor of this grass."

(1941:233) attributed to Father Andrés de Olmos, wild and managed foods transition to the precursors of domesticates, such as *cincocop/cintrococopi*, which Frances Berdan, like Alfonso Caso (1953), sees as teosinte, and finally the sequence ends with maize, *centli* (Berdan 2014:224; Townsend 1992:120).

It is significant that the root *centli* or *cintli*, "dried ear of maize," figures in all but one of the foods of the past eras. It is absent precisely in the first era when giants peopled the earth rather than humans. The myth, while describing a culinary evolution, could possibly retroactively name these past foods in relation to the more culturally salient maize; however, we could also look at the situation differently to see that *centli* foods are those appropriate to human beings. Because each era has a specific kind of people the appropriate foods would likewise vary, but they nevertheless are still placed in the category that essentially means "food." This view would adjust Caso's (1953) observation that the plants cited in this mythology "progressively come to approximate the Mesoamerican Indian's ideal sustenance," maize. This view while accurate in one sense misses the point that each of these foods is the ideal for the "people" who lived in their specific era. Maize is the ideal food for the current era because people of this time are made of it. This perspective diminishes the appeal to a quasi-evolutionary teleology that adheres to the notion that one is moving from imperfection to perfection, when in fact many of these stories suggest the opposite trajectory, in which people move from a state of grace to one of imperfection and blunted abilities. Rather, I am contending that the use of *centli* in each of the human food categories shows it to be a class marker with a general meaning of "food." What this food *is* depends on the nature or perspective of the entity consuming it.

The use of the term teosinte and its cognates to refer to cycads, such as in Hidalgo where this term is used exclusively to name cycads, as opposed to its more common association with *Z. mays parviglumis*, raises the question of whether the references in these mythologies are to the grass or to cycads. As Bonta describes in chapter 8 of this volume, cycads figure in a conceptual system in which they are seen as ancestors or as somehow otherwise related to maize, a fact that further suggests that teosinte in this example possibly refers to cycads specifically, as opposed to *Z. mays parviglumis*. The following provides a contemporary narrative in which this is precisely the case:

> In the days before the arrival of the Spanish, people ate teocintle, but they say that maize was found inside it, but they say that they found maize seeds inside it, instead of [cycad] seeds, and it was said that these were very different from what they had seen before, so they planted them and very beautiful maize seedlings sprouted, and from these many [more were born], but it is the teocintle that in Nahua[tl] mean[s] maize's uncle, and so from this teocintle from here maize was born. They say that the Aztecs ate this and thus they took it from here [Huazalingo]. [orig. at 44:127]. (Bonta et al. 2019:26–27)

Also from the Nahua-speaking town of Tohuaco II, Huautla, Hidalgo we find the conceptually similar terms Chicome-Sintli (seven-cob) and Chicōmexōchitl (seven-flower), which are familiar to Mesoamericanists as names for maize deities. However, in this case they refer to cycads, such as *Zamia loddigesii* and *C. fuscoviridis* (Bonta et al. 2019). In Gulf Coast mythology Chicōmexōchitl is a maize and fertility hero who parallels deities such as the Maya Maize God (Taube 1985), as well as other Gulf Coast maize heroes (Braakhuis 2009:5), underscoring the conceptual slipperiness between cycads and maize.

Teenek Maya

Like the example from the Tenochtitlan world ages, Teenek mythology also presents a series of world creations leading to the present one in which humans now live in an imperfect state after a series of transgressions. This history too suggests evolving foodways as we move from the succeeding creations to the current one. Table 6.2 presents this sequence in five general eras, two of which seem to focus on the Teenek. These are not given dates as precise as the ones in the Tenochtitlan version, so making clear divisions is somewhat more complicated. Nevertheless, as in the Nahua example, giants inhabit the first era and their food is not specified. They are all male and reproduce only through an act of "sinful" sodomy that results in humans who are fed excrement in the ocean. These humans eventually overthrow their father. Real consumption is also absent in the next era and the giants of this period, Lints'i', feed only on the vapors of food and are also unable to reproduce. Because they waste food, the gods send a flood that destroys them and turns them into fish. This imagery suggests a correspondence between this period and the fourth age in the Tenochtitlan version. In the Teenek example, however, greater details are provided for what appears to be the true age of humans. In this regard the surviving person eats a fish and thus commits an act of cannibalism and is transformed into a raccoon as punishment. This act of cannibalism anticipates the next period in which the Aach-Eagle grandmother is also cast as a cannibal. It is at this moment when Thipaak or Dhipaak enters the picture as the adopted grandson of the Aach-Eagle.

Thipaak is said to be the *ehatal*, "soul," the *tz'itziin*, "spirit," and the *ichiich*, "heart of maize" or "embryo inside the seed" (Ochoa 2010:542), but he also takes form in certain zamia that are then understood to be the "spirit of maize" and "Maize Lord" (Alcorn 1984:826; see also Bonta, chapter 8, this volume).[3] Thus as in the case of teosinte, here too we find the convergence of cycads and maize in the figure of the child-maize god, Thipaak, especially in those instances in which he lends his name to a number of cycad terms (Alcorn 1984; Bonta, this volume; Bonta and Osborne 2007; Bonta et al. 2019). The Teenek believed that the "ancestors ate it as maize." Likewise *konlib* or *konlif* is considered to be the "spirit of the milpa," and names certain *Ceratozamia* species, particularly *C. latifolia* or *C. chamberlainii* (Alcorn 1984; Bonta et al. 2019:25; Martínez-Domínguez et al. 2018). In each of these cases these plants are seen as ancestors of maize, as relatives, or as manifestations of deities associated with the origins of maize and culture.

As in the other Corn Master narratives, in his path to bring maize to people, Thipaak's powers are applied against his raptorial, cannibalistic grand-

Table 6.2. Teenek world ages

World Type	World Inhabitants	Type of Food	Reproduction	Punishment/ Defeat
1. Flat Earth	Giants, all male		Sodomy produced new humans who were raised on excrement in the ocean.	Giants overthrown by their children because of sinful reproduction.
2. Pyramid Builders	*Lints'i'* (flat ass) Monkey form: *Mut'in* (stopped-up-ones)	Create Water with magic green stone. Sniff vapors of food, but cannot eat. Actual food is wasted.	No reproduction nor defecation because of third leg. Children made from clay.	Destroyed by flood and turned into fish; their waste of food angers the gods.
3. Post-Flood	Human >Raccoon	Fish, but fish were previous people so cannibalistic.		Transformed into raccoons.
4. Teenek	Powerful humans	Ojox (*B. alicastrum*). No maize.	Produced children. But children are sacrificed to the Aach-Eagle grandmother.	Initially no work and powerful vision, but transgressions result in a life of labor and loss of vision when maize pollen is dusted in their eyes. Fire abandons them.
5. Time of Maize	Mortal humans	Thipaak brings maize	Thipaak, adoptive child of the Aach-Eagle grandmother, defeats her, thus ending child sacrifice.	Teenek look when Muxi' splits maize mountain, ignoring his decree; for this reason they are left with yellow and blue maize.

Source: Based on Alcorn 1984:60–63.

mother. Similar at times to the Classic Maya, Nahua, and Mixtec gods, such as GI, Quetzalcoatl, Lord 9 Wind, or Chicōmexōchitl, Thipaak is a cultural hero at whose command plants grow, and who can transform into tobacco in some stories (Alcorn et al. 2006:603; Hernández Ferrer 2004). Thipaak not only personifies the maize grain or plant, but also embodies fertility in his ability to control the growth of other plants, as well as having a close relationship to thunder in some traditions (Ariel de Vidas 2004:360–363). His evil Aach-Eagle grandmother, in contrast, is associated with the *ojox* or breadnut tree (*Brosimum alicastrum*) (Alcorn 1984; Alcorn et al. 2006).[4] She presides over the world of the creation immediately preceding the current one and wishes to eat Thipaak or otherwise mistreat him. This antagonistic relationship maps onto the Teenek understanding of a dietary evolution, in which they believe that people ate ojox before Thipaak introduced maize and is even played out in how the fruiting patterns of cycads and ojox are seen to impact or indicate the season's maize harvest. On this point Alcorn (1984:354–55) notes that "A good fruit set by ohox (*Brosimum alicastrum*) means that the maize will do poorly while a bad crop of ohox fruit means that the maize harvest will be good. If *tsalaam Thipaak* (*Zamia* sp.?) sets seed then the maize harvest will also be good." Alcorn et al. (2006:608) suggest that "one reading of the Thipaak history could be that through the introduction of maize, an older Brosimum-based religion with a vital female deity demanding human sacrifice coexisted with and was eventually displaced by a hierarchical male-dominated theocracy that required maize products in their rituals and economy."

This mythological background illuminates the significance of the term Thipaak when applied to *Ceratozamia* and *Zamia* species (Bonta, chapter 8, this volume; Bonta and Osborne 2007; Bonta et al. 2006, 2019; Stross 2006). Like the word teosinte/tiusinte, which places cycads in an ancestral relationship with maize in Nahua taxonomies, the term Thipaak also functions in a similar way. These examples evidence just some of the various ways in which cycads and maize are interconnected in Teenek nomenclature.

Recent fieldwork throws additional light on these concepts among the Teenek. Bonta and his associates (Bonta et al. 2019:25) recorded the following recollection from Tancuime, Aquismón, San Luis Potosí in which the narrator speaks to the cycad's role as ancestor. Here *konlib* or *konlif,* initially identified by Alcorn (1984:591) as *C. mexicana* and now known to be *C. latifolia* or *C. chamberlainii,* is considered a friend of Thipaak because

> in the old days it was like a maize cob, this is the maize cob from the old days, and it has stayed around so we could see that it was eaten. The [first] people in the old days did not eat anything, but when God's people disappeared, the [next] people knew to eat the food of today. The konlif is not planted, it just grows from the earth, it is a maize cob from the old days. This maize's spirit was inside a white crystal, and the people would take the crystal to the milpa, and when they were thirsty they put it in the milpa and when they were thirsty they put it in the soil, and before they knew it there was water, so then the men planted these old-time maize cobs. In the old days, God saw that the people did not eat this food and because of this God took it away. In our time God

gave us maize and beans, but God gave the [konlif] crystal to men so that they could survive. [orig. at 44:127]

Konlif is also associated with rain and thunder as indicated in the passage by its water bringing abilities. Elsewhere, Bonta (chapter 8, this volume) notes that *konlif* cycads "are understood to have the power to 'pull down' clouds to provide moisture and are associated with *Maam*, the thunder god." These abilities align them with those attributed to Thipaak himself and suggest that there is a considerable degree of conceptual overlap between these entities.

The etymology of *konlib* is also particularly intriguing and possibly clarifies its shared identity with Thipaak. The key element is the root *kon/m* found as the essential element in *konlib* and its variants, which are typically translated as "plant of sand" or "resident of sand" (Alcorn 1984:591). Bonta (chapter 7, this volume) lists the following variants: *condif, conlif, kombi, kombil, konbi, kondif, konfi, konlib*; however, the root *kon* or ones very similar to it are also found in other regional languages that suggest a possible connection between them. For instance, in Soteapan Gulf Sokean the word *kom* is found in compound with *paampi* (pacaya palm, *Chamaedorea tepejilote*) to name arrowroot or sago (Kaufman 2016a:19; wəəty paampi [NG] = kom = paampi [NG] arrowroot, sago). While this gloss does not provide an exact species identification, since arrowroot and sago are both generic terms for starch produced from cycads, it does confirm a link between the word *kom* and cycads. In other cases *kom* or *kum* (Kaufman 2016b:13) appear to refer to the coyol palm (*Acrocomia mexicana*) and this term is reconstructible to *kuma in proto-Mije-Sokean (Wichmann 1995:352). It is thus possible that Soke *kom* is related to *kon* found in the Teenek term *konlib*, suggestive of a widespread distribution of concepts associating palms, cycads, and maize across multiple language families. Confirmation of this would require further investigation and a fuller understanding of how this term would have been borrowed into the Teenek. Currently, Teenek Mayan and Sokean speaking communities do not abut each other, but the situation might have been considerably different anciently.

Support for this argument is also found in the word *kun*, a term for *Ceratozamia* species found in the Totonac-speaking region of the Sierra Norte (Bonta and Osborne 2007:4). If we extend this line of thought slightly further it is possible that the *kon* root in the Mije deity named Kondoy is also related to these terms since this deity, like Thipaak, is a hero whose story is another instance of the Corn Master narrative.[5] However, *kon* here has been interpreted as "king" and is also found in the possibly even more relevant Mije term for a twinned maize ear *koon-mook*, for "king maize," and in San

Juan Ixcatlán King Maize is described as a "mazorca [ear of maize] fasciated at the tip or divided at the terminal part into two or more ear-like branches" (Lipp 1991:23). Thus, it is possible that one of the sources for *kon* and its cognates in all these contexts is *kon* for "king" or "ruler." Given the prominence of the taxa named and the symbolic associations between twinned cobs and the multiple cones that mature cycads regularly produce, this sense of *kon* might be particularly apt. One of the Teenek terms for *Zamia*, *ahaatik a eem,* "Maize Lord," also presents the use of the word "lord," *ahaatik,* possibly showing that *kon* as "king" would not be unexpected.

If these etymological relationships bear out, then one can string a connecting thread from the cycad-maize spirit of the Teenek, to cycads among the Soke and Totonac, and finally to the deity Kondoy. In this way we find a sustained linkage between cycads, maize, and the Gulf Coast heroes/maize deities that exist across language families, even if both associations are not always found in the same instance, such as in the case of Kondoy, for whom I have yet to find a direct link to cycads. That *kon/kum* also figures in names for food palms such as *Chamaedorea tepejilote* and *Acrocomia mexicana* suggests another line of conceptual overlap between cycads and these palms, presumably because they are culinarily significant and morphologically, if not genetically, similar. Finally, Kondoy, while lacking a specific reference to cycads and palms, nevertheless presents a narrative that closely parallels other Corn Master stories and thus would seem to place him in this paradigm with the caveat that in his story, rulership and military prowess in historical battles are emphasized with little reference to maize.

Thipaak and Cipactli

While *konlib*'s relationship to these similar terms is intriguing, the etymology of Thipaak (or Dhipak) provides a clearer instance of widespread distribution. Thipaak and its cognate terms' appearance across Mesoamerican language families attests to the linguistic spread of a deity who presents similar attributes that tie it to maize, cycads, the earth, fertility, and calendrical nomenclature, among others. Thipaak is cognate to the earth-related crocodilian deity Cipactli of the Nahua and Sipak (Zipacna), a telluric deity described in the *Popol Vuh* of the K'iche' (Stross 2006). Cipactli, "alligator," is also the first day in the Central Mexican calendrical systems and is equivalent to Imix, "waterlily," in the Maya one (Rice 2007:34). It is significant that the Maya Maize God often emerges from waterlily/lotus-skull forms, motifs that are deeply connected to Imix, the first-day sign of the 260-day ceremonial calendar. Likewise, as we will see in the following section, foliation, trees, and anthropomorphic beings often emerge from a crocodilian in Formative period imagery. Like Imix, Alligator is the first-day sign of the 260-day calendar outside the Maya region. Thus, while these day signs do not figure in such imagery per se, the entities associated with them—waterlilies and crocodilians, respectively—seem to inform the iconographies that would correspond to structurally similar situations in which "sprouting" or "emergence" is often the major theme. That is, in the Formative period Gulf Coast imagery, the hero emerges or is part crocodilian while in the Maya area a similar deity emerges from a waterlily-skull complex of imagery. The deities that emerge appear to possess cognate meanings while also reflecting specific regional and temporal differences, just as these calendrical signs are regionally specific.

However, the symbolic richness of this complex does not end here. Building on the Cipactli string of associations, Lorenzo Ochoa (2007:31, 2010) suggested that the term Thipaak is also linguistically related to the word for alligator gar, *zipac,* and because of this connection, he saw this important food fish as also associated with maize. We could extend this line of thought to suggest a linkage also between piscine creatures and cycads, a connection further reinforced by the similarity between gar scales and the scalelike texture of *Dioon* trunks (Carrasco 2020; Englehardt and Carrasco 2020). This piscine-maize-cycad relationship also occurs in the Nahua mythology in which the teosinte-eating humans of the fourth age, upon their destruction, were turned into fish.

The *Popol Vuh,* the sixteenth-century narrative describing the creation of the current age and the origins of the K'iche' people, preserves a similar narrative motif. In this version the Hero Twins, who are personifications of maize, reemerge as catfish after they are burned in the underworld oven of Xibalba, ground, and thrown into a river (Christenson 2003; Girard 1952; Stross 1994:13).[6] As personifications of maize, their remains are processed in kind and at the moment before they take on full anthropomorphic form they are fish people. Perhaps even more relevant is the Hero Twins' defeat of Zipacna. In this case the Twins trick and defeat a creature from the previous age, just as Phillip Arnold (2005) finds evidence of an antagonism between a humanoid figure and a shark or piscine creature, which he also sees as a species of Cipactli.

In Classic and Post-Classic Mesoamerica, Cipactli mythology presents the killing or sacrifice of an aquatic monster that results in the creation of the world. From the body of this creature various things emerge or sprout. In some versions the parts of the creature's body become the things of this world. In this way these ritual primordial sacrificial events are a key component of etiological mythology. We find this event expressed clearly in several central Mexican Codices that depict Tezcatlipoca and Tlahuizcalpantecuhtli battling a fish, which Arnold sees as a shark, or crocodilian (see Figure 6.2). The oscillation between fish and crocodilian is also presented verbally in the *Historia de los Mexicanos por sus Pinturas* (1941:230), which describes the creature as both fish and crocodilian. After the gods created the heavens, they

> made the water and created in it a great fish similar to an alligator which they named Çipaqli [*Cipactli*], and from this fish they made the earth.... Afterwards all the four gods, being united in work, they created from the fish *Cipacuatli* [*Cipactli*] the earth, which they called *Tlaltecli* [*Tlaltecuhtli*], and represent as the god of the earth, extended over a fish as having been made of it. (Phillips 1884:618)

These Post-Classic depictions and early Colonial central Mexican descriptions find a Classic Maya parallel in the story recorded on the south panel of a small platform in Temple XIX at Palenque (Stuart 2005:215). This text describes how one of the patron gods of Palenque, GI, beheaded a Cipactli-like creature called the starry-deer-crocodile, eleven years after Yax Naah Itzamna had enthroned him in the heavens (Stuart 2005:60–77). The key passage reads "1 Etz'naab 6 Yaxk'in is the decapitation of the portal-back-crocodile, the inscribed-back-crocodile. Thrice pooled was its blood... fire was drilled, the sacred work of GI (Jun Ye Nal Chak) was done." This narrative, as Rice (2018) notes, bears a striking resemblance to a passage from the *Book of Chilam Balam of Chumayel* (Edmonson 1986: lines 761–772):

> Then arises
> The great Itzam Cab Ain.
> The ending of the word,
> The fold of the katun:
> That is a flood
> Which will be the ending of the word of the katun.
> ...
> And then will be cut
> The throat of Itzam Cab Ain.
> Who bears the country
> On his back.

In the Palenque version, as in the images reviewed here, we find a central protagonist who kills a crocodilian beast. However, in the Maya version GI

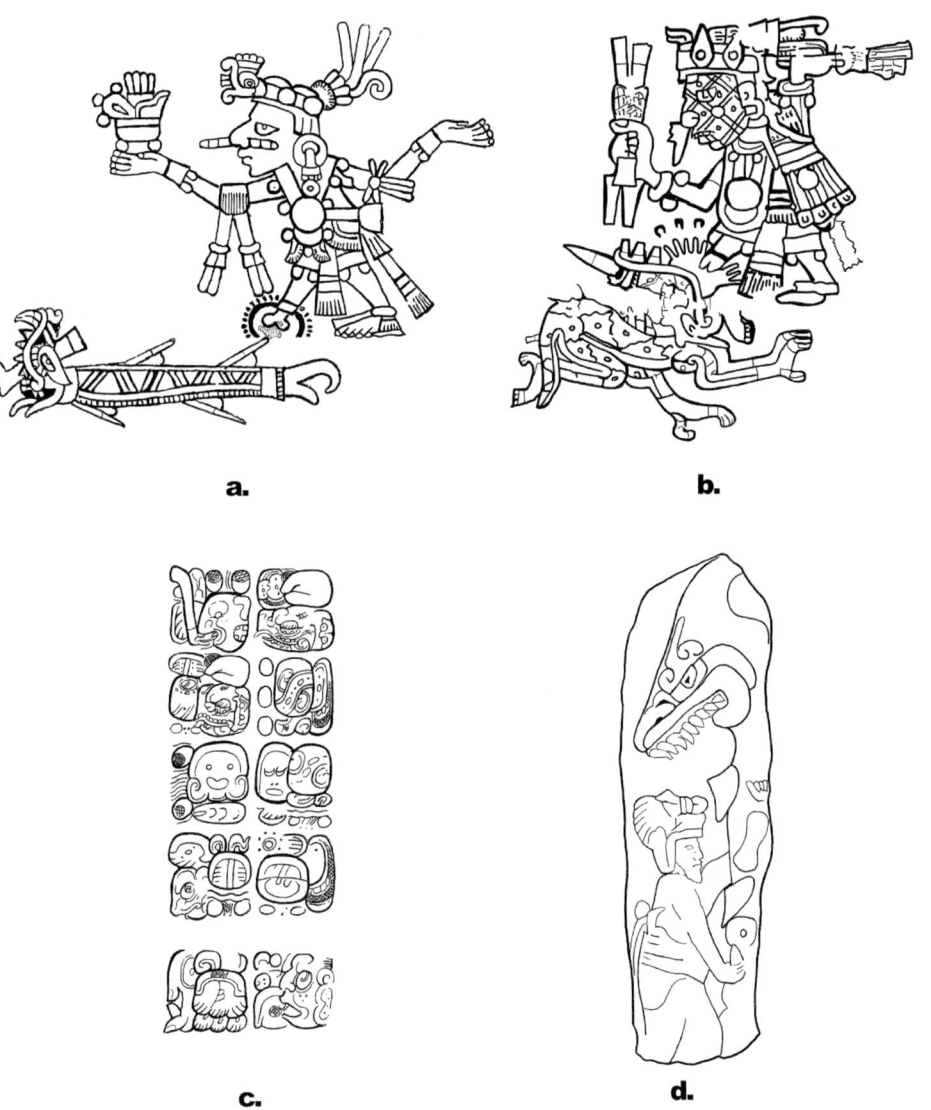

Figure 6.2. Codical, glyphic, and sculptural depictions of Cipactli: *a*, Codex Fejérváry-Mayer, Folio 42. The Aztec world creation story in which Tlahuizcalpantecuhtli (Karl Taube in Arnold 2005:13) loses his foot to Cipactli; *b*, Borgia Codex, plate 27 (note that in *a*, the deity is represented as a piscine, while in the Borgia example it resembles a crocodilian); *c*, Palenque Temple XIX, Southern Platform text; *d*, La Venta, Monument 63 (drawings by author).

is likely more closely related to Ehecatl-Quetzalcoatl (Kelley 1965; Stuart 2005:159, 168), given his birth date on 9 Wind, than to Tezcatlipoca. In these examples we again find a moment when a hero defeats a primordial creature as part of actions possibly leading to the creation of the earth and the establishment of a new period.

These terms and analogic and metonymic relationships thus place cycads into the culinary evolution presented in the above narratives and suggest that they participate in the same symbolic web including maize, fish, and Cipactli, whose specific qualities across various cultural groups introduces crocodilians, waterlilies, and other creatures to the picture. This symbolic web is important to remember in understanding the iconography of Formative and Classic period imagery.

Corn Master Narratives among the Popoluca, Mije, and Xi'iuy

Corn Master narratives similar to the Thipaak mythology are found among other groups in which a child maize god miraculously enters into an adoptive family often two generations removed from his own (Chevalier and Bain 2003:174–175). In some versions he emerges from a pair of eggs, one of which is eaten by the old couple who finds them. In others, such as the Mije story of Kondoy, the second egg hatches into a snake to again form a set of twins. In the Popoluca versions, Homshuk's adoptive parents decide to eat him but he escapes. During his escape the old man dies and the old woman is immolated in a trap that Homshuk sets for her. In other versions he travels to the home of Thunderbolt on a turtle in an attempt to bring his dead father back to life. He eventually defeats the thunder god, sometimes through his own deployment of lightning and thunder, thus showing his close relationship to thunder and rain. In one version, however, he defeats Thunderbolt by "letting himself be cradled in a hammock and then cradling Thunderbolt in like manner. While doing this Homshuk orders a mole to dig a hole beneath the hammock and asks worms to eat up the poles supporting the hammock. Thunderbolt falls and is buried in the earth" (Rodríguez Hernández n.d.a:3, cited in Chevalier and Bain 2003:175). While aspects of these stories differ from one another, their basic narrative symbolism and structure are cognate. As Corn Master narratives, they are also related to those of Thipaak and central Mexico in which cycads play a significant role. When viewed through this lens and in light of the suggestive etymology of *kon* it is possible to again establish a relationship between cycads and a large body of mythology.

This relationship is further substantiated and clarified in a Xi'iuy version in which the role of the hero is taken by a lizard that defeats an old woman who is a known cannibal (Chemin-Bassler 2012). The lizard, in addition to the overall motif of the defeat of a cannibal ogress, would seem to connect this story to the Nahua and Soke versions in which the hero is mocked by iguanas when he goes to fetch water, and in the Popolucan variant when a lizard is sometimes the messenger who warns the hero's mother not to cry when her resurrected husband returns. In the Xi'iuy version, Lizard encounters the woman while collecting firewood and asks to borrow her dress. She declines. Nevertheless the lizard-hero continues about his collection and then digs a pit in which he kindles a fire. He then challenges the woman to jump over the pit. He does so with ease, but she fails the challenge and perishes. Lizard takes her head and womb and dries them in the forest. Several days later *chamal* (*D. edule*) seeds fall from her dried brains and her womb produces *guapilla* (*Hechtia glomerata*) seeds (see Bonta, chapter 8, this volume). While this iteration differs from Mije, Soke, Nahua, and Teenek versions, the structural pattern is strikingly similar. In this version the defeat of the old cannibal produces cycads. The image of seeds emerging from her head/brains is a visual motif that finds extensive expression across Mesoamerica and in some of the earliest artistic presentations of emergence. It also shows how characters within a narrative can change from one setting to the next but retain the same structural logic. We see in this example that Lizard takes a different role than he does in versions from elsewhere, just as Zipacna can both function as an antagonist to maize in the Popol Vuh, but is seen as the hero in the Teenek Corn Master story.

Through these examples, which are indicative of a far richer corpus than can be fully discussed here, we find an etiological story line of intragenerational strife that results in the creation of maize and the present era. The prominent role of cycads in a number of these examples suggests that they were a significant component of such narratives, especially perhaps at an earlier period in Mesoamerican history or in locations in which cycads played a key alimentary role.

Gender Duality and Complementarity and Regeneration

The dioecious nature of cycads also marks them as ancestral plants and as structural equivalents to Ōmeteōtl, the Nahua god of duality who also manifested as the primordial couple, Ōmetecuhtli and Ōmecíhuatl, also known as Tonacātēcuhtli and Tōnacācihuātl. The Teenek recognize cycad gender in terms that assign them as either male, *inik,* or female, *uxum*. Bonta et al. (2019:25) note that one informant explained that the maize plant, which is both female and male, is a result of the fertilization of the female cycad cone by the male cycad cone. In her account, the male cone was conceptualized as a phallus and cycad pollen was equated to semen, while the female cone that opens to receive the pollen was conceptualized as a human-like reproductive apparatus. Maize was thus at some point in the past a creation of this union of dioecious cycad progenitors fused into a single monoecious plant.

The dioecious nature of cycads would certainly have been of interest to Mesoamerican cultures, many of which saw cosmogenesis unfolding from a divine ancestral couple and placed heavy emphasis on gender duality and complementarity (Bassie-Sweet 2000, 2008).

The ritual use of cycads suggests their symbolic role in beliefs and agroecological systems extending into Chiapas in various formerly Chiapanec-speaking (Oto-Manguean) communities. In this case, during the spring, beginning on April 26th and culminating on May 3rd, a pilgrimage is made from Suchiapa to the Cerro Nambiyugua to harvest *D. merolae* leaves, which are used to decorate the cross on the final day of celebration. The original purpose of the ceremony is not entirely remembered; however Pérez-Farrera and Vovides (2006) note that

> upon arrival of the Catholic conquerors, the tradition unified two symbols; the pagan espadaña and the Christian cross. Alternatively, we believe that this was concealment of the true meaning of the pagan tradition by syncretism into the Catholic. Thus the native people were able to keep the original meaning of their custom (which is no longer precisely known) whilst outwardly showing a Catholic practice. Although among various Zapotec groups May 3rd is the time for appeals for rain to Cocijo, the god of lightning, May is the hottest and driest month of the year over most of Mexico and the ritual in Chiapas may probably also be rain-related. We also suspect this may be so since the Chiapanec language is related to the Western branch of the Oto-Manguean language family (Carmack et al. 1996). Although the origin of the ritual is uncertain, we assume it originated in Chiapanec culture since the *Publicación* (herald) is presented in the Chiapanec language, and also because historians recorded that when the Christians arrived, the Chiapanec people adored a plant that "*grew among the rocks.*" (emphasis added)

While May 3 is the day of the Holy Cross, it is especially significant because it also corresponds to planting and the beginning of the rainy season in southern Mesoamerica. In this case it is tied specifically to petitions to Cocijo. The observation that cycads regenerate quickly after fire would naturally also be a powerful allegory for planting fields, which themselves would have just been burned in preparation for planting, and indeed *C. matudae* and *C. mexicana* are both sometimes referred to in Spanish as *tapacapon* or *tapacarbón* for "covers" or "grows on charcoal" (Nicholas Hopkins, personal communication, 2010).

As the above discussion demonstrates, there is a consistent pattern in which what have been understood most frequently as unambiguous maize deities are also associated with cycads. These maize/cycad deities form a conceptual unit that spans the Sierra Madre Oriental, southwest Mexico, and the Gulf Coast and includes speakers from several different language families and ethnicities. Shared elements across these myths, such as child deities, antagonism between the child and its grandmother, thunder, and ovogenesis, especially in the areas around the Tuxtlas and perhaps found obliquely referenced in the raptorial elements in the stories of Thipaak, all suggest either early commonalities likely of local origin (García de León 1976:10–14) or a relatively rapid diffusion across the region, although terminology is not always shared apart from the word Thipaak (Cipactli) and possibly *kon*. This pattern may be compared productively with Classic and Formative period imagery that is thought to reflect narratives similar to the ethnohistoric ones presented above. It is to this material that we will now turn.

The Maize God and Young Hero

Maize deities in historical and contemporary Corn Master narratives, as we have seen, exist across Mesoamerican traditions and share a number of key traits. These deities bring maize and fertility to the current creation and are heroes who defeat entities from the previous age and thereby present paradigms of authority. They are artists and writers, and also closely associated or identical with thunderbolt deities. In some regions their continued presence is manifested by cycads and maize, particularly twin corncobs (Hernández Ferrer 2004:226). Before returning to cycads' role in this symbolic web and as mediators, it is useful to review the Classic period maize god narrative and its relationship with the Gulf Coast folklore, since it is from this perspective that many of the theories about Formative Gulf Coast art and ideation have been developed. In this vein, I begin with Classic period interpretations of the Maya Maize God, which have often served as a baseline for ones formulated for the Formative period or traditions outside the Maya region. From this point, I move to Formative period iconography with specific examples that return to creatures, such as Cipactli and crocodilians, and cultural heroes that would appear to be Formative period equivalents to the entities discussed in the previous section. Finally, through the connections between cycads and these entities established previously, I end with the possible role a maize-cycad iconographic category might have played in Formative period art.

Scholarship on maize deities in iconography spans more than a century[7]; however, Karl Taube's (1985, 1996, 2000, 2004) work on this subject was and has continued to be particularly stimulating, and has resulted in the isolation of the Tonsured Maize God (TMG; Figure 6.3), which Nicholas Hellmuth had identified earlier as the Principal Young Lord (cited in Taube 1985:172). Additionally, Taube proposed that this deity corresponds to Hun-Hunahpu, the father of the Hero Twins (Coe 1973, 1976), in the *Popol Vuh*, the sixteenth-century narrative describing the creation and the origins of the K'iche' people (Christenson 2003; Tedlock 1985), thereby putting this Classic period visual record into conversation with this colonial mythological narrative. In iconography the Maya Maize God's iconographic features and his emergence or sprouting from seedlike elements or earth imagery naturally pointed to his identity with maize, the single most important crop by the Middle Formative period and certainly thereafter.

While the *Popol Vuh* has often served as the main mythological reference for these narratives, Edwin Braakhuis (2009, 2014) finds that the Gulf Coast Corn Master or origin-of-corn mythologies better parallel the Classic period imagery. These stories, as we have seen, present regional differences, ones that speak to specific ancient foodways and agroecological systems that would have been and are reflective of the Gulf Coast and its various ecological environments. Finally, this Gulf Coast folklore reveals that TMG is likely better compared with the versions of maize-god-hero or Gulf Coast Maize Hero (GCMH), rather than with Hun-Hunahpu of the *Popol Vuh*. This readjustment allows us to tie this imagery to the cycad-maize-ancestor complex described above and to better understand the nature of this entity and its specific relationship with maize at a time before the conquest.

In these Gulf Coast traditions we find figures—Thipaak, Homshuk, Chicōmexōchitl, and Kondoy, among others—that taken together present a fertility hero associated with maize, the arts, thunder, fertility, and cultural authority. Like Oswaldo Chinchilla (2011), Braakhuis (2009) sees both the TMG and GCMH as encompassing "the domains of sustenance, water and rain, the dead, kingship, and such arts as writing, dance, and music" in addition to maize. In this way, he prefers to see the TMG as a Classic period Maya

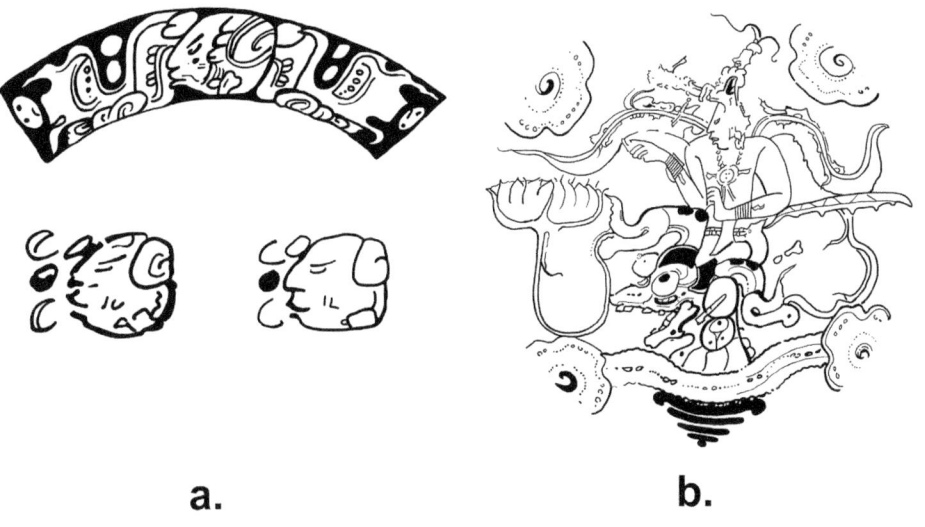

Figure 6.3. Tonsured Maize God: *a*, various versions of the glyphic name of TMG likely read Jun Ixim (drawings by the author after Taube 1985). Two versions of the TMG's emergence: *b*, detail of a Classic period Maya painted plate, TMG sprouting from the waterlily-skull complex; and *c*, an unprovenanced capstone from the Northern Yucatan present him seated on a skull and jaguar-skin cushion within skeletal portal jaws (drawings by author).

manifestation of a maize/fertility god–cultural hero represented in numerous Gulf Coast mythologies.

Within the current discussion his interpretation provides a more robust framework for understanding specific iterations of visual narratives than do attempts that tie the TMG primarily to Hun-Hunaphu and the *Popol Vuh* narrative, recourse to archetypes, or distant cross-cultural comparisons, such as the Puebloan ones often invoked (Taube 2000). It also allows us to

Table 6.3. Characteristics shared between Tonsured Maize God and Gulf Coast Maize Hero

Tonsured Maize God	Gulf Coast Maize Hero
Maize deity associated with fruit trees, youthful	Maize deity associated with fruit trees, youthful
Creator god (young form of Itzamna)	Creator god (Xochipilli as Tonacatecuhtli)
Lightning axe in forehead	Inventor of lightning / son of lightning
Dancing	Dancing
Music (rattles, drum, carapace drum)	Music (rattles and carapace drum)
Games (acrobat, possibly ball player)	General playfulness (including ballgame)
Writing	Writing (Chicōmexōchitl)
Aquatic habitat of flowers, herons/cormorants, and music	Lagoon paradise, aquatic paradise of the Old Thunder God, Tlalocan-Tamoanchan "Lord of Tlalocan"
Attributes of kingship (headband, throne, ceremonial bar)	

Source: After Braakhuis 2009:table 1.

see a mythological continuity from as far north as the Teenek Maya of Tamaulipas through the entire Maya region. Indeed, the origin-of-maize story from the Pokomchi (Mayers 1958:3–11) bears striking similarities to these Gulf Coast examples as does the Ch'orti' Kumix story (Braakhuis and Hull 2014), which extends iterations of this mythology into Honduras. Table 6.3 presents Braakhuis's comparison of the TMG with the GCMH. Classic period inscriptional and iconographic data make these connections stronger (see Carrasco 2010).

Thus, if the Classic Maya Corn Master narrative is indeed cognate to the Gulf Coast ones, we can use this mythology to better understand the significance of this entity at an earlier historical moment and account for imagery that has always appeared outside the kin of the *Popol Vuh*. The aspect of his character that most fascinates here is his relationship to broader concepts of fertility and hybridity with other species beyond maize. As in Table 6.3, Braakhuis (2009, 2014) shows that the TMG and GCMH both possess associations with fruit trees, as well as other flora, with one of the best examples found on a carved vessel (K4331) now in the Dumbarton Oaks Collection (Figure 6.4).

On this vessel the Maya Maize God exhibits features beyond those normally associated with maize (Braakhuis 1990; Martin 2006:154–155). In this scene cacao (*kakaw*; *Theobroma cacao* and possibly *bicolor*) pods hang from his body, which has the effect of merging maize and cacao. The hieroglyphic

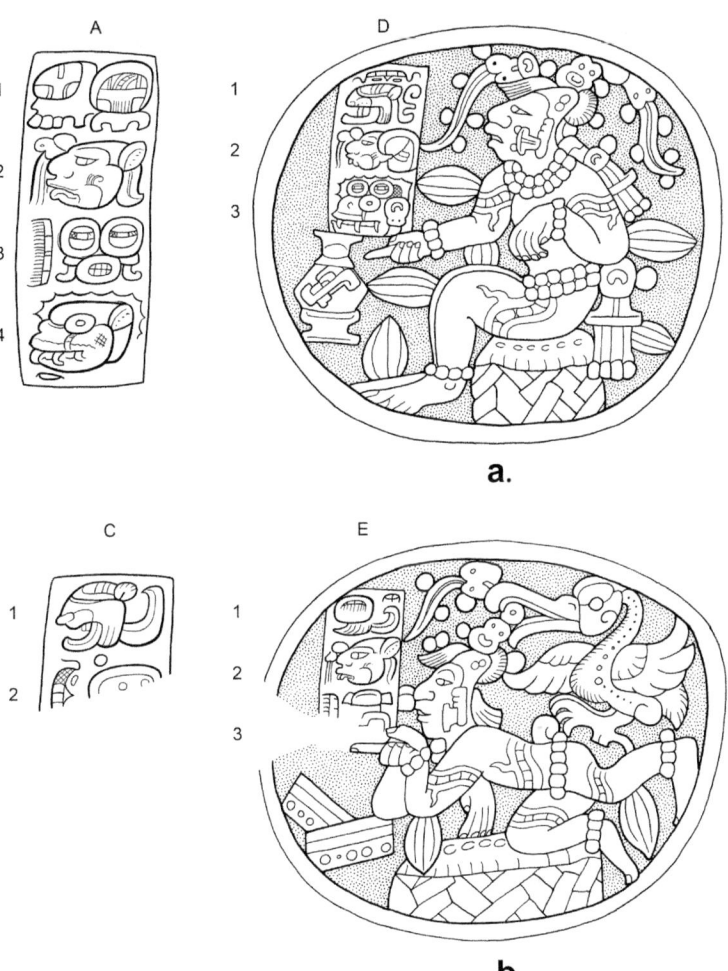

Figure 6.4. Maya Maize God as cacao tree: *a*, the seated god points to a vessel; *b*, the Maize God points to an open codex. The text names the human protagonist as a "Maize Tree" (A2, D2). Unprovenienced stone bowl (K4331), Early Classic period, The Dumbarton Oaks collection, Washington, DC (drawing courtesy Simon Martin).

caption glosses this figure as *iximte'* (A2, D2, E2), which Simon Martin (2006:156) translates literally as "maize tree"; however, there are terms in Maya languages in which the compound *iximte'* occurs. For example, in Tzotzil and Yucatec Maya *iximte'* and *iximche'*, respectively, name the ojox or breadnut tree (*B. alicastrum*); there is no attested use of the term to name maize. What are we to make of this visual and terminological mix?

To the extent that we understand it, this scene presents several layers of complexity. First, it depicts a deity that is normally regarded as a maize god

a.

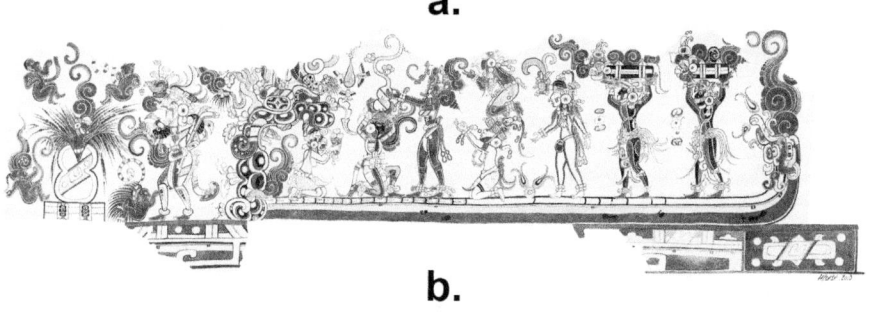

b.

Figure 6.5. Maya Maize God emerging from bottle gourd: *a*, Kerr vessel 5761 (photograph courtesy Justin Kerr); *b*, San Bartolo Sub-1A North Wall mural (San Bartolo Mural, illustration by Heather Hurst, ©2004).

bearing cacao fruit. Second, the caption records a term, *iximte'*, that seems more related to a third kind of plant rather than to either maize or cacao. Thus, in this case *ixim*, like the Nahuatl word *cintli* examined previously, appears to carry a broader semantic range than simply "maize." This image complicates the certainty with which we can say the TMG is conceptualized solely as a maize god.

Another significant example of interspecies hybridity is the maize god's birth from a split gourd in both the San Bartolo Murals and a sculpted vessel (K5761) (see Figure 6.5). In each of these images a maize deity emerges from a split gourd, a gourd which is perhaps introduced in an adjacent image that shows it being brought out of a mouth or from an animated cave, respectively. Thus, in addition to better connecting the ethnographic traditions discussed in this chapter and volume to this rich material, the point of making this detour into the Classic Maya Maize God iconography and the presentation of several examples in which Maya maize deities include other species in their representation is that the Maya conceptualization of this deity has often come to inform the interpretation of Formative period deity images. The result is that sprouting deities in Formative period Olmec art have often been taken to represent maize nearly exclusively. In turn this has further reified the notion of a singular "Maize God," as opposed to a

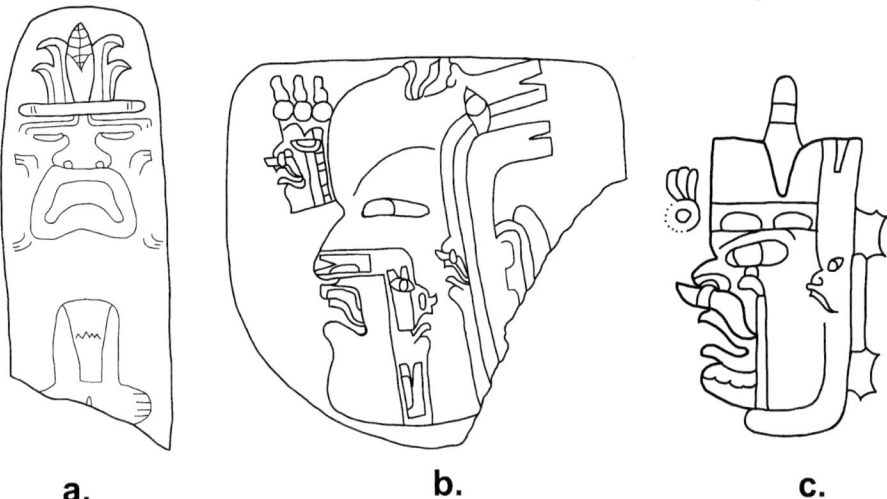

Figure 6.6. Olmec Maize God: *a*, El Sitio celt; *b*, unprovenanced jade plaque in the Museo Nacional de Antropología (MNA-10-9656); *c*, detail of celt from La Venta Offering 1942-c (drawings by author).

generalized fertility deity and possibly, as is argued here, hidden the role of other significant taxa and processes, as well as other contributing symbolism, such as human gestation (Tate 2012).

Indeed it was from the headwaters of Classic Period Maya iconography that interpretations of Formative period imagery flowed (Looper 1993; Reilly 1991; Taube 1992, 1996). In 1996 Taube brought together many of the earlier discussions of maize-related imagery to advance an argument for Formative period vegetal imagery in which he identified a number of figures as representations of a maize deity or aspects of this deity, one of the most prominent of which he dubbed the Olmec Maize God (see Figure 6.6) (OMG; GII in Joralemon's system) (Coe 1968:111; Joralemon 1971:59–66; Taube 1996, 2004:27). In most of these interpretative regimes, maize came to be the dominant lens through which Olmec deity images were understood in the sense that the representation came to be seen as isomorphic with the maize plant, although even among the Maya this deity appears to embrace a broader notion of fertility.

Matthew Looper (1993) also noticed that there were distinct differences in sprouting imagery that might reflect the contrasting growth habits found between monocotyledons and dicotyledons (a single versus double seed leaf, cotyledon). While this observation was never extensively developed, it isolated a real iconographic difference and pointed to the complex nature of these sometimes deceptively simple images. The question remains: Was this simply a component of artistic license, illustrating a generic plant form in which these different morphological features were used to signal differing plant attributes, or was it meant to be diagnostic? Offering support to Looper's observations is the Chol word *ña'al, discussed at the beginning of this chapter, indicating that even in the Maya cultural context, "maize" terminology covered a sufficient semantic range to include other flora as well as animals.

Scanning even the limited range of examples of the Olmec Maize God in Figure 6.6 shows that it is a complex entity that is at once phyto- and zoomorphic and possibly even based on spontaneously aborted human embryos and fetuses (Tate 2012). Early and Middle Formative period imagery draws from a diverse range of taxa and phenomena to conceptually and visually construct deities (Reilly 1991; Stross 1994; Tate 2012). This fact alone complicates explanations that posit an isomorphic relationship between maize and specific deities in Formative period imagery and that do not take into adequate consideration the ecological, social, and symbolic processes that Mesoamerican deities personify (Hunt 1977; López-Austin 1992).

While discussions of this Formative period Gulf Coast material often continued to rely on the *Popol Vuh,* to illustrate the significance of the maize god in the region's cultures and to show the parallel life cycles of maize and humans, Taube did invoke Foster's (1945:180) observation that the contemporary Gulf Coast Sierra Popoluca believe that Homshuk passes "from childhood through maturity to old age each year during the cycle in which the maize sprouts, grows tall, ripens, and then withers." Nevertheless, it is somewhat ironic that, despite being within the cultural region of the Olmec, the Gulf Coast maize mythologies we briefly reviewed in the previous section have played a relatively minor role in analyses of Formative period imagery apart from a few, significant exceptions (Braakhuis 1990, 2009; Chevalier and Bain 2003; Chinchilla 2011; see especially Tate 2012:61–64). Likewise the differences between these iterations have not been considered to reveal significant information about corresponding iconographic differences across regions.

In light of the clear connections between these Formative period images and later Classic period examples, Taube (1996:68), citing Hastorf and Johannessen (1994:436), raises an interesting conundrum toward the end of this seminal article that still remains unaddressed even after a quarter century:

> Early Formative lowland diet raises interesting questions regarding the role of corn among the Olmec of Early Formative San Lorenzo.

However, it is important to distinguish between the dietary, cultural, and economic roles of this food. In a cross-cultural discussion of maize in the pre-Hispanic New World, Hastorf and Johannessen (1994:436) note that "corn appears to have had a special significance in many cases before it became a staple, thus indicating that its unique cultural importance did not stem primarily from any dietary prominence." Moreover, the storage capabilities of dried maize, particularly when smoked, must have made it an especially attractive tribute food even in Early Formative times. As a readily stored item, maize would have obvious advantages over abundant but difficult to preserve aquatic fauna.

In this passage and elsewhere (Taube 2000:298–299, 2004:6)[8] he observes, as others have since, that despite not being the dominant food during the Early Formative, maize had still apparently become symbolically important. Likewise, Taube and others (2000; Arnold 2000:117–118, 2005) note that maize motifs and referents did not become common in Olmec art until the Middle Formative period was underway, which appears unsurprisingly to stand in positive correlation with this grain's increasing role in the Formative period diet. Returning to the above passage he also suggests that maize held specific advantages over other staples. From its symbolic significance he immediately turns to what seems an expedient solution that maize would have been easier to store than aquatic resources. In the first place, given both the great abundance of aquatic resources and their durability when preserved correctly (Arnold 2000, 2009), this assumption that aquatic resources couldn't be stored is debatable. Moreover, it is unclear how this relates to the central issue of maize's symbolic importance prior to its culinary dominance. What we have seen in the preceding discussion is that the category of *cintli*, "maize," extends beyond maize itself. If this is the case, then we could imagine, as Bonta has also suggested, that cycads as a primordial foodstuff presented an original "*cintli*" model into which maize eventually grew to become the paradigmatic form.

The archaeological evidence also shows the dominance of other resources in the Early and Early Middle Formative period diet.[9] Researchers have observed that the early significance of maize might be tied to its sugary stalks, which could have been fermented and consumed as part of feasting or rituals (Iltis 2000, 2006; Smalley and Blake 2003); however, isolating the date of the first use of maize in a form typical of Late Formative and Classic period cultures is still undetermined (see, however, Kennett et al. 2020), as is detecting the precise moment when it became a symbolically significant plant and why. In any case, I do not contest Taube's and others' skillfully crafted iconographic genealogies or the various reasons advanced for maize's eventual preeminence. While it is not possible to assign precise dates, I do, however, want to focus on this conundrum and the initial moment when such iconography emerges, at a time seemingly before or just at the increasing alimentary dominance of maize and in light of substantial evidence that maize was but only one component of a larger system of botanical forms that eventually came to crystallize around it.[10] In this I want to revisit Arnold's (2005:3) question of "what was Olmec iconography depicting for the half-millennium prior to ca. 700 BC?"

Arnold identifies marine resources in his analysis of piscine imagery, which he also associates with Cipactli imagery and terminology, as discussed previously. However, here I focus additionally on terrestrial resources and make the case that the mythologies cited previously suggest that both cycads and ojox were also components of relic foodways that came to play a significant role in mythology and accordingly also likely contributed to Formative period iconography. As foodways changed in the Middle Formative period so too did mythologies, and in such ways that earlier practices were often framed as antagonistic to novel ones or were somehow otherwise modified to fit regional foodways, as we saw with particular clarity in the case of the Xi'iuy, in whose version of the Corn Master narrative cycad seeds are produced through the defeat of the cannibal ogress.

When we turn to representations of the Formative complex of fertility/maize gods we find amalgamations similar to the Classic Maya ones discussed above. These Formative period instances also complicate identifications that make one-to-one correlations between a specific taxon and deity representations (Carrasco 2020; Englehardt and Carrasco 2020). A particularly instructive set of images depicts a composite entity created from crocodilians, phytomorphic forms, and humanoid components. The scenes in Figure 6.7 present a series of hybrids with a crocodilian. The Arroyo Pesquero Celt presents God II, the Olmec Maize God, holding a bundle or possibly a serpent bar. A crocodilian head replaces his lower torso and legs. Similarly, a stone scepter representing a bundle presents a sprouting deity whose body is formed from wrappings. Like the Arroyo Pesquero example, a reptilian, possibly crocodilian, head replaces his legs and feet. On Izapa Stelae 25 and 5, a crocodilian forms the trunk and roots of trees. In yet another example from an unprovenanced sculpture from the Soconusco, dubbed the Young Lord, both crocodilian and piscine imagery are found inscribed on his thighs. When read together we arrive at a conceptual picture in which trees or a figure arises from a crocodilian. More specifically, as discussed previously, we could read this entity as a kind of Cipactli, an interpretation given greater

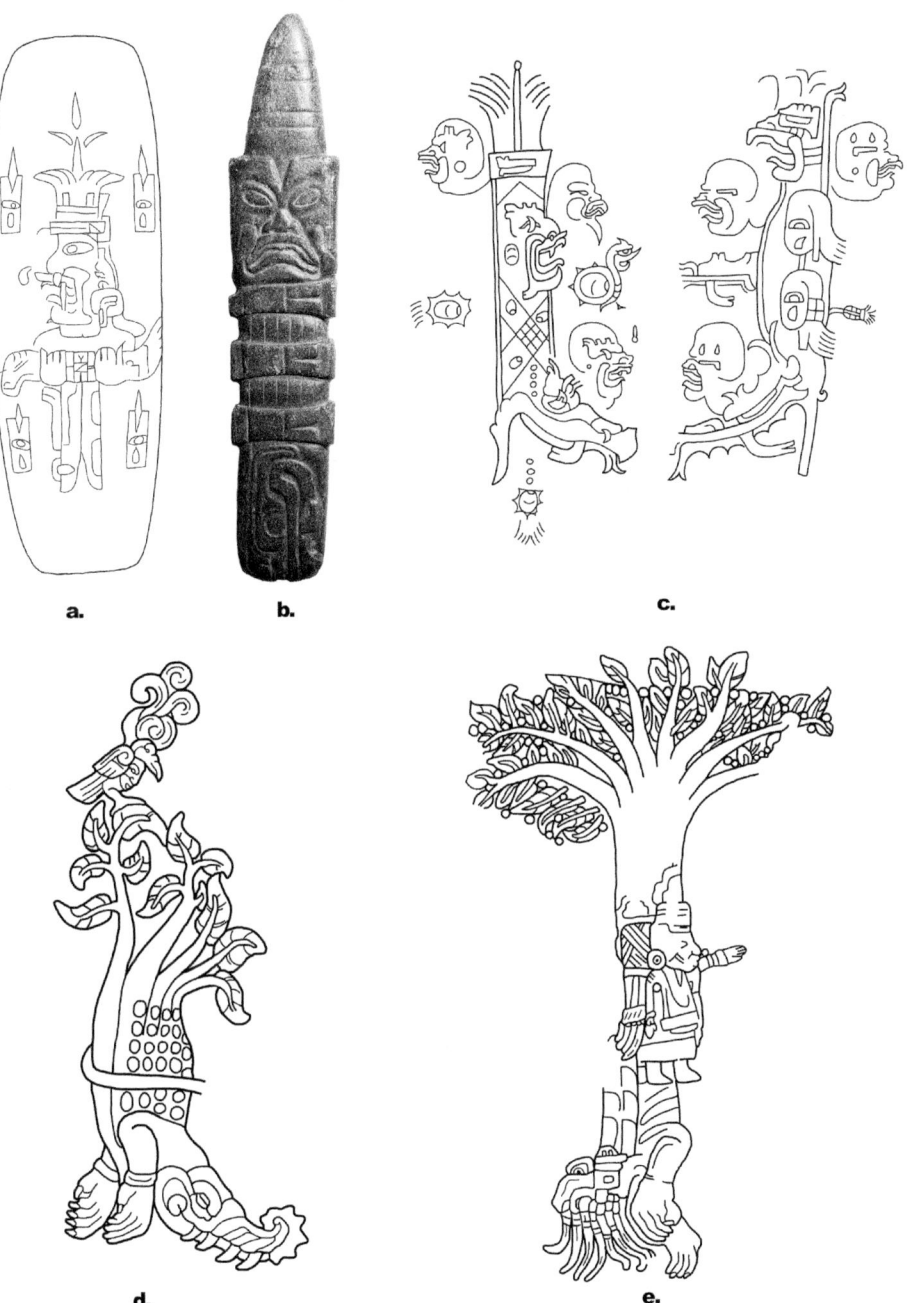

Figure 6.7. Flora emerging from crocodilian: *a*, Arroyo Pesquero celt (drawing by author); *b*, Ojoshal Scepter, Cardenas, Tabasco (Museo Regional de Antropología Carlos Pellicer Cámara, photograph by the author); *c*, Green Stone figure "Young Lord," detail of inscription on thighs (drawing by the author, after Reilly 1991 and consultations with photographs); *d*, Izapa Stela 25 (drawing by author); *e*, Izapa Stela 5, detail of central tree (drawing by author).

Figure 6.8. Equivalent Central Mexican (Alligator, *left*) and Maya (Imix, *center* and *right*) day signs (drawings by author).

strength in the visual pairing of crocodilian and fish presented on the Young Lord, which parallels exactly the two versions of Cipactli creatures presented in the codical examples in Figure 6.2.

In these examples we find a structure that through one line of visual evidence suggests a maize deity, were we to rely primarily on previous interpretations and the cleft head of the figures on the Arroyo Pesquero Celt and stone scepter as the key taxonomic element. However, if we broaden the scope to include the larger composition then we see a pattern in which a hybrid figure seems to denote a range of taxa that have a particular relationship with crocodilians, a relationship that is structurally similar to the one found between the maize god and the skull-waterlily complex in Classic Maya art. This is a particularly interesting and systemic contrast because it exactly parallels the difference between the crocodilian and Imix waterlily imagery of the day signs of the central Mexican and Maya 260-day calendrical systems respectively (Figure 6.8). Through this lens we can see that the Arroyo Pesquero celt imagery is nearly exactly the same as that presented in the Maya Maize God sprouting imagery seen in Figures 6.4 and 6.5. It also suggests that in Formative period iconography the crocodilian serves a similar role as the waterlily-skull complex in Classic Maya imagery. The significant point, especially in the present argument, is that, whereas the TMG nearly always is the one emerging from the waterlily-skull complex, in Formative period imagery there is a range of taxa sprouting from the crocodilian, some of which are not maize at all. What is more, through the example of the Young Lord we find that this crocodilian is structurally equivalent to a piscine creature, perhaps a shark (Arnold 2005:7; Stross 1994) or gar fish (Reilly 1991). This connection is strengthened by Arnold's (2005:27) observation that the foliation sprouting from the Arroyo Pesquero figure's head is strikingly similar to the tailfin of the piscine entity on the Young Lord (see Figure 6.9). Again, it is useful to recall that gars, sharks, crocodilians, and a maize/cycad deity can all be kinds of Cipactli.

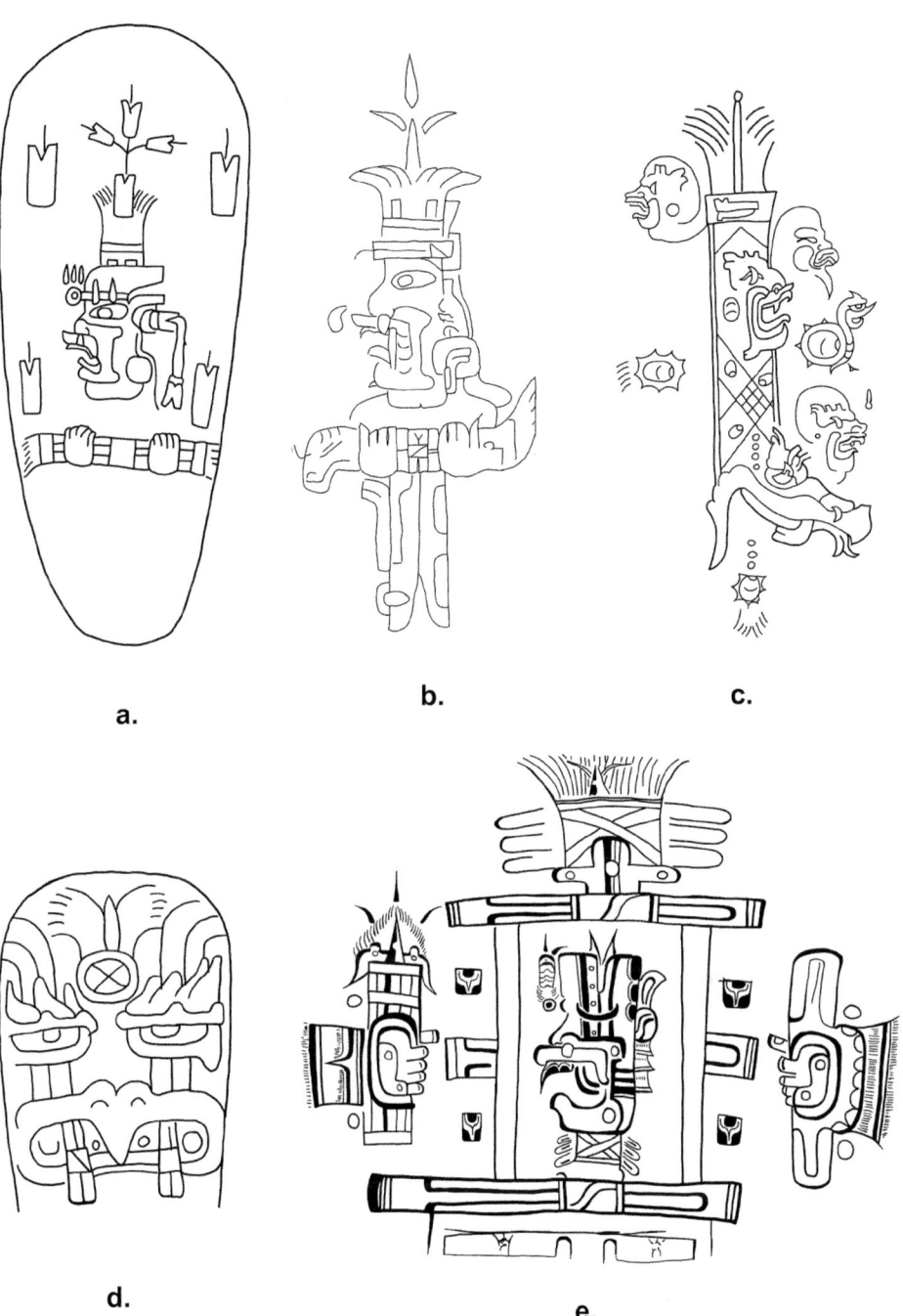

Figure 6.9. Piscine-monster tails substituting for sprouting foliage (adapted from Arnold 2005:fig.17): *a* and *b*, Arroyo Pesquero celts; *c*, Green Stone figure "Young Lord," detail of inscription on thighs; *d*, unprovenienced incised serpentine celt; *e*, Chalcatzingo Vase (drawings by author).

Finally, returning briefly to the Classic period Maya, the Berlin Tripod vessel and the reliefs on the sides of K'inich Janaab' Pakal's sarcophagus both present scenes of humans turning into trees (Figure 6.10). On the vessel, these figures show the head of the ancestor transformed into a base of the tree and their fingers turning into roots just as the teeth of the crocodilian on Stela 25 also become roots. In contrast, on the sarcophagus the named ancestors sprout as specific trees from the cracked ground marked with earth signs (*kab*). These figures are both ancestors and in their foliation resemble deities of fertility. In these images ancestors sprout as a variety of trees (Braakhuis 2009:15; Grube and Gaida 2006:128). Yet, their features sometimes also mirror those of the Maize God. This association with fruit trees reminds one that as Thipaak fled his cannibalistic stepmother, fruit trees sprang up behind him (Alcorn et al. 2006:603). Furthermore, as discussed previously, figures can sprout from bottle gourds, or as seen here from seed elements that are not identical to those of maize. Thus, even within Classic period Maya art the Maize God shares features with a broad range of taxa. While there is insufficient space here to fully unpack this, it is interesting to note the resemblance between these fruit tree ancestor figures and the crocodilian trees discussed previously. If we consider Cipactli creatures as a kind of ancestor yet also as a source for new life, we might find an even more striking conceptual bridge between this set of imagery. The differences we see between these sets of imagery, like those found among iterations of Corn Master narratives, are another instance of transformation in mythological structure that seems keyed to specific regions and periods.

Cycads in Formative Period Iconography

In this final portion of the chapter I turn specifically to Formative period Gulf Coast imagery that possibly makes specific reference to the cycad-maize iconographic category. The egg or seedlike stone sculpture from Tenaspi Island in Lake Catemaco provides a compelling example of a Gulf Coast specific Corn Master mythology encoded in Formative period imagery (Figure 6.11). This sculpture likely references the stories of ovogenesis that are a distinctive element in Homshuk, Kondoy, and Sintiopiltsin (lit. ear-of-corn-god-son-honorific [Chevalier and Bain 2003:177]) Corn Master narratives. In these stories this child-deity is the source and precursor to actual maize while at the same time seemingly being maize himself. As examined previously, cycads play a similar role in a number of these stories in such a way as to make them structurally equivalent to maize—a kind of maize from a previous creation. Therefore, we might consider this sculpture in light of

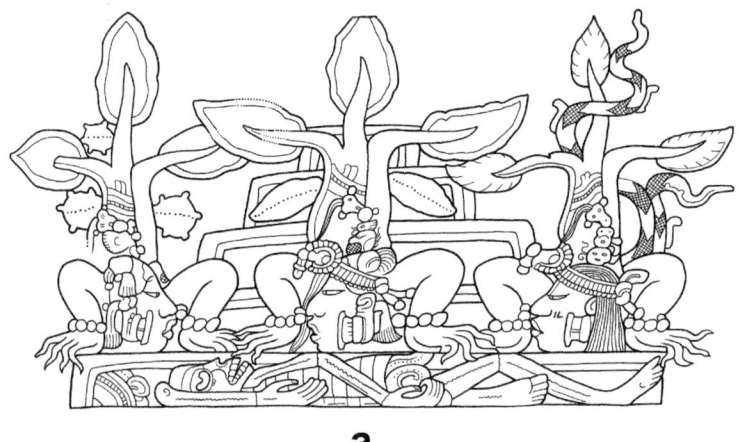

Figure 6.10. Scenes depicting ancestors sprouting/growing as trees: *top,* Berlin Tripod Vase; *bottom,* K'inich Janaab Pakal's sarcophagus, detail of sprouting ancestor, Temple of Inscriptions, Palenque (drawings courtesy Simon Martin).

this expanded understanding of cycads' role in the Gulf Coast Corn Master narratives.

The Tenaspi Egg depicts a human face wearing a tripartite headdress lightly carved onto one side of an oval stone (Blom and La Farge 1926:22). The face rests on a day sign cartouche followed by the number six. However, the day sign itself is either too eroded to read or was originally painted. Earspools flank the face, and the headdress presents a large central diadem from which a fan-shaped element extends back over the top of the stone.

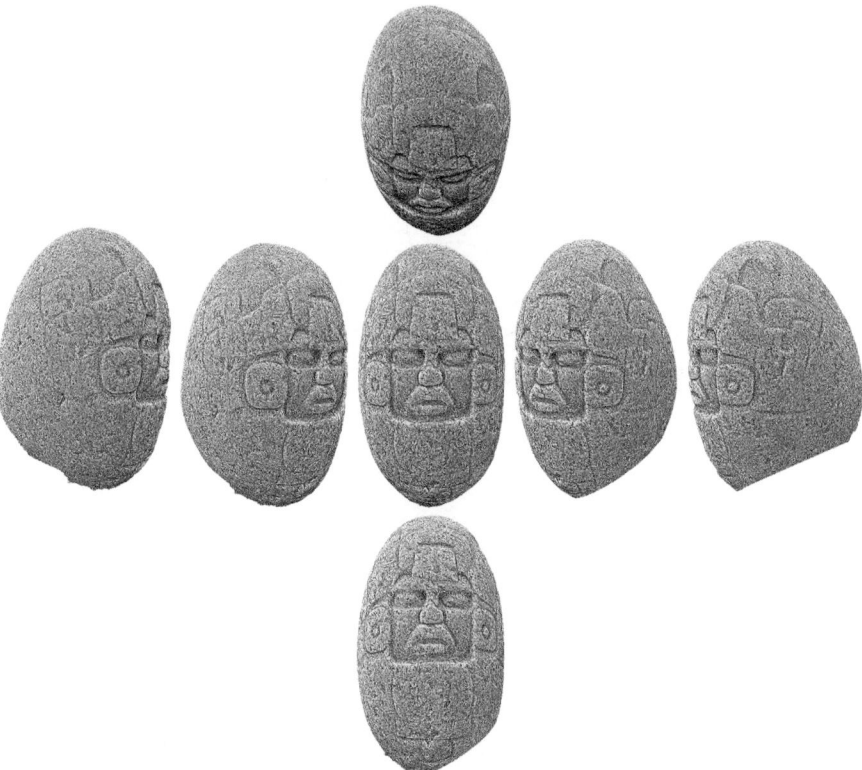

Figure 6.11. Late Formative period stone sculpture "Tenaspi Egg," reportedly from Tenaspi Island, Catemaco, Veracruz (photogrammetric model by author).

The lateral ends of the headdress terminate in knots. The elements are not specific enough to tie to exact iconography, but overall it resembles other tripartite headdresses often associated with rulership and foliation.

The emergent face and the egg shape of the stone not only recall the stories of ovogenesis presented previously but also greatly resemble cycad seeds (Figure 6.12), and the initial eruption of the radicular end of the embryo through the hard shell-like sclerotesta gives the impression that the seed is hatching. Although these formal resemblances might appear to us coincidental, in Mesoamerica cultures likeness often played a key role in creating meaningful associations. Indeed, Chevalier and Bain (2003:180–182) enumerate a list of forms including eggs, progeny, fruits, seeds, and stones, among others, that are currently viewed as symbolically related based on formal or functional similarities.

If we turn to the *Ceratozamia* spp. endemic to the Tuxtlas there are additional points of interest to consider in relation to the Corn Master narratives

Iconology of Cycads, Maize, and Fertility Deities in Mesoamerican Art · 203

Figure 6.12. *Dioon spinulosum* seed germination (photographs by P. Barry Tomlinson, courtesy Montgomery Botanical Center).

and patterns of semblance. In *Ceratozamia* species seeds are held in pairs on the sporophylls in the female strobilus (Figure 6.13). In this, not only does each seed resemble an egg, but they are also twins. It is unlikely that this characteristic would be overlooked given the preponderance of additional data speaking to cycads' close association with maize.

Finally the parallel between the morphology and growth habits of cycads and Olmec Maize God (Olmec God II) iconography presents a striking resemblance worth considering in detail. Formative period fertility iconography often presents figures with sprouts emerging from their cleft heads. These scenes have often been identified, as discussed previously, as Formative depictions of the Maize God. This emerging "mazorca" often resembles a cycad strobilus growing from the apex of the stem. Thus, while previous interpretations of Formative period phytomorphic deities have often looked exclusively to maize as a source for sprouting imagery, the formal qualities of God II more closely resemble the physical attributes and the growth habits of cycads.

One of the finest examples of this iconographic similarity between cycad strobili and what has been considered a maize cob is found on a jadeite sculpture discovered underwater at the site of Arroyo Pesquero (Wendt et al. 2014). Figure 6.14 presents this object next to the female strobilus of a *C. latifolia*. Notice that the forked elements on the sculpture that have been interpreted as split kernels of maize resemble the two prominent spinelike

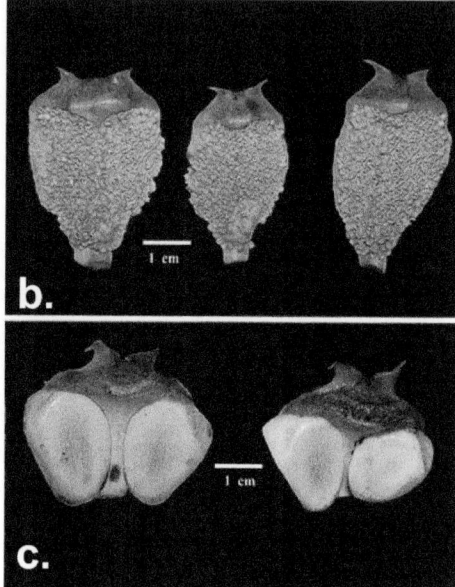
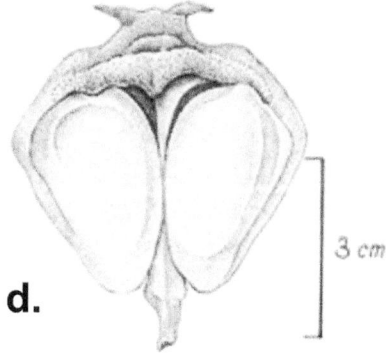

Figure 6.13. *Ceratozamia mexicana*: *a*, megasporophyll; *b*, detail of pollen strobili/ microsporophyll; *c*, detail of ovulate strobili cross section (photographs from Lilí Martínez-Domínguez et al. 2018:112, fig. 11c–d, e); *d*, *Ceratozamia vovidesii*: drawing of ovulate strobili cross section (drawing by Edmundo Saavedra).

horns on the sporophylls of the strobili. These are in part diagnostic of the genus *Ceratozamia*. Each sporophyll holds two seeds mirroring the twin eggs found in many of the ovogenesis narratives. Wendt has speculated that this object was perhaps attached to a staff. One wonders if this sculpture then would be an implement that would invoke the generative powers of the maize ancestor, while itself also being formally suggestive of maize. Other objects, such as the Ojoshal Scepter (Figure 6.14c) present similar scenes that provide additional details that connect the idea of the scepter or staff to sprouting imagery and crocodilians. In yet other examples (Figure 6.14d–e), the female seed-bearing strobili of *Dioon edule* resembles the iconography of

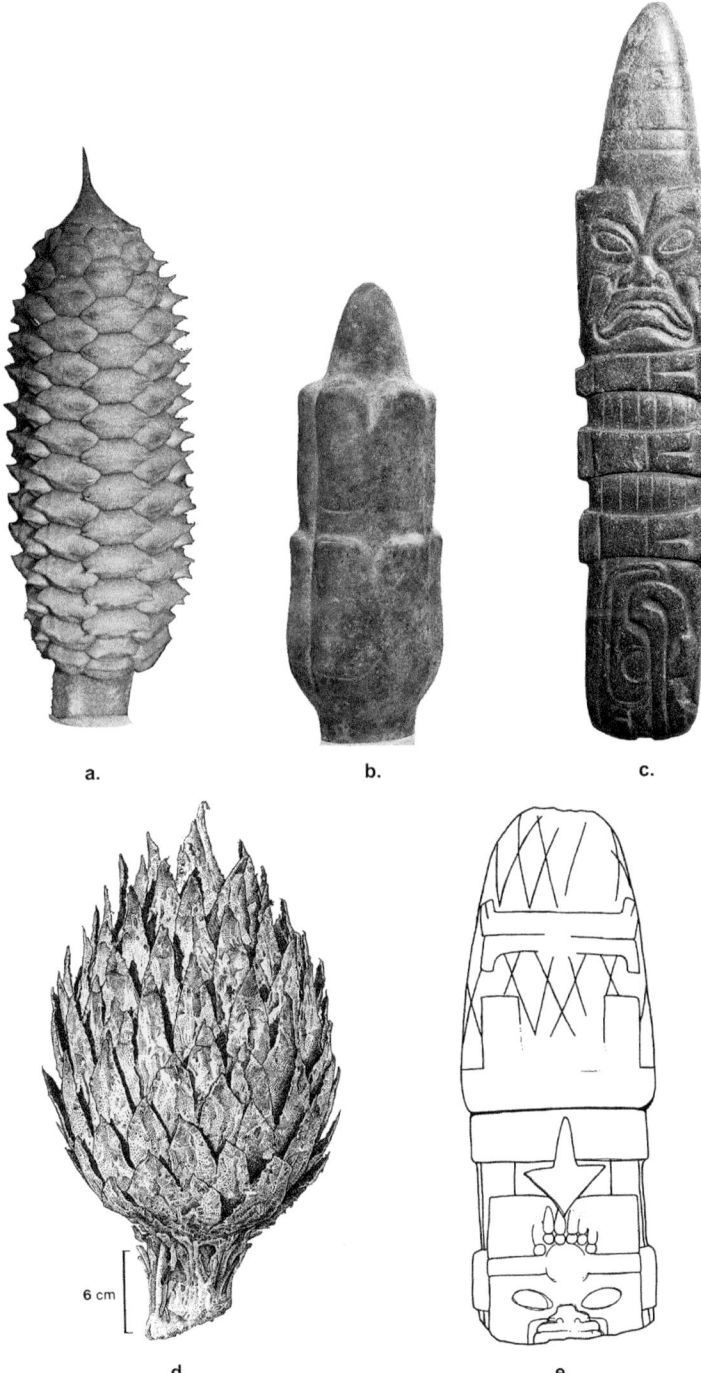

Figure 6.14. Similarities between cycad strobili and Olmec imagery: *a, Ceratozamia latifolia* female strobilus (photograph by author); *b,* jadeite sculpture, Arroyo Pesquero, Veracruz; *c,* Ojoshal Scepter from La Encrucijada, Tabasco (photograph by author); *d, Dioon edule* female strobilus (drawing by Edmundo Saavedra); *e,* Vegetal Bundle motif (drawing by author).

a corn fetish, as well as bundles and "torches" common in Middle Formative iconography.

At this time these formal similarities remain speculative; however, we might imagine a situation in which cycad seeds were considered within the same category as eggs because they contained food and represented generative powers. Seeds and eggs in many ways could have easily been regarded as flora and fauna equivalents and also related to other items within this category such as skulls or stones. Also the cycad, at least in the case of Thipaak, is equated with Cipactli creatures, possibly because of the texture of its trunk and the fact that the strobili's sporophyll resemble scales. Therefore the cycad is a kind of crocodilian that produces eggs in the form of its seeds; however, these seed-eggs must be processed (in nixtamalization) like maize. Thus the role of eggs in maize stories would seem to identify the cycad as a kind of maize from a previous creation that presents or produces "eggs," in some cases twin "eggs," that must be processed in the same manner as the maize of this creation. Indeed, as we have seen and as others describe in greater detail (Alcorn 1984; Bonta, chapter 8, this volume; Bonta et al. 2019), cycads are believed to be the ancestors of maize, either because they are a couple similar to the primordial couple, or because they are one of the embodiments of a deity who brings maize to people.

Given the archaeological, ethnographic, and linguistic data presented here and in the other chapters of this volume that demonstrate the close association of cycads and maize, it is unlikely that these homologies and analogies were lost on Formative period Mesoamericans. To the contrary, the early evidence of cycad use suggests that this close symbolic relationship originated in the Late Archaic or Early Formative period and then was adapted and elaborated in various regional forms and in response to the eventual dominance of maize. The sprouting element in Formative period imagery, particularly that of the Olmec Maize God, could stand for the emerging maize plant, the mazorca, or the cone of the cycad, among other possibilities. These images reference all these instances, thus making the attempt to isolate a true referent futile and ultimately undermining the subject's symbolic richness. Indeed, these resemblances create an Indigenous category, including maize, cycads, fish, and crocodilians, among others, that functioned as a core component of Pan-Mesoamerican mythology. Within these symbolic and food systems it would appear cycads played something of a mediating role (Figure 6.1, Figure 6.15).

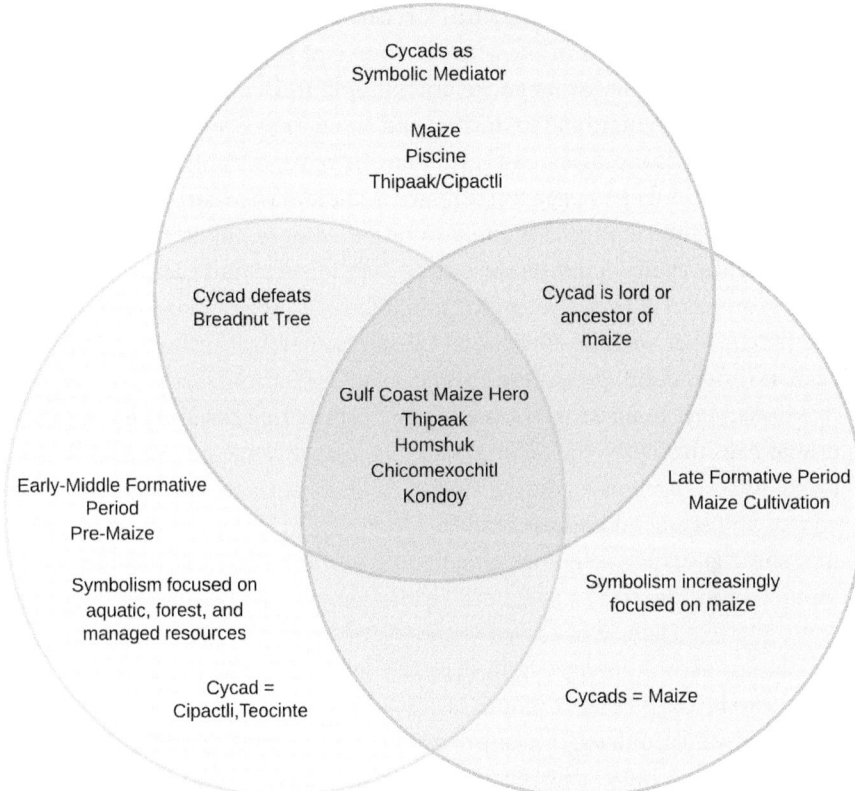

Figure 6.15. The relationship of agroecology, symbolism, and mythology (created by author).

Conclusions

Through these examples I wished to accomplish two related goals. First, based on the patterns and symbolic links between maize and cycads as expressed in shared terminology and mythology, I traced their affiliations to broader symbolic categories (Carrasco 2015, 2020) that include gar fish, sharks, crocodilians, and the ojox or breadnut tree, among others. Second, through this we found that Indigenous classificatory schemes present semantic ranges that show Mesoamerican taxonomies to be predicated along different axes of meaning. Many appear to be based on the alimentary importance of a plant and morphological and processual similarities, or they have been brought together through specific roles within core Mesoamerican mythologies. In some cases the saliency of certain plants made them particularly suited to represent these larger cultural patterns or as sources for terms that could serve as meta-labels for other taxa with which they stood in morphological, functional, and/or symbolic relation, although it is not always possible to discern the original source of the terms. Identifying the cycad-maize category in language, mythology, and symbolism has allowed us to reconsider Formative period art and thought in a new light. It permits us to conduct iconological analyses that move beyond iconographic ones that rely on iconic isomorphism between depiction and object. Through this lens I have suggested that Middle Formative imagery possibly depicts a botanical category that would have included both maize and cycads and could be thought of as a centli category that includes a range of taxa beyond that of maize proper and is evidenced in the terms that include the root "centli," such as those found in the Tenochtitlan sequence of world eras.

Mesoamerican artists illustrated a vast array of plants and animals. For example, in the Proto-Classic murals at the site of San Bartolo, sabal palms, pines, aroid, and bottle gourds, among other species, populate the mythological landscape and adorn figures. Moreover, in these murals anthropomorphic figures are seen emerging from botanical items, such as the child/infant "sprouting" from a split bottle gourd, which is a theme found repeated some 600 years later on a carved vessel (Kerr vessel 5761). As these instances clearly reveal, not only are flora depicted to give a sense of place—such as directional trees on page 1 of the Codex Fejérváry-Mayer or the use of cacti to indicate Chicomoztoc in many examples—but botanical forms also merge with humans to illustrate fundamental philosophical propositions that saw human and plant life cycles as mutually analogous. Thus, despite the highly detailed depictions of specific plants, Mesoamericans did not conceptualize representation as a window onto an objective reality nor wished to emphasize the same visual priorities and distinctions that modern science prioritizes. In Mesoamerica, depictions of the environment and the fruits of nature served multiple roles. In particular, such illustrations often presented key theological and philosophical ideas about a world in motion and cyclic change (Chevalier and Bain 2003; Maffie 2014). The sprouting, fruiting, and maturation and eventual death or consumption of plants manifested a prime example of such dynamic change, one which could also function as a metaphor for human life. Through a cultural logic in which like elements share a common identity, scientifically discrete taxa were combined into hybrids or amalgamations, or, perhaps better put, placed into higher order categories that contained each of the depicted elements and that potentially allow us to discern the parameters of this more abstract category, as I hope to have done in this chapter.

In the case at hand, we find a symbolic set, including maize, fish, and cycads, among other elements, that appears to have been established in the

Formative period and continues into the present in a variety of permutations. These items appear also to be tied to specific temporalities in myth so a proper food in one era ceases to be so or transforms into another in the succeeding period. In this way we see that, from the perspective of the world ages discussed here, each era had its own "people" who then were destroyed and became something else. Specifically I have suggested that cycads served as mediators between these ages with a foot, as it were, in both the previous moment as a kind of ancestral food or guardian of maize and the present, since cycads are still consumed or used in ritual. In this way—like the equivalency found in incense which is both a food for those to whom it is offered and a ritual offering to the devotee—cycads' position seems dependent on the particular perspective from which they are examined. By acknowledging the significance of cycads I have also suggested that it would be productive to reexamine Formative period imagery that has traditionally been read as primarily maize related. Again, my view is not that we should see cycads in the place of maize, but rather we should think in terms of a category inclusive of both and probably other items as well that have yet to be isolated.

Notes

1. "Corn Master" is a Popolucan expression used along with Jomxuc (Homshuk) and Mok Santu (maize saint) to refer to an entity that brings maize to people. I use the term Corn Master here so as to not favor one particular Indigenous language term for this entity who is found throughout Mesoamerica and beyond.

2. The likelihood that maize was not processed as *nixtamal* (boiling dried kernels in lime or wood ash) raises an intriguing possibility that nixtamalization originally arose as a preparation to detoxify cycad seeds, as is still done in Honduras and Hidalgo, Mexico (Bonta 2010a). The transference of this preparation technology would have been critical for initiating Early Formative period population growth that led to the establishment and expansion of such sites as San Lorenzo on the Gulf coast and Tlatilco and Tlapacoyo in the central highlands, among others. Nixtamalization both enhances maize's nutritional value and reduces endemic mycotoxins by as much as 94% (Staller 2018:183). Therefore, while it is impossible to prove with current evidence, the adoption of cycad processing techniques for maize might have been critical for one of the major culinary developments in Mesoamerica (see Cibrián-Jaramillo et al., this volume, Figure 2.5).

3. The concept of the "spirit of maize" is also found among the Ch'orti' Maya in the gendered terms *ijben winik*, "male maize spirit," and *ijben ixik*, "female maize spirit," and is also likely found personified in the cultural hero Kumix, a child deity who, like Thipaak, defeats beings from the creation before the present era (Hull 2009). Alan and Pamela Sandstrom (1986) document similar spirits of maize and other plants in the paper cutout figures of the Huasteca.

4. The motif of antagonism between a grandmother and young boy extends from the lowland tropical forests of South America (Lévi-Strauss 1969–1981[1964–1971]) through

Panama (Sherzer 1990:162–202) into Mesoamerica. This motif continues into the US Southeast in narratives that describe the origin of maize as resulting when a young man kills his adoptive mother after witnessing her create corn from her body. After her murder and subsequent burial, maize sprouts from her grave (Witthoft 1946).

5. Miller (1956:196) suggests several alternatives for the etymology of Kondoy: Kondoy seems to have been "a nickname and not properly a name if we accept the etymology of the Mixe. According to one interpretation, it comes from '*kon*' or '*konk*,' chief or strong man, and from '*oí*,' good. Another explanation relates the name to the legend that tells how the Zapotecs lit the Zempoaltépetl in a futile attempt to burn Kondoy and his soldiers, causing it to derive from '*kon*,' chief and '*toi*,' burned." [Kondoy parece haber sido "un apodo y no propiamente un nombre si aceptamos la etimología de los mixes. Según una interpretación, viene de '*kon*' o '*konk*,' jefe u hombre fuerte, y de '*oí*,' bueno. Otra explicación relaciona el nombre con la leyenda que narra cómo los zapotecas encendieron el Zempoaltépetl en un inútil atentado de quemar a Kondoy y sus soldados, haciéndolo derivar de '*kon*,' jefe y '*toi*,' quemado."]

6. Alcorn notes that the Teenek likewise believe that the people and giants known as Lints'i' drowned in a flood and became fish (Alcorn 1984:60). These fish were seen as the ancestors of the next era of humans who, because they ate these fish in a fit of hunger, were transformed into raccoons. Thus, fish, like corn, is seen as an ancestor of people. The blood-fish connection is also found at El Tajín (Rex Koontz, personal communication, 2018).

7. In the 1980s interpretations of phytomorphic deities in Classic Mesoamerican iconography (Fields 1991; Hellmuth 1982 [cited in Taube 1985:172]; Taube 1985) returned to iconographic work conducted in the late nineteenth and early twentieth centuries on codical and post-Classic imagery of maize-related deities (Dieseldorff 1922; Goodman 1897; Schellhas 1897, 1904; Seler 1902–1923, 1963, 1976; Spinden 1913) to renew scholarship on the central symbolic importance of this plant. Early work had focused by necessity on codical representations, and some of the first identifications were of a maize deity that was what Schellhas termed God E in the Madrid and Dresden Codexes.

8. Taube (2004:6) also notes that "although maize is documented at Mokaya sites, it probably was not the primary staple. The ears of recovered specimens are small and relatively unproductive, and chemical analysis of Mokaya human bone collagen reveals that type C-4 pathway plants, such as maize, were not a significant part of the local diet (Blake et al. 1992; Clark and Blake 1989:389)" (see also VanDerwarker and Kruger 2012).

9. The archaeological evidence also shows the dominance of other resources in the Early Formative period diet (e.g., *Agave* spp., *Brosimum alicastrum* L., *Manihot esculenta* Crantz, *Persea americana, Sabal* spp., *Theobroma cacao, Xanthosoma* spp., and cycads among others (Brown 2010). For a recent discussion of the various subsistence strategies of the Early to Middle Formative periods in Coastal Oaxaca, also see Hepp (2019). Sheets and colleagues (Sheets et al. 2011) have demonstrated the cultivation of manioc beyond the household garden, and excavations at Cuello and San Andrés (Pohl et al. 1996; Pope et al. 2001) provide evidence of its cultivation in the Formative period. For example, Borstein (2001) found that maize was not used until the end of the Early Formative at Laguna de los Cerros.

10. Phillip Arnold (2005:3) asks this same question: "In fact, such a reconsideration has already begun; Taube (2000:298–299) recently observed that corn motifs and referents did not become common in Olmec art until the Middle Formative period was underway. An intriguing question, therefore, is what was Olmec iconography depicting for the half-millennium prior to ca. 700 BC?"

7

Tracking *Teosinte* across Mexico and Northern Central America

Mark A. Bonta

Teosinte, Cycads, and the Sacred-Maize-Ancestor Concept

In modern scholarly literature, "teosinte" typically refers to maize's genetic ancestor, *Zea mays parviglumis,* a grass from the Balsas drainage basin of Guerrero state in western Mexico (Matsuoka et al. 2002). However, in parts of Mesoamerica, "teosinte" is also a local term for other cereal and forage grasses as well as for cycads (Wilkes 1967). The Nahuatl word is pronounced and spelled in numerous ways, combining the root *centli* (*cintli, sintli,* etc.) and the prefix *teo* (*teu, tio, tiu, tau, teoc,* etc.). *Teo* derives from *teotl,* god, and broadly signifies the divine or consecrated nature of the root to which it is affixed in many Classical Nahuatl words. *Centli* is the ear of maize. Thus, *teosinte* signifies "sacred maize ear," although it also suggests "old," "of the ancestors," or "wild" (Bonta 2010b; de Molina and de Spinosa 1966).

At first glance, the striking superficial resemblance of the male cone of certain cycad species to the maize cob suggests that the name *teosinte,* when applied to cycads, is derived simply from this appearance. However, as evidence in this chapter and elsewhere in this volume indicates, cultural connections between the two are far deeper and more intricate than appearances alone. The name itself was the first reason I began to investigate the maize-cycad connection around 1999 (and the ultimate reason for this volume; Bonta 2010b), by tracking its occurrence as a Honduran term both for forage grasses and for cycads to try to determine when, where, and ultimately why cycads and maize have been culturally intertwined. I refer to the most common iteration of this relationship as the sacred-maize-ancestor (SMA) concept, explicated in broader detail in chapter 8 (this volume). Here, I wish to focus particularly on making sense out of "teosinte" and closely allied terms. The guiding questions of this chapter are where the term and concept originated, and how they came to be applied to both cycads and grasses.

Postconquest *Teosinte* in Mexico

In terms of the available literature, the story of *teosinte* commences in the Nahua-dominated city of Huayacocotla, Veracruz, in the Sierra Madre Oriental. In the 1560s, Spanish naturalist Francisco Hernández obtained the first written postcolumbian record of *teosinte* from this location. His Nahua source described a plant known to local people as *teocintli* or *tepecentli*:

> Del TEOCENTLI o tepecentli, o sea mazorca de maíz de monte.
> Es un arbusto de raíz redonda parecida a una piña de pino, de donde nacen tallos cilíndricos, lisos, verdes con blanco, hojas que ocupan la parte superior del tallo, poco más o menos de un palmo de largo, de un dedo de ancho, gruesas, duras, con nervaduras rectas, semejantes a hojas de palma, y fruto oblongo parecido también a una piña y que nace en la punta del tallo. La raíz se come cocida con maíz; es dulce, astringente, y detiene aplicada las diarreas; el fruto se come crudo. Nace en lugares montuosos de Hoeiacocotla. (Hernández 1946:866)

> Of TEOCENTLI or tepecentli, that is, ear of wild corn.
> It is a bush with a cylindrical root, in appearance like a pinecone, from which cylindrical trunks bearing smooth, whitish green leaves extend that occupy the upper part of the trunk, more or less a palm's length, a finger's breadth, thick, hard, with stiff nerves, similar to palm leaves, and oblong fruits similar to a pinecone that are born at the tip of the stem. The root is eaten cooked like maize; it is sweet, astringent, and can prevent diarrhea; the fruit [*presumably the sarcotesta*] is eaten raw. It grows in brushy places in Hoeiacocotla. (translation and comment by the author)

In the twentieth century, some maize experts suggested that the plant described in this text was a cycad, though other commentators suggested distinct identifications (Hernández 1946; Kempton and Popenoe 1937; Wilkes 1967).

Recent research has revealed that the only plant known by these two names in present-day Huayacocotla is indeed a cycad: *Ceratozamia fuscoviridis* (Bonta et al. 2019). The second name (*tepecentli*) mentioned by Hernández translates as hill-maize ear (*tepe* = hill) and is the most common Nahua and Totonac name for *Ceratozamia* species in the Sierra Norte de Puebla, to the south of Huayacocotla (see below). In other Nahua areas I have visited, sacred hills, which often have cycads growing on them, are conceptualized as the origin places of maize (Bonta et al. 2019; Diego-Vargas 2017). Thus, the two names suggest that cycads were seen as wild and sacred maize ancestors at least by 450 years ago.

No other unequivocal pre-1800s reference to *teosinte* (including any spelling variant) has come to light in Mexico, and it is not found in postcolumbian codices as a term for maize ancestors. A group of sixteenth-century texts concerned with the Nahua origin stories of the Cinco Soles (del Paso y Troncoso 1903; Florescano 1997; Wilkes 1967) do mention *cincocopi* and related terms that have been established with reasonable certainty as referring to the *Zea mays* genetic ancestor and related grasses—defined as the food the ancestors ate before they had maize, but without explicit reference to a divine quality. But other than Hernández, writings in the herbals, codices, and other sources from postcolumbian Mexican culture (e.g., Gates 1939; Sahagún 1577) do not contain obvious references to cycads. In large part, this is probably because cycads are not native to the Valley of Mexico and other parts of the Mexican central plateau where the Mexica and other Nahuatl-speaking groups that formed polities such as Tlaxcala and Tenochtitlan were based, and where the majority of later Spanish documentation of Nahuatl language and culture was gathered.

Overlap between native cycads and Nahuatl speakers occurs in the Sierra Madre Oriental as well as the Gulf Lowlands from southern Veracruz north to San Luis Potosí, with the highest cycad diversity in the Huasteca region. Figure 7.1 (see also Figure 8.1) shows the Huasteca as the assumed origin area of the term *teosinte* (discussed in more detail below).[1]

Teosinte Grasses in Mexico and Central America

The story of *teosinte* is different in Central America. Speakers of Pipil, a language derived from Nahuatl and brought to the region from Mexico 500 or more years ago (Kaufman 2001), apply the term *teosinte* (with numerous phonetic variants) to wild cereal grasses in the genera *Zea*, *Tripsacum* (gamagrass), and *Setaria* (foxtail millets), used as human food and as forage for livestock (Kempton and Popenoe 1937; Sutherland 1986; Wilkes 1967). These terms entered Spanish usage centuries ago and are commonly attached to toponyms found in the historical record since the late 1500s (Hernández 2008) to the present in Guatemala, El Salvador, western Honduras, and Nicaragua, mostly on the Pacific slope. Kempton and Popenoe (1937) were the

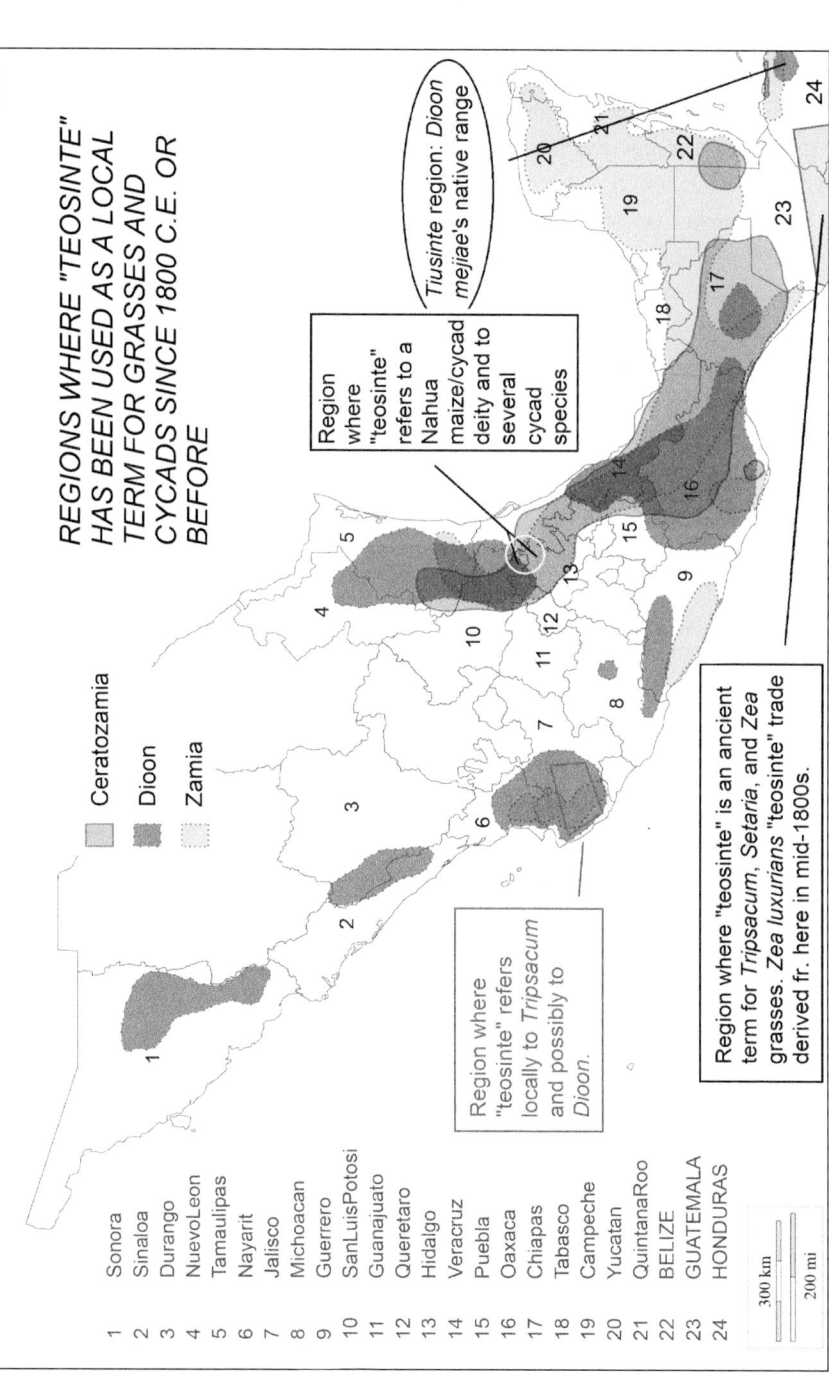

Figure 7.1. Teosinte regions in Mexico and northern Central America (map by the author).

first of the modern maize origin researchers to point this out in the 1930s after working in Guatemala, and later researchers have been able to locate wild populations of these grasses in the region by finding toponyms that incorporate *teosinte* (e.g., Haynes and Bonta 2003).

Wilkes (1967) corrected the misconception, often encountered in research on Mexican wild *Zea*, that *teosinte* is a widespread autochthonous Mexican term for wild grasses. What happened is that the Guatemalan *Zea luxurians* entered the world market (as *Reana luxurians* in the botanical literature, then as *Euchlaena*) as a popular forage grass in the 1800s, under the trade name *teosinte*. It was imported to Mexico, among many other countries, and thus became the term that scientists, government officials, and farmers used. It was later reapplied to the native *Zea mays* varieties, such as *Z. m. parviglumis* in the Balsas Valley, during research into maize's origins. Nowhere in Mexico is *teosinte* an original term for these grasses (e.g., CONABIO et al. 2008; Mondragón Pichardo and Vibrans 2005; see also the discussion of *Tripsacum teosinte* in Jalisco, below). In Guatemala and elsewhere, no ethnobotanical study has yet shown that there is still a sacred-maize-ancestor concept attached to the *teosinte* grasses, and the people who utilize the term have not retained any Indigenous languages.

Teosinte Cycads in Central America

The botanical discovery of another *teosinte* "sacred-maize-ancestor" in an isolated part of northeastern Honduras can be credited to Paul Standley and Louis Williams (1950), although Honduran natural historians knew of the species prior to that date (Bonta 2007). Evidently unaware that the term was also used for cycads in northeastern Mexico, Standley and Williams wrote that

> In Honduras this Dioon is known commonly by the strange name of "teosinte." This term belongs properly to a grass of the genus *Euchlaena* [*synonym of* Zea; *in this case they are referring to* Zea luxurians]. This true "teosinte" has been found at various localities in Mexico and Guatemala and is known also from western Honduras. How a name so well fixed came to be applied in Honduras to another plant so different as Dioon is a subject about which it is futile to speculate. (Standley and Williams 1950:38; comment by the author)

They named it *Dioon mejiae*, after the Honduran naturalist Isidoro Mejía, in whose garden they had encountered the plant.

In northeastern Honduras, the name *teocinte* is applied uniquely to this cycad and to no other plant. (Bonta et al. [2006] established *tiusinte* as the standard term for ethnobotanical usage to avoid confusion in the literature, and to better reflect local pronunciation.) Unusually for a well-known local plant species, *Dioon mejiae* has no other name derived from any of the various non-Nahuatl languages originally endemic to the region (Lenca, Pech, Tawahka, etc.). The Tolupan term *tiñuc* appears to derive directly from the local pronunciation variant *tiucsinte*.

For centuries, a local ethnic group now known as the Nahoa de Honduras has inhabited the area where *tiusinte* is found (Bonta 2009). They long ago lost their native language, but presumably spoke a Nahuatl dialect. Though archival evidence is fragmentary, the terms *teocinte* and *tiusinte* were already incorporated into place names on Indigenous land titles in the Nahoa de Honduras region by the 1700s.

How, then, did "teosinte" end up as a term for cycads, rather than grasses, at opposite Atlantic-slope peripheries of Mesoamerica—northeastern Honduras and northeastern Mexico—but virtually nowhere in between?

Given the comparative richness of cultural interrelationships between cycads and maize in northeastern Mexico compared with northeastern Honduras, and the diversity of Indigenous names for cycads in the former region, compared with the single Indigenous name in the latter (see chapter 8 in this volume, and Bonta et al. 2019), it is virtually certain that Nahuatl speakers from northeastern Mexico were the first to apply the term "teosinte" to Honduran cycads. According to a letter by Hernán Cortés (1963[1526]), and to other contemporary sources from the early 1500s immediately post-Contact, northeastern Honduras at that time contained an enclave of speakers of a Mexican language, as well as speakers of other, non-Mexican tongues—what are known today as Pech, Lenca, and so forth (Davidson 1991). Those Mexican language speakers, whose direct descendants are most likely the Nahoa de Honduras, are culturally distinct and geographically distant from the Pipil speakers of El Salvador and Guatemala who applied "*teosinte*" to grasses.

Available data are insufficient to establish whether the *teosinte* term and SMA concept were imported and applied to northeastern Honduras' native *Dioon* cycad before or during the Spanish colonial period. Nahua traders and other migrants from a longtime homeland somewhere in northern Mexico moved widely across Mesoamerica and beyond, in distinct settlement phases and streams, from long before the establishment of the Mexica (Aztec) empire to after its fall (MacLeod 2010). Accompanying the Cortés expedition to northeastern Honduras in the 1520s, for example, were thousands of Mexica, and the Spanish were in the habit of awarding land and other concessions to them in the areas they conquered.

Further clarifications are necessary. No wild *Dioon* cycads occur in the home area of the Central American *teosinte*-as-grass region (see Figure 7.1)—a region that extends as far south as Nicaragua, and possibly to northern Costa Rica. Indeed, no wild cycad species other than the extremely scarce *Zamia herrerae* is known to occur there. Conversely, in the *tiusinte* region, no wild *Zea* species has been recorded (Sutherland 1986), although *Tripsacum* or *Setaria* may occur (the botany of the region is relatively poorly known). *Dioon mejiae* is a range isolate and phenotypically rather distinct even from the nearest *Dioon* species in Mexico (Haynes and Bonta 2007; Moretti et al. 1993). It is the only one of its eighteen-member genus found outside of Mexico. While it is certain the species is native to northeastern Honduras, it is possible that the many present-day *D. mejiae* populations straddling ancient, long-distance trade routes were seeded or transplanted by immigrants of Mexican origin centuries ago from relict populations in the upper Aguán Valley. This hypothesis has some support from Mexican and Honduran analogues: Indigenous and non-Indigenous people alike uproot and transport *Dioon* cycads long distances to have local sources of leaves for ritual purposes (Bonta et al. 2019), and if, as is likely, the pollinator insects are transported as well, in the cones (presumably inadvertently), viable populations and easily accessible cone sources also result. The extraordinary alimentary importance of abundant *Dioon* cones as food in resource-impoverished northeastern Honduran pine forest landscapes, combined with such local terms for *Dioon* populations as "fincas" (farms) and "plantaciones," also point toward ancient origins of *tiusintales* (cycad groves) in cultivation (Bonta et al. 2006).

Although the term *tiusinte* leads me to believe that Nahua speakers from Mexico imported cycad knowledge to northeastern Honduras 500 or more years ago, the technically difficult food preparation methods suggest that the ethnic origin of at least the women who implanted cycad food culture in northeastern Honduras was not Nahua, but rather Xi'iuy (Pame). In northeastern Mexico, Nahua and Pame speakers formed a mixed culture for several centuries (Gutiérrez and Ochoa 2009), which would explain this seeming incongruity. Indeed, the complicated techniques used to detoxify the cycad seeds and make them edible, among other cultural factors, are identical between Xi'iuy culture and northeastern Honduran mestizo and Nahoa culture, but exclusive to those two groups (Bonta et al. 2019; Tristan Martínez et al. 2020). Similar practices are not found among any other northeast Mexican ethnic group that utilizes cycads, and this extends to other beliefs discussed in chapter 8. The chance that two identical detoxification techniques evolved in separation is effectively nil—one must have influenced the other.

Why didn't the term *teosinte* become a common name for cycads between the northeastern Mexican origin area and northeastern Honduras? In northeastern Mexico, *teosinte* and closely related names (among many other terms) are applied to sacred-maize-ancestor cycads in all three local genera (*Ceratozamia*, *Dioon*, and *Zamia*), so it stands to reason that the Nahua-speaking diaspora would have applied these terms to cycads they encountered along the way. And this may be the case; for example, what is apparently a *Zamia* was at one time apparently known as *teosinte* somewhere in northern Guatemala (Standley and Steyermark 1958). Otherwise, south of the extended *teosinte* region in northeastern Mexico described below, the record is silent. There are several possible explanations. The group transporting the name could have been quite small and did not linger anywhere long (i.e., as traders, refugees, accompaniers of Spanish conquistadors, etc.). Perhaps they traveled around the Yucatán Peninsula by water to avoid contact with local Maya groups. Or perhaps the term existed at one time but subsequently died out. In any case, even in the destination area of Caribbean littoral Honduras, where there are six endemic cycad species, only *Dioon mejiae* was given the name *teosinte*, and if any other local names for it once existed, they have been lost. The other five cycads are referred to by completely different, non-Nahuatl-derived names related to manioc, and are used as poisons and rarely as food—in the latter case, detoxification methods are typical of those used for cycads in the Caribbean basin, and completely unlike the northeastern Mexico-derived methods for *teosinte* (Bonta 2007).

Teosinte Extended: Sacred Tamale, Sacred Leaf

A diverse collection of Nahua terms for cycads in the Huasteca SMA source area and south as far as central Veracruz can be found in Bonta et al. (2019), Diego-Vargas (2017), and Vite-Reyes (2012) (see also Table 8.2 in this volume). A related term for a *Dioon* cycad, *tiotamal*, is recorded from central Veracruz around Xalapa, used by local mestizos in what was at one time a Nahuatl-speaking area. It is applied to the southernmost population of the widespread and widely consumed northeastern Mexican *Dioon edule* cycad. English-language commentators, probably mistakenly, assumed it meant "Uncle Tamal" (Chamberlain 1906). Much more likely is that the term originally meant "sacred tamale" (*teo-* is often pronounced and spelled *tio-*), a variation of *teosinte* as used to the north. Another cycad term, *tionishuatl*, currently used by Nahuatl speakers in northern Hidalgo, means "sacred leaf" (the leaf is often used in cycad/maize deity celebrations). These additional terms suggest that the maize ear/cycad cone *cintli* homology is only one component of the sacred connection of the cycad plant to the maize plant.

TEOSINTE IN JALISCO

One further set of instances of the local use of *teosinte* remains to be clarified but may indicate the Nahua transport of the term from northeastern to western Mexico. Specifically, in Jalisco, where *Zea* grasses do not appear to have had this local name prior to the introduction of *Zea luxurians* from Guatemala (see above), several "teosinte/tiosinte" toponyms exist. The situation is confusing: Ramírez and Alcocer (1902) list "teoxintli" as the local name for "*Zea luxurians*," which may be either the introduced species or a misidentified native grass now known as *Zea perennis*—in either case, this reference dates from the period when the trade name had already entered common usage. The "teosinte" toponyms, however, predate the introduction of the Guatemalan species.

A cycad—specifically, the native *Dioon tomasellii*—is a candidate for the source of these toponyms. Unfortunately, local ethnobotanical knowledge of this arborescent species is depauperate, despite several prominent wild populations (Bonta et al. 2019), as is local understanding of the original meaning of toponyms such as Tiosintal. The only "Teocinte" toponym that still applies to a viable community originates from the name of a pre-1950 hacienda, which was named for a watering place called El Teocinte on the main trade route along the Pacific Coast.

Another possibility exists. According to Collins (1921), "teosinte" was a local name for *Tripsacum* (gamagrass) in Jalisco. In his intriguing account of *Tripsacum* in Coscomatepec, Veracruz, a few local people knew terms for plants that outsiders used. They told him that the outsider name was "teosinte," but the grass's "true" name was *Zacate Guatemalteco* (Guatemalan grass), meaning it had almost certainly been introduced from Guatemala. But the situation was different on the western side of the country:

> We found later that the application of the name teosinte to Tripsacum was wide-spread in Mexico; for example, near the Ampara mines in the State of Jalisco there is a section of the country known by the name of "Los Teosintes." The natives say that the place is so called because of the abundance of teosinte growing in that locality. Here again the plant to which they refer proved to be Tripsacum, and no Euchlaena [*Zea teosintes*] could be found in the region. (Collins 1921:347)

As we found in western Honduras (Haynes and Bonta 2003; see also Sutherland 1986), *Tripsacum* gamagrasses are the most common and well-known "teosintes." While these were apparently also introduced to Mexico from Guatemala with that name, it is entirely possible that the term preexists this introduction, at least in Jalisco. Only clear references from before the mid-1800s will clear up this conundrum.

Conclusions

In summary, there are two separate bona fide origin areas for the term *teosinte* (including spelling and pronunciation variations) and closely allied terms that predate the modern trade in *teosinte* forage grasses that begin in the mid-1800s:

(1) a region stretching from southwestern Guatemala (and possibly neighboring Chiapas) to Nicaragua, and possibly as far as Costa Rica, along the Pacific Slope, and barely onto the Atlantic Slope in western Honduras (Río Ulúa Basin) and northern Nicaragua (Río Segovia headwaters), involving only grasses.
(2) a region encompassing the Sierra Madre Oriental of northeastern Mexico, from central Veracruz state at Xalapa northward to southeastern San Luis Potosí, involving several cycad species.

There is a secondary *teosinte* zone, derived from number 2 and exclusively involving the single wild cycad species *Dioon mejiae*, in northeastern Honduras, with no evidence of overlap or even meaningful cultural contact vis-a-vis cycad or grass knowledge with number 1.

There is a fourth *teosinte* zone restricted to Jalisco, involving an unknown plant species that is not a *Zea*.

How can we reconcile all this? Given that by far the most complex SMA concepts and accompanying lexicon, regardless whether the term *teosinte* is used or not (for example, the term *chamal* was also widely transported [see chapter 8, this volume]), derive from northeastern Mexico, it is a safe bet that *teosinte* originated there in the multiethnic mix of Teenek, Nahua, Totonac, and Xi'iuy found in the Huasteca and bordering regions. Given that Teosintle itself is a Chicomexochitl-like maize/cycad deity in northern Hidalgo Nahua culture (Bonta et al. 2019; Diego-Vargas 2017), and that maize there is considered a type of cycad, it would seem highly likely that multiple groups of Nahua speakers transported at least the outlines of the SMA to diverse regions, and even to places where no cycads are located, like southwestern Guatemala. Grasses were not exempt, of course—because maize is still seen as a type of cycad in some cultures, other grasses, as well, would presumably have been placed in the same conceptual group. However, we are still lacking ethnographic and ethnohistoric evidence of *teosinte* grasses conceptualized as sacred-maize-ancestors, beyond the etymology of the name itself.

What, in essence, is the cycad-grass *teosinte*? As the next chapter shows in greater detail, in its purest form, it is one name for a deity manifesting itself as a cycad that provides strength and protection to the milpa, and through the union of its dioecious female and male manifestations, continually births the monoecious maize plant. Thus, the "sacred" prefix "teo" refers not to a distant temporal origin, but to a continuous, cyclical rebirth of maize and milpa under the guardianship of the cycad deity.

Note

1. Figures 8.1 and 8.2 (this volume) show the broader context of sacred-maize-ancestor concepts applied by local people to cycads across Mexico and northern Central America.

8

The Sacred-Maize-Ancestor Concept

An Ethnographic Perspective

Mark A. Bonta

Rapidly fading Mesoamerican beliefs in the sacred nature of cycads as ancestors and protectors of maize are found from the Northeast Mesoamerican periphery in the Huasteca region of San Luis Potosí state all the way to the Southeast Mesoamerican periphery in northeastern Honduras (Figures 8.1 and 8.2). Intensive field research undertaken in Honduras and northeastern Mexico has revealed a rich panoply of beliefs and suggested numerous avenues for explanation of the origins and significance of the SMA (sacred-maize-ancestor) concept (Bonta et al. 2019). This chapter explores the principal SMA beliefs among the Teenek, Nahua, Xi'iuy, and Totonac and outlines the little we know about them elsewhere across the region, particularly from the Zoque-Popoluca, Mazatec, and Chontal de Oaxaca ethnic groups. From the ethnographic evidence, I hypothesize that the SMA concept originated in the Gulf Lowlands between eastern San Luis Potosí and northern Veracruz sometime prior to Spanish conquest, spreading westward into the Sierra Madre Oriental, southward along Gulf of Mexico trade corridors and across the Continental Divide to Pacific coastal Oaxaca, and by sea, land, or both to northeastern Honduras. It seems nearly certain that the SMA concept did not reach the Sierra Madre Occidental or Pacific Coast anywhere north of Oaxaca, and it is also absent from the Mexican high plateaus where wild populations of cycads are not found.

The SMA is best iterated among first-language Teenek and Nahua speakers of the Huasteca who still practice traditional milpa agriculture. It is progressively *mestizado*, shorn of sacred meaning and surviving solely through words and practices without an anchor in Mesoamerican beliefs, in Indigenous communities that are losing their languages, and in mestizo communities. Nevertheless, the SMA has not been exhaustively researched, and it

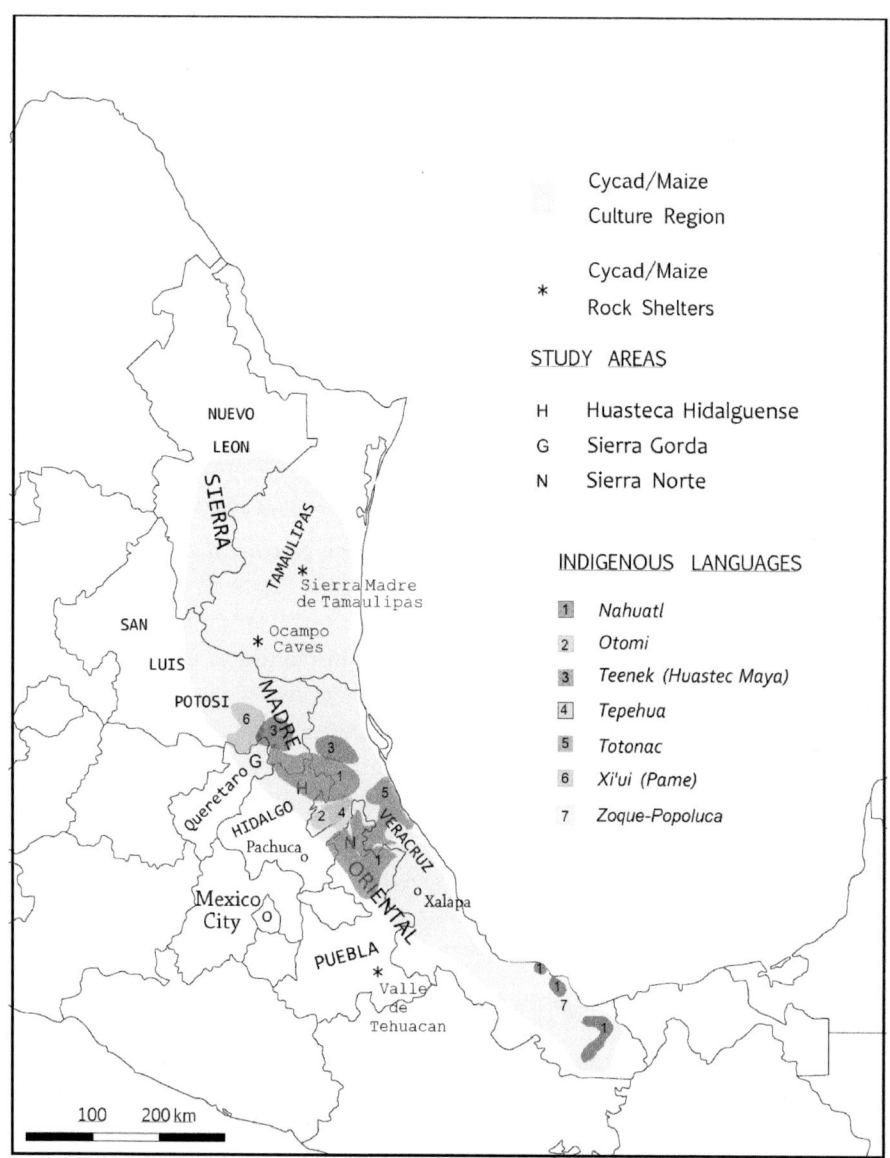

Figure 8.1. Indigenous language areas in northeastern Mexico within the cycad/maize culture region. The area outlined in light gray contains over 20 cycad species connected strongly (in Indigenous language areas) or loosely to the SMA (sacred-maize-ancestor) concept.

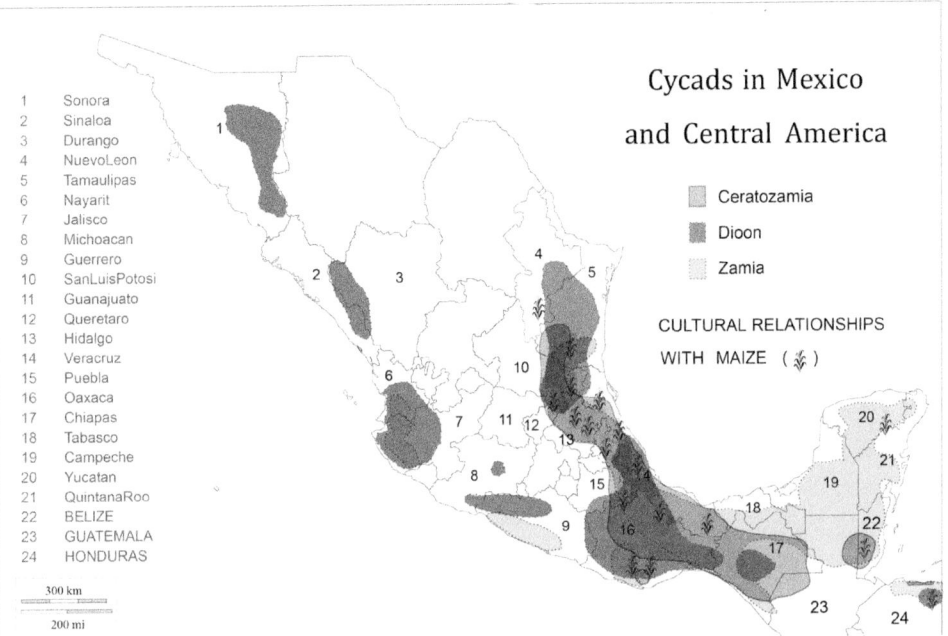

Figure 8.2. Principal locations for the SMA (sacred-maize-ancestor) concept related to cycads in Indigenous and mestizo regions of Mexico and northern Central America. Figure adapted from Bonta et al. (2019) under Creative Commons license.

is quite possible that Indigenous groups elsewhere still maintain complex beliefs and practices relating cycads to maize.

Teenek of San Luis Potosí

A single Mayan ethnic group is found in northern Mexico, having separated from a southern Maya ancestral group several millennia ago (Sandstrom and García Valencia 2005). The Teenek (Huastec Maya) once controlled a large polity including parts of present-day Tamaulipas, San Luis Potosí, Querétaro, and Veracruz states. Today, the Teenek are concentrated in numerous municipalities in the northern Veracruz lowlands and in the lowlands and Sierra Madre foothills of San Luis Potosí. Buried in Janis Alcorn's (1984) extensive Huastec Maya ethnobotany treatise are significant references to cycads that indicate the extraordinary place that cycads hold in Teenek culture, that is, in close association with Dhipak.

Dhipak and Konlif

Teenek believe that Dhipak (Thipaak), whom some authors suggest is related to the Nahua Sipactli (Cipactli), a Mesoamerican deity associated with crocodilians or fish (Stross 2006), was a boy who discovered maize and gifted it to the Teenek people (Alcorn et al. 2006). Many stories about Dhipak exist—in a single village in Aquismón municipality, for example, each elderly interviewee had a different Dhipak story (Bonta et al. 2019; Diego-Vargas 2017). Stories vary in the degree of inclusion of Roman Catholic motifs. Commonly, Dhipak, now embodied in the maize plant, defeated his evil grandmother (Miim), embodied in a cycad or in another plant eaten in an earlier era, such as *Brosimum alicastrum* (Alcorn et al. 2006). In the villages of Aquismón, the principal cycad associated with Dhipak is *Ceratozamia latifolia*, known locally as *konlif* (sand-plant) but also bearing many other names, most related to maize ancestry (see Table 8.1).

In the Sierra Teenek villages ensconced in the Sierra Madre Occidental karst foothills of western Aquismón, cycad knowledge is disappearing rapidly under the onslaught of evangelical Christianity hostile to maize festivals and other Catholic-linked celebrations. Nevertheless, some elderly Teenek still conceptualize *konlif* as the life-force (*manu, tsubal, ejatal em*), king (*rey de las mazorcas*) and flesh (*la carne del maíz*) of maize and the milpa. Thus, *C. latifolia konlif* cycads, while only rarely used for food or other purposes, are left to grow wherever they are found, and are favored as maize-guardians around the periphery of the milpa. They are understood to have the power to "pull down" clouds to provide moisture and are associated with Maam, the thunder god, in this capacity. Their presence in and around a milpa not only confers strength to maize plants, but also acts as a trap for predators, by attracting maize pests with their tasty cones and young leaves and thus luring them away from maize. *Eumaeus* moth caterpillars (*gusano de konlif*), the main consumers of cycad leaves, are seen in a good light because their frass is understood to help fertilize the milpa.

Konlif has an ancestral, familial, and dominant function vis-à-vis maize, as witnessed by the plethora of local names (Table 8.1): *ahaatik a eem* (maize lord); *compañero del maíz* (maize companion); *amigo de Dhipak* (Dhipak's friend); *ayudante de Dhipak* (Dhipak's helper); *hermano de Dhipak* (Dhipak's brother); *tío de la mazorca* (maize ear's uncle); *mamá de las mazorcas; madre del maíz* (maize's mother).

Among the Sierra Teenek, each vegetative part of *konlif* is homologous to maize, but with a difference: maize is seen as the result of the sacred union of the *inik konlif* (male plant) and *uxum konlif* (female plant). The male cone,

Table 8.1. Teenek cycad-maize terminology

Teenek name[a]	Alternate names	Spp.[b]	Translation	Part referred to
ahaatik eem	ahaatik a eem	CL	maize lord (ajātic, Christ the Lord; ēm, maize)	whole plant
aján		CL	maize cob	cone
amigo de Dhipak		CL	friend of Dhipak	whole plant
ayudante de Dhipak		CL	Dhipak's helper	whole plant
bo'jor		CL	maize cob	male cone
chukmal		CL, DE	hairs	tomentum on female cone
chum		CL, DE	point	tip of female cone
compañero del maíz	compañero de la mazorca	CL	maize's companion	whole plant
dhókob		CL	leaf	leaf
ejatal em		CL	the life of maize (ejattalāb, soul, life)	whole plant
ekmito'tol		CL	outer shell	female cone scales
elotito de konlib	elote de konlib	CL	tiny maize ear	male cone
el rey de las mazorcas	el rey y carne de las mazorcas	CL	king (and flesh) of maize ears	whole plant
espíritu del monte		CL	spirit of the forest	whole plant
flor de Dhipak		CL	Dhipak's flower	male cone
flor de konlif		CL	konlif's flower	male cone
hermano de Dhipak		CL	Dhipak's brother	whole plant
inik konlif		CL	male konlif	male plant
inik tzamaal		DE	male tzamaal	male plant
jojo	jobal, jojobal	CL, DE		tomentum on female cone
k'id		CL	spine (q'uīth, spine)	spine on leaf
konlif	condif, conlif, kombi, kombil, konbi, kondif, konfi, konlib	CL	plant of sand	whole plant
la carne del maíz		CL	the body of maize	whole plant
madre del maíz		CL	mother of maize	whole plant
mamá de las mazorcas		CL	mother of the maize ears	whole plant
manu		CL	strength/power of maize	whole plant
Miim	mim	CL	grandmother (mim, señora)	whole plant

(continued)

Table 8.1—Continued

Teenek name[a]	Alternate names	Spp.[b]	Translation	Part referred to
mim konlif		CL	grandmother of *konlif*	whole plant
o'k		DE	head	female cone
o'tlab	*o'tol*	DE	shell	sarcotesta
olote de konlif	*olotito*	CL	*konlif*'s ear	male cone
palma del Maam	*palma de Nuestro Señor*	CL	*Maam*'s palm	whole plant
planta del Diablo		DE	Devil's plant	whole plant
planta dueña de la tierra		DE	plant that owns the land/earth	whole plant
po'jodh		CL	powder	pollen
po'jodh tzamaal inik	*pojostlil tsamal inik*	DE	male *chamal* powder	pollen
t'ichol		DE	membrane	endotesta
t'uul		DE	the body	sarcotesta
thimaloon poko		CL	wild palm	whole plant
Thipaak	*Dhipak*	CL	*Dhipak* deity	whole plant
tío de la mazorca		CL	maize's uncle	whole plant
toro		CL	bull	male cone
ts'een Thipaak		CL	Sierra *Dhipak*	whole plant
tsakam Thipaak		CL	little *Dhipak*	whole plant
tsakam way'		CL	child maize ear	cone
tsalam Thipaak	*tzalam-thipac*	CL	*Dhipak*'s shade	whole plant
tsubal		CL	life of maize	whole plant
tzamaalib	*tsamal, tsamalib, tsamay, tzama[a]l, tzamalib, tzamay*	DE	plant of cold land	whole plant
ushum konlif	*ushu konlif, uxum konlif*	CL	female *konlif*	female cone
ushum tzamaal		DE	female *tzamaal*	female plant

Source: Data adapted from table S1, additional file 1 (Bonta et al. 2019) unless indicated otherwise.
[a] Names in Teenek also incorporate Spanish and Nahuatl words.
[b] CL = *Ceratozamia latifolia*. DE = *Dioon edule*. CL entries may also refer to other local species of *Ceratozamia* as well as to *Zamia fischeri*.

called *toro* (bull), *bo'jor* (maize cob of konlif), *elotito de konlif,* and *olote de konlif,* is seen as a phallus that creates and releases semen (pollen) to *uxum konlif*. Though it is unclear whether cycad insect pollination is known to local people, the mechanism of the opening of the *ekmito'tol* ("outer shell"—the female cone scales) covered with *chukmal* (tomentum, their hairy covering) to allow the pollen entrance, is likened to the penetration of the vagina of *uxum konlif*. Maize, a monoecious plant, is thus understood to originate "mystically" through the union of dioecious male and female cycads. The evidence can be found in the tiny, maize-kernel-like cycad seeds before they mature.

In the Teenek communities I have visited, no other plant is understood to have such central importance in milpa origin, strength, and renewal. It is difficult to understand how it has been overlooked by so many studies on the origins of maize culture. The oversight may be attributable to the general lack of knowledge of cycads by the outsiders who have asked the questions, and because *konlif* is also sometimes called simply Dhipak, thus like Teocintle and Chicomexochitl (see below), it is possible that researchers have misunderstood informants, assuming they are referencing either maize or a disembodied deity.

In summary, among the Teenek, the *konlif* cycad is a fundamental component in milpa culture, seen by many as what human beings ate before maize, and by most people as a mystical plant evoking positive feelings. Its close cycad relatives, the zamias, including *Zamia fischeri*, not thought of as *konlif* but known to be related to it, have a much more limited role in Aquismón that includes shamanistic uses enshrouded by secrecy (Bonta et al. 2019).

Tzamaalib

The third Teenek cycad type, *tzamaalib*, "cold-land plant," is *Dioon edule*, a species that grows sparsely in the Huasteca lowlands and Sierra Teenek but thickly in the Sierra Gorda highlands a day's trek to the west. From the shortened form *tzamaal* is derived the Nahuatl word "chamal," which by the 1600s or before had become the standard term for shrubby or arborescent *Dioon* cycads throughout northeastern Mexico (Bonta et al. 2019).

Tzamaalib, however, evokes intensely negative sentiments among Sierra Teenek of Aquismón old enough to recall the hunger that forced their families to walk west to the Sierra Gorda of nearby Querétaro to harvest cones for making food. *Tzamaalib*, like *konlif* and other cycads in the region, is deadly without the proper processing, but even though *konlif* was also eaten (and still is, to an extremely limited extent), its toxicity does not affect its revered status. Mention of *tzamaalib*, however, brought tears to the eyes of informants who associated it with the misery they had suffered as children—and with the memories of sometimes mortal intoxication episodes in their communities when starving families could not wait the requisite time to cook the seeds sufficiently.

Other than variations of the term "tzamaalib," *planta dueña de la tierra* (plant [that is the] owner of the land) and *planta del Diablo* are the only other terms we collected that refer to *Dioon edule* in Teenek culture. The intricate SMA concept associated with *C. latifolia* is largely absent for *D. edule*. Maize associations are limited to the types of cycad foods that people prepared (tamales, tortillas, etc.) that are also made from maize.

Xi'iuy

Northern Pame, who refer to themselves as Xi'iuy, among other names, appear to be the descendants of peoples who inhabited northeastern Mexico far longer than the neighboring Nahua and Teenek people with whom they interacted for many centuries. In particular, the three groups cohabited in communities in the Sierra Gorda mountains of Querétaro such as Jalpan, which might explain the intermingling of so many nearly identical cycad beliefs and practices, including the SMA (Chemín-Bassler 1984; Gutiérrez and Ochoa 2009; Ordoñez Cabezas 2004).

During the Spanish colonial period, Xi'iuy people were one of the Chichimeca nomadic groups associated with hunting and gathering and often hostile to Spanish incursions. A succession of Roman Catholic missions, particularly in the late 1700s under Fray Junípero Serra, attempted to restrict them to valley-bottom settlements and sedentary agricultural practices, but in general they remained dependent to a large part on wild animals and plants, particularly the *dameu* (*chamal*) that grew in vast concentrations in the Sierra Gorda and the mountains to the north in San Luis Potosí, known today as the Pamería. Xi'iuy concern for *chamal* is mirrored in the plant motifs on the restored plaster facade of the Tancoyol de Serra mission church in Querétaro, apparently created by Xi'iuy artisans in the 1700s (Bonta et al. 2019; Zapata Avendaño 2018). Misinterpreted previously as maize or *Acrocomia* palms, the numerous *chamal* cones and leaf crowns portrayed, alongside agaves and other native wild plants, afford some indication of how important sustenance from cycads was in the primarily semidesert and arid tropical scrub conditions of their homeland. This would not have been seen in a favorable light by Spanish missionaries, who regarded the use of edible wild plants—particularly the poisonous ones—as just another indication of Xi'iuy barbarism (Chemín-Bassler 2012).

The clearest indication of the SMA in Xi'iuy culture are the terms *maíz del monte* and *maíz gordo* (forest maize and fat maize; Bonta et al. 2019, Tristán 2012; Tristán Martinez et al. 2020). The term *dameu* and its variants I suspect are derived from the Nahua *chamal* or directly from Teenek *tzamaal*. For indications of the significance of *dameu* to the least acculturated Xi'iuy of southern-central San Luis Potosí around Santa María Acapulco, some of whom remain fluent Xi'iuy language speakers, we are indebted to anthropologist Heidi Chemín-Bassler (1984, 2000, 2012).

Dameu is thought of as a miraculous plant in Xi'iuy culture, though use and appreciation of it have faded substantially, particularly with rapid modernization, as it gains the inevitable negative tinge associated with dire poverty and suffering (Tristán Martinez et al. 2020). Nevertheless, even among mestizos with no need for a food to supplement their diets when maize stores run low, what they call *chamal* is held in high regard, to the extent that some people even box up *chamal* tamales to send to relatives living in the United States. Numerous components of the harvest and food preparation cycles are similar or identical to the treatment of *tiusinte* (*D. mejiae*) in Honduras. Specific elements of seed detoxification, including the use of ash rather than lime, are unique to Xi'iuy *dameu* culture and Honduran *tiusinte* culture but to no other cycad cultures, indicating a tangible historical connection between the two, as mentioned in chapter 7 of this volume, and discussed below.

Complementarity of maize and cycads is evidenced through beliefs that cycad productivity and maize productivity alternate, such that in years one harvest is abundant, the other is not. Maize is often planted in areas where *dameu* grows, resulting in a form of maize-cycad intercropping also seen commonly in Honduras. The most important foods are made with cycad meal as well as with cornmeal (tortillas, tamales, atoles, breads), with tamales being the primary form of sustenance. Among the Xi'iuy of Santa María Acapulco, the most complex cycad foods recorded anywhere in Mexico and northern Central America are found, including tamales stuffed with meat, grasshoppers, garbanzos, and other ingredients (Chemín-Bassler 2000). Whereas in many places in Mexico the fact that cycads were one of the foods that helped former generations survive famines is seen in a negative and embarrassing light, in the Pamería and Sierra Gorda this is often still viewed as a point of pride and an important part of cultural heritage. The contrast is striking between the Teenek and Xi'iuy, who together compose the most socioeconomically marginalized populations in their respective regions (Xi'iuy in the Sierra Gorda of Querétaro and Pamería of San Luís Potosí; Teenek of the Huasteca potosina just to the east). The former and the latter groups have customarily harvested *chamal* from the same populations.

On the Origin of Cycads

I have not conducted cycad research among first-language Xi'iuy speakers, so I am not aware of a rich SMA culture paralleling that of the Teenek or Nahua. However, versions of a story about the origin of *dameu* collected by Chemín-Bassler (2012) and by Gibson et al. (1963) suggest connections to Nahua and Teenek stories and more widely to maize-ancestor stories throughout the Gulf Lowlands.

In "El chamal, alimento divino de los Pames-Xi'iui de San Luis Potosí y Querétaro," Chemín-Bassler (2012) relates that this "divine food" is still prepared by Xi'iuy women and is locally considered "de lujo" (a luxury). Most notably, she relates a story told her by an Indigenous ex-leader of Santa María Acapulco about the origin of *chamal,* according to the informant's grandfather. It tells of an elderly woman, an outsider, who arrived in the community around 300 AD, when there was no more Heaven-sent manna to eat, and not yet any maize. Local people knew that the woman was evil and had killed her own children, ripping out and eating their hearts—and that she intended to do the same in Santa María Acapulco. After her first night in the community, she encountered a lizard that complimented her on her multicolored dress, the likes of which had never been seen before, and asked to borrow it to gather firewood. She declined, so the lizard went off to gather firewood and, upon returning, dug a deep hole and kindled a fire in it. The lizard then challenged the old woman to jump over the hole, which it did with ease, whereupon the woman fell in and was burned up. The lizard then retrieved her head and womb and took them to the forest to dry, then later retrieved the rest of the body parts and did the same. But, to the lizard's surprise, a few days later, *chamal* seeds fell out of the dry brains, and *guapilla* seeds from the womb [*guapilla* seeds are eaten when *chamal* runs out; the species referred to is presumably *Hechtia glomerata*, a bromeliad]. From other parts of the body, semen came out that produced many men.

Chemín-Bassler's informant then related that God had sent the evil old woman to us, the poor, because he did not want us to die of hunger, and ever since, *chamal* has been considered a delicious "maize of the poor." The informant mentioned that even when maize is available, Xi'iuy still prefer *chamal* tamales. The belief in the superiority of cycad food to maize is elsewhere found only among the most traditional Nahoa de Honduras and mestizos in the Honduran *tiusinte* region (Bonta et al. 2006).

Another version of this story (Gibson et al. 1963:Text I.–The Ogress, 126–130) was related in the Xi'iuy language by 16-year-old Librado Gabriel Correa to linguists in 1952. With a basket of flowers, an "ogress" would lure children to a cave where she would cook and eat their internal organs. One time, a boy managed to escape on the pretense of needing to leave the cave to go to the bathroom. The ogress later searched high and low for him. Birds that she asked told her they had seen the boy during plowing time, or else during the time of green corn [*elotes*], or during the harvest. Finally, she met

Table 8.2. Xi'iuy *Dioon edule* cycad terms

Xi'iuy term	Alternatives	Translation
chamal		[widespread term derived from Teenek *tzamaalib*]
dameu	*Dameaō, dameaw, namEo',*[a] *nameu*	[derived from "chamal"?]
Ilye'e dameaw kon garbants		*chamal* tamales with garbanzos
kanaw namew	*kanau nameu*	*chamal* head [e.g., female cone]
maíz del monte		forest maize
maíz gordo		fat maize

Source: Data adapted from table S1 in additional file 1 (Bonta et al. 2019) unless otherwise noted.
[a] Gibson et al. (1963).

some ants whose clothes she admired, and they told her they jumped over a fire pit to make their clothes turn color. They then arranged for the striped lizard to jump over to change his colors and lured her to the edge of the pit. They pushed her in, and as she burned alive, her brains exploded and went to become *namEo'* (*chamal*) in the hills; her eyes became *guapilla*.

Although the stories do not explicitly locate the origin of maize in cycads, they are nevertheless highly significant. They suggest a link to Teenek conceptions of the evil grandmother in the Dhipak stories and in other cultures' accounts. Other common elements of Mesoamerican maize-origin stories—the small boy, birds, lizard, and ants—are also present (see Alcorn et al. 2006). The importance of the head, and particularly the brains, as the ultimate origin of cycads, may be related to Huasteca beliefs in the cycad cone as a head, and features prominently in Xi'iuy cycad terms (Table 8.2) (the term *cabeza* is widely used to refer to the cone; see references in Tables 8.1–8.4) and the capacity for neurotoxic cycads to make one go "crazy in the head." Given that cycads have been consumed for at least 4,500 years in northeastern Mexico (Bonta et al. 2019; see also Englehardt et al., this volume), it is presumably true that knowledge of how to prepare the carbohydrate-rich but highly toxic starch in the seeds diffused from one culture group to another at distinct points in time (knowledge of how to eat the sarcotesta required no special insight, because it does not require detoxification).

Nahua

The Nahua of northern Hidalgo and immediately adjacent San Luis Potosí are the largest concentration of this populous ethnic group in Mexico, and many lowland Huasteca communities, though acculturated to dominant Mexican mestizo culture in various aspects, still contain many bilingual Nahuatl-Spanish speakers and a few elderly monolingual Nahuatl speakers. Other than a healthy respect for cycad toxicity, there is not evident dislike for any of the local cycad species as is evinced for *tzamaalib* among the Teenek. In terms of the SMA, the Nahua are equal to the Teenek in possessing a rich set of beliefs and practices interweaving maize and cycads, even though they have not inhabited the region nearly as long as the Teenek have.

Nahua elsewhere in the region also connect cycads to stories about the ancestry of maize. This is directly evident through terms for cycads, oblique references to wild maize and child maize heroes, and other indications, but fine-grained ethnobotanical research is sorely lacking. It does appear that the SMA concept survives in some form or another among the Nahua of the northern Veracruz Gulf Lowlands (part of the Huasteca), the Sierra Madre Oriental in Veracruz (e.g., Huayacocotla; see chapter 7, this volume), and among the Gulf Nahua of the Gulf Lowlands of southern Veracruz. There appears to be overlap with mestizo terms as well as with Zoque-Popoluca and Totonac terms and concepts, as discussed below. One particularly intriguing cycad name is *"maiz de los chaneques,"* a term used in the mixed Nahua/mestizo area around Catemaco, a southern Veracruz city heavily associated with shamanism. It is one of the names for *Zamia furfuracea*, a cycad that grows almost exclusively along sandy beaches. *Chaneques* are the Nahua version of a widespread Mesoamerican trickster spirit, Hispanicized as *duendes*. They are often cast as mischievous child sprites.

Cycads in the Huastecan Nahua Milpa

The Nahua relationship to *chamal* (*D. edule*) is limited to the few places where traditional Nahuatl-speaking people overlap with native populations of this cycad in northern Hidalgo and southeastern San Luis Potosí (Vite-Reyes 2012). However, the *teocintle* and to a lesser extent the *tzompoyo* cycads have numerous names and associated beliefs and practices related to maize. The species are primarily *Ceratozamia fuscoviridis* at higher elevations, associated closely with the local maize-cycad deity Teocintle (also known as Chicomexochitl or Chicomecintli, among other names: see below) in places such as Huazalingo, Hidalgo, and *Zamia loddigesii*, associated with SMA beliefs quite like the Teenek *konlif* vis-à-vis the power and strength of the milpa. Other *Ceratozamia* and *Zamia* species are also involved to a lesser extent.

Literature concerning these and other Nahua maize deities in the Huasteca and in southern Veracruz, while not citing cycads explicitly, refers to wild maize, the spirit of maize, the master of maize, maize ancestors as children, and other concepts that strongly suggest ethnobotanical research could unearth connections to cycads similar to what we have found in Teenek and Nahua communities. These maize deities include Centeuctli from northern Hidalgo (Sánchez Jimenez 2005); Cintektli or Piljcintektli from Texquitote, San Luis Potosí (Hernández Vaca 2010); Sindiopi from the Minatitlán region, southern Veracruz (García de Léon 1968); and Tamakasti (Law 1957) or Tamakastsiin (López Austin 1992) from Mecayapan, southern Veracruz.

The Nahua community of Tohuaco II, in Huautla, Hidalgo manages communal milpas that include swidden plots in deep forest on hills where *Zamia loddigesii* grows. Because farmers know that this *sintli de cuautitla* (forest maize) resprouts vigorously from subterranean stems, they are not troubled to protect the cycads when they slash and burn a forest plot during the milpa cycle. But once the new field is prepared, cycads poking up through the ash layer are left to grow and are conceptualized as the *alma de cintli* (also, *alma del maíz*, soul of maize; see Table 8.3). *Cunelcintli* (maize ear children—i.e., seedlings) mature into the *cintli i nana* (maize ear mother) and *padre del elote* (maize ear father), male and female cycad plants, and then become Chicomexochitl (seven-flower) or Chicomecintli (seven-ear), terms for the maize deity (e.g., Rosas Guerra 2016; van't Hooft 2008) reflecting the fact that both male and female zamias produce numerous cones on the same plant. Female cones are known as *elotsintli* and *pilelotsi*, among several other names, referencing their appearance as tiny maize ears.

As the strength or force anchoring the milpa, zamias are believed to confer unique power to maize ears, enlarging them and making them more pest resistant. Andrew Vovides (personal communication, 2018) has suggested that this belief is connected to the fact that cycads do indeed fix minerals such as iron that strengthen maize plants. This opens the possibility that observed maize-cycad intercropping—seen in areas where no complex SMA concept has survived, and as distant as Honduras—may have a tangible benefit for maize (Bonta et al. 2006; Bonta et al. 2019)

In Huastecan Nahua culture, the terms Chicomexochitl and Chicomecintli refer to the twinned ancestral and divine aspects of maize and of cycads, such as *Ceratozamia fuscoviridis*, which is principally known by that most portable of Nahua terms, *teosintle*. The riot of cycad names that also apply to maize, and vice versa, were explained in simple terms by one informant as follows: maize is a type of cycad. In other words, modern maize is the descendant of "hill maize"—"wild maize"—and is thus a form of the maize-cycad plant fit for human consumption. Cycads were the maize eaten by the ancestors and were transformed into maize.

Table 8.3. Mexican maize-cycad terminology derived from Nahuatl

Nahuatl name[a]	Variations	Spp. (R)[b]	Translation	Refers to
alma de cintli		CF-ZF-ZL (H)	soul of maize ear	whole plant
alma del maíz		CF-ZL (H)	soul of maize	whole plant
chamal		CF-C (H), DA (N,T), DE (G,H,Q,S), DP (P), DS (Y), DT (J)	[from Teenek *tzamaalib*]	whole plant
chamalillo	*chamal chico, palmilla chamalillo*	C-CC-CF-Z-ZF (H,S)	little *chamal*	whole plant
chicomesintli	*chikomesintli*	CF-ZL (H)	seven-ear	male plant
chicomexochitl	*chicomexochil*	CF-ZL (H)	seven-flower	male plant
cintli		ZL (H)	maize ear	male plant
cintli cuautitla	*sintli de cuautitla*	CF-ZL (H)	forest maize ear	male plant
cintli i nana		ZL (H)	mother of maize ear	female plant
cuacintli	*cuacintle kuasintli*[c]	CF (H,NV)	forest maize ear	whole plant
cunelcintli		ZL (H)	maize ear children	seedlings
elote	*elote de monte*	CF-ZL (H)	forest maize ear	whole plant
elotillo		CF-ZL (H)	tiny maize ear	cone
elotsintli	*elolcintli*	ZL (H)	maize ear	cone
eloyo		ZL (H)		whole plant
maíz de los chaneques	*maíz cimarrón*	ZFu (SV)	maize of the chaneques (forest spirits), wild maize; used by mestizos and Nahuas in Catemaco region	whole plant
padre del elote		CF-ZL (H)	father of maize ear	male plant
pileloltsi	*pilololcintli*	CF-ZL (H)	small maize ears	male cones
tatatlcintli		ZF-ZL (H)	father of maize ear	male plant
teocinishuatl	*teocinhuatl tiozin[is]huat tiosin[is]kuatl*	CF (H)	sacred maize leaf	leaf
teocintle[d]		CF-ZF-ZL (H,S)	sacred maize ear	whole plant
tepexi		CT (P)	hill seed, hill maize ear, hill leaves	whole plant

Nahuatl name[a]	Variations	Spp. (R)[b]	Translation	Refers to
tepe[t]zintle	*teocintli, tepecintle, tepecintli, tepetlcintle*	CF-CT (CV,NV,P)	hill maize, wild maize. See Table 8.4 for Totonac use.	whole plant
tío del maíz		CF (H)	uncle of maize	whole plant
tiotamal	*quiotamal*	DE (X)	sacred tamale	whole plant
tzompoyo	*guachompoyo*	CF-ZF-ZL (H)	evil head, evil hair	whole plant

Source: Data adapted from table S1 in additional file 1 (Bonta et al. 2019) unless otherwise noted.

[a] Names incorporate Spanish and Teenek, and in some cases are also found in non-Nahua mestizo communities.

[b] Regions and species involved. C: *Ceratozamia* spp.; CC: *Ceratozamia chamberlainii*; CF: *Ceratozamia fuscoviridis*; CT: *Ceratozamia totonacorum*; DA: *Dioon angustifolium*; DE: *Dioon edule*; DP: *Dioon purpusii*; DS: *Dioon spinulosum*; DT: *Dioon tomasellii*; Z: *Zamia* spp.; ZF: *Zamia fischeri*; ZFu: *Zamia furfuracea*; ZL: *Zamia loddigesii*. (CV): central Veracruz highlands; (G): Guanajuato; (H): Hidalgo; (J): Jalisco; (N): Nuevo León; (NV): northern Veracruz Gulf Lowlands; (P): Puebla; (Q): Querétaro; (S): San Luis Potosí; (T): Tamaulipas; (X): Xalapa, Veracruz; (SV): southern Veracruz Gulf Lowlands; (Y): Yucatán.

[c] Flores Martínez 2010.

[d] See chapter 7 (this volume) for spelling variants.

This divine ancestry, the central SMA concept, explicitly or implicitly, is what can be detected in cycad traditions from northeastern Mexico all the way to Honduras. But in the most traditional areas of the Huasteca, the Nahua concept still includes the deity, the *teotl,* Teosintle, the maize-cycad god who embodies both plants together. In communities such as San Juan Huazalingo, San Miguel (Saint Michael) is the syncretized, Roman Catholic form of this deity, and the festival dedicated to him, Michaelmas, prominently features floral arches that interweave *teosintle* cycad leaves (*Ceratozamia fuscoviridis)* and dried maize cones to symbolize this dual "spirit" (Diego-Vargas 2017).

The term *tzompoyo* is also used widely in Hidalgo to refer to cycads, and often interchangeably with *teosintle;* informants profess ignorance of its original Nahuatl meaning. It appears to mean "evil head" (de Molina and de Spinosa 1966; Karttunen 1985: *tzom* = "something hairy"; Karttunen 1992: *poyotl* = evil, rotten, decayed, bad), thus relating it to Xi'iuy and Teenek ideas about *chamal.*

In summary, what we know of Xi'iuy, Teenek, and Nahua beliefs and practices concentrated on the commonest cycads in their surroundings strongly suggests that there has been considerable interchange of SMA culture among the three groups, and it seems that any of the three, or another group that has disappeared, could have been the originators of the SMA concept.

MAIZE AND CYCADS ELSEWHERE IN MEXICO

Chamal

Use of the term *chamal* is widespread in mestizo culture across tropical and subtropical Nuevo León and Tamaulipas, and southward in Guanajuato, Querétaro, San Luis Potosí, and Hidalgo (Bonta et al. 2019). Species involved are the very closely related *Dioon angustifolium* and *D. edule*. Tamales, tortillas, and other foods are still eaten or were eaten in the past across this region, but otherwise, no explicit mestizo SMA tradition seems to exist. The same condition applies to practices associated with cycads elsewhere that the name "chamal" has been recorded: Puebla, the Yucatán, and Jalisco. It is possible that in these latter places, the term "chamal" arrived with the modern landscaping trade.

Totonac and Tepehua

Totonac and Tepehua, related languages, are spoken in conjunction with Spanish, Nahua, and Otomí (Ñuhu) in the Totonacapan region of northern Veracruz, the Sierra Otomí-Tepehua of eastern Hidalgo, and in the Sierra Norte de Puebla. Superficial field research among Tepehua and Otomí in Hidalgo revealed no connections between cycads and maize (Bonta et al. 2019), but there are clear connections between maize and cycads among Totonac speakers in both the Gulf Lowlands and the Sierra Madre, in close conjunction with Nahuatl terms and concepts, and apparently also among Tepehua of the Sierra Norte (e.g., Flores et al. 1992). In Papantla, *Zamia loddigesii* is or was understood to be the origin of maize. The term *tepetzintli* is used there as well as in the Sierra Norte by both Nahuas and Totonacos, but the true Totonaco names are variations of "kun." Mújica Vélez (2007) says that "our [Totonac] grandparents" called *tepetzintli* "kin lihuaykan," which he translates as *nuestra carne,* "our flesh." The conjunction of maize ancestry and people as embodied cycads has close parallels in Teenek and Nahua conceptions from the nearby Huasteca.

Zoque-Popoluca

The Zoque-Popoluca (Soteapanec, Sierra Popoluca) of the region around Soteapan in southern Veracruz appear to also preserve the SMA concept, given certain names preserved in the literature. Table 8.4 lists these, with the

Table 8.4. Cycad terms relating to maize in Totonac, Zoque-Popoluca, Chontal, and Mazatec

Ethnic group (Region), name/s	Translation: Species
TOTONAC (SIERRA NORTE, PUEBLA AND TOTONOCAPAN AREA, NORTHERN VERACRUZ)	
amigo del maíz	friend of maize: *Zamia loddigesii, Zamia vazquezii*
kin lihuaykan[a]	our flesh (*Ceratozamia* and *Zamia* spp.)
kun (kone, kene)[b]	*Ceratozamia* and *Zamia* spp.
tepetzintle, tepetzintli[b]	hill maize, wild maize–used by both Nahua and Totonac speakers to refer to *Ceratozamia* and possibly *Zamia* spp. in Ixtepec, Santiago Ecatlán, Papantla, and elsewhere
ZOQUE-POPOLUCA (VERACRUZ)	
cahua	[apparently a Nahuatl word]: *Zamia loddigesii*
ikiapacinti[c]	[Nahuatl: rain-maize ear]: *Zamia loddigesii*
maíz de coxca[c]	[Nahuatl: possibly from *cocoxca*, sick person + maize]: *Zamia loddigesii*
maíz de los antiguos[c]	maize of the ancestors: *Zamia loddigesii*
pek mok[c]	old maize: *Zamia loddigesii*
poua	[apparently a Nahuatl word]: *Zamia loddigesii*
CHONTAL DE OAXACA	
comida antígua	old-time food: *Dioon merolae*
comida de los abuelos	grandparents' food: *Dioon merolae*
comida de gentil	old-timers' food: *Dioon merolae*
gentil[d]	old-timer?: *Dioon* sp.
la-fané-tejuá	ancient food: *Dioon merolae*
lan-zi-lé (lan-zi-li)	[no translation]: *Dioon merolae*
maíz viejo (palo maíz viejo)	old maize: *Dioon merolae*
MAZATEC (OAXACA)	
tush-kjù	seedling-guardian: *Dioon purpusii, Dioon rzedowskii*
ya-tuj-cho (*Yahtuchó, ya-tuj-cho-chu*:female; *ya-tuj-cho-shi-i*: male)	tree-seedling-guardian: *Dioon spinulosum*

Source: Data adapted from table S1 in additional file 1 (Bonta et al. 2019) unless otherwise noted.
[a] Mújica Vélez 2007.
[b] Deance Bravo y Troncoso 2012; Caballero Zamora 2009; Escobar Fuentes 2006; González Torres 2007; Mújica Vélez 2007.
[c] Blanco Rosas 2006.
[d] Zarate Escamilla 2007.

provenance of several terms almost certainly Nahuatl. "*Pek mok*" (old maize) and "*maíz de los antiguos*" seem to be sufficient evidence for the SMA, and the overlaps with Nahuatl (e.g., *ikiapacinti*) follow a familiar pattern. As with Xi'iuy, Teenek, Tepehua, and Totonac to the north, the sharing of concepts and terms with Nahua in the same or nearby communities exists. The Zoque-Popoluca also have a maize deity, Homshuk (Elson 1947; López Austin 1992), which may be related to cycads, though an unambivalent source has not yet been encountered to establish the linkage.

Cycads and Maize in Oaxaca

The diffusion of the SMA concept from the Huasteca and the Gulf Lowlands to Oaxaca left its mark, though apparently not among the many Zapotec-speaking groups in the state, who classify cycads as palms (see Bonta et al. 2019)—Zapotec terms such as *jango* and *yazn-goag* mean "spiny palm tree" or "spiny leaf," for example. The only indication of maize compatibility is the widespread knowledge in cycad regions that *Dioon* were formerly eaten in place of maize, along with other "famine foods" like the roots of plantains (*Musa*). People mentioned to me that tortillas once made from *Dioon holmgrenii* seeds in Santiago Textitlán had "always" been mixed with maize meal. But otherwise, unlike in the Huasteca, the abundant and significant use of cycad leaves for syncretic Roman Catholic festival decorations does not appear to have any maize connection per se.

Other Oaxaca culture groups we have researched that have cycads but not any indications of the SMA concept include the Chinantec, Mixtec, and Mixe, while the extent of cycad knowledge among the Amuzgo, Cuicateco, Huave, Triqui, and several smaller groups is unknown.

Mazatec

Intriguing terms in the Mazatec language indicate some sort of significance for *Dioon* in the Mazatec milpa. In San Bartolomé Ayautla and Santa Catarina, Oaxaca, the term *tush-kjù*, "seedling-guardian," has been recorded for *D. purpusii* (Table 8.4). The common term for *D. spinulosum* in Cerro Tepezcuintle, Tuxtepec, is *ya-tuj-cho*, "tree-seedling-guardian" (Table 8.4). Cycads as guardians or shepherds of the milpa is an SMA belief found elsewhere only in the most traditional parts of the Huasteca. The referenced Mazatec speakers live in northern Oaxaca in and near the Gulf Lowlands, thus not far from Gulf Nahua and Zoque-Popoluca communities that connect cycads to maize, as discussed above.

Chontal de Oaxaca

The Chontal de Oaxaca, who inhabit both the Sierra Madre del Sur and the Pacific Coast lowlands of Oaxaca and speak a language isolate, draw a clear relationship between maize and cycads. This is most evident in a story about *maíz viejo* ("old maize," a cycad) collected by Miguel Pérez-Farrera in San Pedro Huamelula (Bonta et al. 2019), where an unidentified *Dioon* species, referred to as *gentil*, is considered sacred (Zarate Escamilla 2007). The story came from the informant's grandfather, who related that the *palo maíz viejo* was created for people who had nothing to eat, who would make tortillas from it. The one condition imposed was that one should eat the plant only when there is nothing else left to eat. If one eats it but there are still maize grains to plant, then the cycad becomes jealous and causes harm (presumably, intoxication). The grandfather's older relatives had told him that it was the *antecesor* (ancestor or antecedent) of maize. Parallel to this, Zarate Escamilla (2007) relates that *gentil* produces maize grains instead of cycad seeds during times of "necesidad" (i.e., when hunger or famine looms).

During several days of ethnographic research among the highland Chontal, I documented a set of intricate beliefs (Bonta et al. 2019) about cycads (*comida antigua* or *comida de gentil*) focused on the old-timers' or grandparents' food (*la-fané-tejuá*), what was eaten before maize (today, only the nontoxic sarcotestas are still consumed). In San Lucas Yautepec, beliefs in the carcinogenic qualities of cycad pollen and knowledge of insect pollinators indicate a particular knowledge of cycads I have not encountered anywhere else.

Western Mexico, Southern Mexico, and Belize

In the entire region where wild *Dioon* and *Zamia* cycads are found in western Mexico (Sierra Madre Occidental and Pacific Coast) from Sonora in the north to Guerrero in the south, I have found no indication of the SMA. In Chiapas, Tabasco, and the Yucatan Peninsula, there is little indication of the presence of the SMA, despite diverse traditions of cycad use among Indigenous groups. Notably, this includes a nearly complete lack of association among Maya groups.

One potentially notable exception is the *chak waj*, red maize, among the Tojol-ab'al Maya of Las Margaritas, Chiapas (Méndez Pérez 2012). *Chak waj* is one of the four recognized colors of maize, and the author's brief mention of it appears to come from a single informant. The red-seeded cones of zamias seem candidates for this type of maize, rarely eaten, placed in the eaves to protect the home, and associated with medicinal uses such as childbirth. While these may very well be attributes of *Zea* maize with red grains, they are also commonly associated with cycads as well, given that the color red (which among the Tojol-ab'al is related to human blood) is incorporated into cycad names in other Maya languages (Bonta et al. 2019). Specifically, the Lacandón Maya term for a local zamia translates as "red spine." Most compellingly, among the Yucatec Maya, the standard and widespread term "chak waj" refers to *Zamia loddigesii*, though we translated it as "red tortilla" in that case.

The only modern unequivocal reference to maize for a cycad that I have come across in Maya communities is the Mopán Maya name "corn palm" for *Zamia decumbens* in Belize (Calonje et al. 2009). While potentially indicating a deeper connection, it also may be analogous to the term *maicito* (little maize) used to refer to *Zamia obliqua* in Colombia (Bonta and Osborne 2007), a nod to the strong morphological similarity between maize cobs and male *Zamia* cones but without any tangible connection to the SMA.

The Tiusinte Region

In northeastern Honduras, beliefs and practices surrounding *Dioon mejiae* parallel those of the Xi'iuy people for *Dioon edule* (Bonta et al. 2006; Bonta et al. 2019; see chapter 7). Only echoes of the types of SMA beliefs found among the Teenek and Nahua are detectable. Beliefs that have translated well into practical measures are still found, such as the custom of leaving cycads in maize fields or planting maize in cycad groves (but with no explicit belief in cycads being the "strength" or "power" behind the milpa), or the idea that in good maize years, cycad cones are scarce, and vice versa. Nowhere in Honduras is maize thought of as a type of cycad, or vice versa, nor are there any Nahua speakers, even among the Nahoa de Honduras, who could have preserved the literal meaning of *tiusinte*. In foods, however, maize and cycads are seen as highly complementary, with cycads and maize eaten together during Holy Week, at the height of the dry season when maize stores start to become scarce. *Tiusinte* is what helps tide impoverished families over until the first maize harvest. The same types of foods are made from maize and from cycads, and in some cases meal from the two are combined. No connections to Mesoamerican beliefs are imputed to any of this per se.

Conclusions: Summary of SMA

What remains of the sacred-maize-ancestor concept is clearly an extremely impoverished version of what it was a century ago, and given that it has little practical value to modern agriculture, it has been cast aside or ignored by local people and outsiders alike as they strive to raise maize yields, simplify or abandon the milpa, and in some areas, jettison local foodways altogether. The survival of the outlines of the SMA is probably secure in Mexico, in the form that it is in Honduras, as a vague set of folk beliefs with no reference to a deeper cosmology and a series of recipes for cycads and other material practices. However, it is also possible that cycad-eating and all knowledge about the plant will die out altogether, as it apparently already has in Jalisco and Tamaulipas. Further research should focus on the Teenek, Xi'iuy, Nahua, Totonac, Mazatec, Zoque of Veracruz, and Chontal de Oaxaca, judging by what we know so far, but it should not be assumed that even local people concerned about rescuing the spiritual concepts guiding milpa creation are familiar with ancient SMA beliefs and practices.

The question of ethnic origins of the belief can be resolved parsimoniously by the spread of Nahua speakers across Mesoamerica and beyond, carrying the SMA concept from an origin area in the northeastern Mexican Gulf Coast lowlands. This would date the SMA in any given area outside the Gulf Lowlands to the last two millennia. Nahua (Yuto-Aztecan language family) contact with and SMA diffusion to Oto-Mangue speakers (Oto-Pamean: Xi'iuy and Mazatec), Maya (Teenek), Chontal de Oaxaca (Huamelula and other locations along a principal ancient route through highland Oaxaca into the Isthmus of Tehuantepec), Totonacan, and Mixe-Zoquean is more convincing than independent inventions among all these separate groups. An alternative hypothesis is a much older and deeper origin for the SMA, perhaps connected to the origin and diversification of maize itself—a potentiality explored throughout this volume.

Beyond the spread of the SMA by Nahua speakers throughout much of Mesoamerica, the major takeaway from this chapter is that cycads, long overlooked in this role, are the maize heroes, the maize deities of so many interconnected cultures in the Gulf Coast Lowlands and Sierra Madre Oriental. This places them squarely in the most central of roles within Mesoamerican cosmology relating maize origins to human origins. Many critically important avenues of exploration open with this insight, and some can be explored by ethnobotanists in tandem with archaeological and iconographical approaches championed by Carrasco, Englehardt, and other authors in this volume. Collaboration with Indigenous communities and recording of cycad narratives are the most important labors, anywhere that maize-hero narratives and maize deity ceremonies still exist alongside cycads. It is worthwhile remembering that ethnographic collaborations were what first sparked interest in the cycad-maize relationship, but what has been achieved thus far is but a tiny fraction of what remains to be investigated.

Epilogue

What Have We Learned and Where Are We Going?

Dennis William Stevenson

Cycads often reveal themselves when one least expects it. For example, one day in the early 1980s, I found myself deep in the Colombian jungle with my friend and mentor Knut Norstog, and with Sergio Sabato, a colleague from the Universita di Napoli and founding member of the Naples cycad group. We had been searching for weeks in the southwest of the Department of Antioquia, west of Medellín, for the elusive *Zamia wallisii,* a species not seen in more than a century after its original description by the German botanist Alexander Braun in 1875. Because no extant specimens could be found in any of the world's herbaria, many had doubted its very existence. In fact, we had experienced a world of trouble looking for it—including frightening encounters with guerillas and paramilitaries—and were only rewarded when, while clearing a path through the underbrush, my arm was painfully scratched by a spiny plant. Annoyed, I whacked it with my machete and the leaf that fell to the ground was—lo and behold!—*Zamia wallisii*. It was as if the plant had to tell us, "here I am—right in front of you!"

These plants are indeed all around us and have played significant roles in human societies for millennia. As is evident from even a cursory review of research topics, annotated bibliographies, and scholarly work in the last half century, interest in cycads has transcended time, space, and cultures. For example, archaeological data suggest cycad use by the ancient Etruscans (DeLuca 1990), and a wealth of evidence clearly illustrates the use of cycads by aboriginal Australians, throughout Asia, as well as in Africa and the New World (Beck 1992; Bonta et al. 2019; Bradley 2005; Hayward and Kuwahara 2012; Khuraijam and Singh 2012; Kira and Miyoshi 2000; Osborne et al. 1994, 2007; Patiño 1989; Smith 1951; Smith 1982; Thieret 1958). These examples reveal a larger pattern of independent development of interest in

Figure E.1. The author with colleague and noted cycad expert Hiep T. Nguyen examining a handsome specimen of *Cycas pectinata* during fieldwork in Vietnam, ca. 2006 (photograph by Damon Little).

these plants in a diversity of religious practices, artistic traditions, and foodways throughout human cultures. Yet, despite what at first glance may appear to be an impressive array of research cited above, cycads—particularly their use and significance in human cultures—remain curiously understudied and relatively unknown to larger audiences outside the comparatively small circle of cycad researchers and enthusiasts.

This volume is here to introduce these plants to a wider academic public, to scratch more arms with cycad spines, and to encourage more people to get "bitten" by the metaphorical "cycad bug." It brings together a wealth of new research and exciting observations on multiple aspects of human-cycad relationships in Mesoamerica and the Caribbean over this region's long history. As presented here, this area is a microcosm for studying the relationship of cycads to human endeavors around the world. As I see it, all elements of the cycad-human interface are present in Mesoamerican and Caribbean societies over time: art, religion, and food.

Now I have to confess, although I have always been interested in these aspects over the course of nearly a half century of studying cycad biology,

I never had time—with teaching, fieldwork, and other commitments—to explore the wealth of information available concerning the "utilitarian" and "symbolic" facets, as the editors call them, of human-cycad interactions. I would rely on limited information from presentations at meetings and conferences, in textbooks such as Norstog and Nicholls (1997), and from my own fieldwork (Figure E.1). One particularly evocative example from personal experience is the case of Native Americans in the Chocó region of Colombia, who, as I learned through my work in this region, have a poem thanking the gods for *chigua* (a native term for cycads, in this case referring to *Zamia obliqua*) as a food source (Figure E.2). On the whole, however, my own research focused far more on the biological aspects of cycads, and less so on their cultural uses. In part, this omission was due to trying to keep up with new approaches in my own field of plant biology and a lack of easy access and time to pore through journals and books in archaeology, anthropology, and other disciplines that focus on the place of plants in cultural systems. The current volume provides exactly what I wanted and needed. Of course, such reciprocity works both ways; that is, the chapters of this volume dedicated to cycad biology, natural history, and population genetics will—or *should*—be appreciated by other disciplinary audiences (e.g., in anthropology or archaeology), because they provide equally valuable data and perspectives.

I am the Chigua grown
on the shire of the sea
After the formation of the world
and I put forth shoots from the earth
I asked of myself
what food would I be?
Cook the Chigua well
watch out for the emetic effect
take care lest happen to you
what happened to Tenerio.

Figure E.2. Indigenous Colombian poem on cycads (transcribed by Victor Patiño, translated by Julia Morton); *right,* photo of *chigua* (*Zamia obliqua*) in front of a family's house in the Chocó region of Colombia (photograph by Knut Norstog).

I concur with the editors that the natural starting point for this book is Calonje, Vovides, and Gutiérrez-Ortega's chapter 1, which provides a thorough yet succinct review of cycad biology for a broad audience to set a context. Understanding the taxonomy and morphology of these plants paves the way for understanding cycad diversity in terms of not only species diversity but also their structure, and life histories and reproductive biology as well as biogeography, habitats, and ecology. This contribution is followed by the population genetics data presented in chapter 2 by Cibrián-Jaramillo et al., concerning molecular markers that enhance our understanding of genetic changes through time, and potentially allow for the identification of cycad remains in archaeological samples—as Englehardt et al. consider more thoroughly in chapter 5. Comparatively analyzing the genetic evidence against other data sets should also give us some ideas regarding the possible movement of species by Indigenous people through time. Such an integrative approach is clearly a potentially powerful tool for biologists and anthropologists alike.

Another interesting point raised in many chapters is the use of cycads as food and the preparations for detoxification, which vary from place to place and perhaps suggest separate histories. This is well represented in Jaime R. Pagán-Jiménez's contribution concerning the use of *Zamia* in the Caribbean. Using the characteristics of starch grains for species identification of archaeological samples is a breakthrough also applicable for studying and analyzing species diversity and in contributing to phylogenetic analyses. Whereas I vaguely knew of the use of cycads for food in the wider circum-Caribbean world, as presented so superficially in Stevenson (1987, 1991), I had no idea of the extent and the thorough documentation that was available and is now in one place. That said, cycads do not appear to be a staple food in most cultures; rather, as mentioned above, they appear basically reserved as a specialized, ritual, or famine food in this region, as elsewhere.

Pagán-Jiménez's chapter also allowed me to put a personal experience in a broader historical context. The use of cycad as a famine food is something that persisted in Jamaica during World War II, according to an elderly lady who, when visiting me, recognized the *Zamia erosa* plant I had in my house and subsequently related to me how her father and neighbors had grown them as a starch source during the war. At the same time, she told me they were toxic and not to try it myself. Of course, people around the world have successfully dealt with the issue of detoxification, and enough studies have been done now on the processes as well as the successful extraction of toxic compounds to allow researchers to understand that, for example, "Guam Dementia" was actually a biomagnification issue involving fruit bats (flying foxes) and not the detoxification of cycads by consumers (see Cox and Sacks

2002). This all highlights the need to integrate information from as many sources and disciplines as possible to create a more complex and nuanced understanding of the world around us.

Fortunately, such a perspective is presented and advanced in this volume, through the integrative combination of data from various chapters ranging from the history of cycad research in Mesoamerica and the Caribbean to genetics to ethnobotany to Indigenous agriculture. A very interesting and clear instance of such integration is evident in the sequence of chapters 3, 4, and 5 (by Pagán-Jiménez, VanDerwarker, and Englehardt and colleagues, respectively), in which the reader can appreciate the combination of a historical perspective with an agricultural perspective, and with an archaeological perspective—while all three contributions speak to cultural data as well. Such synergy provides a holistic picture and, although this fact should come as no surprise, serves to confirm the integrated, dynamic whole of Indigenous lives and cultures, as well as hinting at what we can learn from them in contemporary society.

Similarly, the place of cycads in art and religion is important and often overlooked, but no more. Renewed interest in recent years has led us to integrate considerations of art and religion with cycad studies in many parts of the world where cycads grow natively. Is it just a coincidence that maize and cycad seed cones look very similar, and maize and cycad seeds are starch rich and storable—and that both must undergo detoxification through nixtamalization? Did Indigenous cultures consider these plants gifts from the gods? Did cycads represent specific things (e.g., maize) and/or abstract concepts such as longevity, fertility, and/or annual rebirth—and do they still? We are just now beginning to assess, investigate, and understand these questions, as seen in the various integrative approaches presented in chapter 6 by Carrasco and chapters 7 and 8 by Bonta. Their results not only move our knowledge forward but simultaneously challenge us and generate new questions—as scientific research should—which is in any case a key goal of this volume, as the editors state in their introduction. Whereas I had been tangentially aware of cycads in art, as mentioned in my introductory paragraphs, I had only a vague concept of their religious aspects and significance. Fortunately, that situation has been greatly ameliorated by these chapters. Presumably, it will encourage others to conduct similar studies in other parts of the world.

An interesting and unique aspect of cycad studies presented in this volume is that, unlike many food plants, almost all cycads are in danger of extinction in the wild. However, and to return to the beginning, as it were, as Calonje et al. point out in chapter 1, there is a potential remedy that incorporates Indigenous conceptions to provide alternative sustainable conservation strategies and to generate new knowledge. This chapter's section on conservation presents an excellent example of the use of an integrated, eclectic approach to solve a tangible, real world problem. I would consider this an important lesson learned.

Finally, we might now ask ourselves: have we done enough to save the cultural aspects from going extinct? How deep is our understanding of those issues, and do we have enough data and experience globally? Can we apply our methodologies and lessons learned to other areas of the world? The present volume will certainly provide a model for doing so. The important point here is that this volume as a whole successfully presents an integrated approach of cycad biology and genetics—"hard" sciences, perhaps—with social science data and methodologies derived from ethnobotany, archaeology, ecology, agriculture, and history, among other fields. This conjunction, in turn, allows us to appreciate how discrete disciplines and approaches can lead to a bigger picture and a more complete, hopefully more consciously reflective understanding of the world around us and our own history and relationships with the natural world.

In that sense, this book is, like this epilogue, not intended to be an "ending," but rather a new beginning in our studies and understanding of the relationship between humans, their cultures, and cycads. It most certainly sets the stage for students and the rest of us. Just as cycads regenerate with new growth after fire or after the bulldozers pass, so too will our knowledge and interest expand with the advent of the conjunctive methodologies and insightful observations presented in this publication. Moreover, I would hope that this volume will facilitate worldwide efforts toward the conservation of these ancient, endangered, CITES-listed plants, as well as their habitats. Such preservation can be achieved only through bringing together diverse groups in pursuit of a common interest—one shared by scientists, stakeholders, and local people who live among these plants. Although such aspirations are indeed lofty, it would not be an overstatement to say that this volume is one of the first that explicitly attempts to unite these sometimes disparate groups and to introduce these fascinating plants to a wider audience beyond the small but zealous circle of cycad scientists and enthusiasts.

On a personal note, this book has allowed me to reflect on my decades of experience in cycad research—a figure that approaches 75 years when including my formative work and that of my mentor Knut Norstog. When I was an undergraduate at Ohio State University in the late 1960s, Knut came to Columbus to give what turned out to be a fascinating talk on *Zamia* reproduction, and I was hooked. Later, as a postdoc at Harvard and the Fairchild Tropical Garden in Miami in the 1970s, I worked closely with Knut,

Figure E.3. *Zamia stevensonii*, the cycad species named after the author, in habitat and displaying its characteristic white flush (photographs by Greg Holzman).

who had by then retired to south Florida. I was also fortunate to accompany Knut during fieldwork throughout Florida and Latin America, particularly Mexico and Colombia, in the 1980s.

My experiences in cycad research began with basic taxonomy and structural botany in an attempt to provide a classification (see Stevenson 1992). This work basically involved field research and global exploration. Perhaps the most rewarding aspect of this for me was not so much finding new species, but rather "rediscovering" supposedly "lost" species, such as *Zamia wallisii*, my painful encounter with which is related in the introductory paragraph above. Such exploration was tremendously exciting, in many ways even more so than the recent privilege of having a new species named after me (Figure E.3)—although given the exquisite emergent foliage of this *Zamia*, I am indeed deeply honored to have this named after me by two of my close colleagues (Taylor Blake and Holzman 2012). Further, such exploration paved the way for other avenues of research and collaboration. As plant science continued to evolve over the past half-century, I was fortunate enough to be able to learn new approaches via friends and colleagues, which I was subsequently able to apply to my own interests in cycad systematics and biology. It started with cytology (the study of chromosome number and morphology), and rapidly expanded to include phytochemical and other cutting-edge methodologies—all while continuing field work and producing cycad floristic treatments, which provided a vehicle for new species discoveries throughout the Neotropics, and indeed worldwide.

For example, when I started researching cycads in the 1960s, there were 87 described species; there are now more than 360, with more than 80 in the *Zamia* genus alone. Or contrast the original 1959 treatment of *Macrozamia* by the noted Australian botanist Lawrie Johnson, in which 14 species were identified, with the 38 species described by Ken Hill in the 1998 *Flora of Australia* volume on gymnosperms. Examining the history of *The World List of Cycads* (see Calonje et al. 2022 for the most recent version) over the years reveals a similar trend in other large genera such as *Cycas, Encephalartos,* and *Zamia,* as well as in relatively smaller ones such as *Dioon* and *Ceratozamia.* These discoveries and revisions all started with field work and exploration. During such research, botanists rely on earlier flora works to identify specimens. The inability to properly identify what we see thus leads not only to the description of new species, but also to new flora treatments, in a constant cycle of discovery, revision, and verification that starts in the field. Although in my career some of the longest lasting and most widely used publications are precisely these flora treatments, their revisions can occur in relatively short time frames—as the fourfold increase in identified cycad species over my career has shown. A further personal example is the current diversity of *Cycas* species in China: when I published the Cycadaceae section in the *Flora of China* with my colleague Chen Jiarui (Jiarui and Stevenson 1999), we recognized 14 species out of 60 total known for the genus at that time. Over the past twenty years, field research has built on and revised our original, and today there are 23 recognized species of *Cycas* in China, out of 120 for the genus.

With advances in cladistic methodology and systematics, a basis for reconstructing the genealogy of the cycads using these new data emerged, which was eventually further augmented with the advent of new methods and techniques in molecular biology, systematics, and the evolution of development (EvoDevo), among other fields. An excellent example of the use of molecular biology in the interface of conservation and systematics is the development of DNA barcoding in cycad identification for CITES enforcement when identifying morphological features have been removed for smuggling or making products (Sass et al. 2007). Yet to achieve effective results with this technique, one needs to be aware of not just how to pipette, but also of issues and questions from other disciplines in both the sciences and the humanities (as this volume advocates)—for example, what cultural uses or demands would necessitate cycad smuggling, or what products might be made with endangered species and how?

These were—and remain—exciting times indeed, and I saw no reason to just sit on the sidelines. Of course, to understand cycad evolution through time, I had to have forays into paleobotany. Knut and I saw the advances in technology and as a result we could incorporate them in our own research in meaningful ways. Another fascinating departure from my own research agenda and into the fields of molecular biology and genomics occurred via my participation in the New York Plant Genomics Consortium. The founding projects of this consortium were focused primarily on the genomics of cycads and on the neurotoxin β-N-methylamino-L-alanine (BMAA), a nonprotein amino acid that naturally occurs in the coralloid roots of cycads and which has been implicated in Guam Dementia and Amyotrophic Lateral Sclerosis (ALS, or Lou Gehrig's Disease) (Brenner, Stevenson, and Twigg 2003; Brenner et al. 2003; Brenner, Stevenson, Chiu, et al. 2007; Brenner, Stevenson, Katari, et al. 2007; Cibrián-Jaramillo, de la Torre-Barcena, et al. 2010; Cox and Sacks 2002). We discovered that BMAA interferes with the nitrogen fixation cycle normally associated with cycad roots (as VanDerwarker notes in her chapter 4) and results in plants not being able to respond to light when exposed to it (Brenner et al. 2000). In the lab, BMAA caused thale cress (*Arabidopsis thaliana*, the favored experimental species for plant molecular biologists) to germinate with abnormally long hypocotyls (hypocotyledonous stem or "below seed leaf") and cotyledons (seed leaves) that did not turn green or simply did not open at all—just as with the bean plant grown in a closet for a middle-school science class. The question thus arose as to how cycads—which produce BMAA naturally—are able to turn green and grow properly? Is there something in their genome that facilitates their proper growth? A potentially exciting solution to this query may be found in a BMAA insensitive thale cress mutant (BIM409) that perhaps parallels cycad genomics in allowing this mutant to germinate successfully, and it may even have biomedical applications. For example, if BMAA causes both ALS and interferes with the growth cycle of some plants, could the treatment and/or cure for ALS in humans be found in the cycad genome or that of BMAA-insensitive mutants of other plants? An ongoing project on whole genome sequencing, focused specifically on *Cycas rumphii* and *Zamia furfuracea*, may provide clues that assist in unraveling this mystery. Again, exciting times, and an excellent illustration of the inherent potential of research on cycads—as well as a very good reason to appreciate these plants, and to reflect on what we might lose should we continue to drive them to extinction.

Much of what was accomplished in the past—by me and others—was serendipitous, to say the least, such as in the case of *Zamia wallisii* described above, or in the documentation of insect pollination in a cycad. In that case, I was studying seed cone development in *Zamia furfuracea* while Knut was studying pollen development and ultrastructure at the Fairchild Tropical Garden during the 1970s. I was complaining about weevils inside the seed cones, and Knut was noticing them on the pollen cones and emerging from them—serendipity, to be sure! The rest is history; after that, wind pollination was no longer the default concept for cycad reproduction (Norstog and Stevenson 1980; see also Norstog 1987; Norstog and Fawcett 1989).

In truth, many of my field and research experiences reflected such happy coincidences—or blind luck, if you prefer. I recall several instances in my fieldwork, a further two of which I will share here. One was trying to collect *Zamia paucijuga* from Las Tres Marías Islands in the Pacific Ocean off the coast of Nayarit, particularly the island of María Cleofas, in the 1980s. Although now a tourist destination, the islands were then used as a prison—with the ominous, Kafka-esque name of the Islas Marías Federal Penal Colony—which operated from 1905 until 2019. Needless to say, permits for collecting there were not officially available. Not to be dissuaded by this difficulty, we hired a fishing boat and went after dark, snuck on to the island from the beach, and collected a few plants before returning to the mainland port of San Blas. The heady days of youth—or at least a youthful middle age! I would of course never do something like that today—nor would I recommend it!—but I might like to return there now as the prison is closed and the islands have been declared a UNESCO heritage site. I even hear that ecotours are forthcoming.

Another example stems from some fieldwork in Venezuela with my friend and colleague Aldo Moretti, also in the 1980s. We were out looking for specimens of *Zamia lecointei*, but we had been unsuccessful in our search for weeks and were therefore feeling dejected. While stopping for refreshment in a small town, we noticed a single *Zamia lecointei* planted in a pot in front of a cottage. We stopped to look at it and the owner came out to greet us. We asked about it, and, after cleverly checking our credentials by asking if we knew of the cones, which were not on his plants, he told us that he himself had brought the plant from the forest near his birthplace some 100 km away. He then proceeded to accompany us to the locality, where we found all of the *Zamia lecointei* specimens we could have hoped for. I was reminded of this story when reviewing Cibrián-Jaramillo and her colleagues' chapter 2, specifically as they mention the possibility of people moving these plants across the landscape (a potentiality also explored by Bonta et al. [2019] and Englehardt et al. [2020]). In this case, it was evident that this plant held special meaning for our host, and was important to him culturally, perhaps as a reminder of his home. Or maybe he simply liked the way it looked?—a sentiment shared by many cycad enthusiasts and collectors.

There are two serious lessons that I think can be taken from these anecdotes, amusing as they may be. The first is that local people, citizen scientists, and those who live among cycads want to be involved, and to help, with both field research and preservation—we just need to learn to ask the right questions, and to listen carefully. As mentioned above, meaningful dialogue between scientists, stakeholders, and locals is key to both cycad research and conservation. Although this volume is overtly academic in nature, I feel it will have a wider appeal beyond "specialists," and it may prove useful in beginning such engagements. In addition, having previously worked with the majority of the contributors, I know that their own investigations are predicated on and reflect such a commitment to openness, sharing, and community involvement—a commitment that also comes through in the pages of this volume. The second lesson here is, as in all sciences, to question, observe, and trust what you see—not necessarily what you are told. This is a lesson I took from Knut, an inspiring mentor both in the field and in the laboratory, and that for me made all the difference. It is easy to imagine the contributors to this volume stimulating the next generations of cycad researchers in a similar manner as Knut did for me all those years ago, exhorting them to continue exploring, questioning, and pushing the envelope in the constant cycle of research, discovery, and revision—reminiscent of the flora treatments mentioned above—that is ubiquitous in all intellectual inquiry.

Although it is clear that much remains to be done if we are to achieve a deeper and more complex understanding of cycad history, biology, culture, and human-cycad relationships, the present volume begins to show us the way. Combining the integrative approaches and innovative methodologies advanced in the contributions assembled here with the "traditional" botanical and biological methods for the study of cycads (with which I am most familiar)—as well as developing techniques and emerging research on plant molecular biology, "EvoDevo," paleobotany, and the genomics and "-omics" approaches described by Cibrián-Jaramillo et al. and Englehardt and colleagues in their respective contributions—provides a tantalizing and eminently worthwhile path for the future. As I detail above, there are numerous potentially fruitful avenues for cycad research in the future, from possible medical applications to the comparison of genomic and archaeological data to detect human movement of these plants. I share the contributors' enthusiasm and excitement, and I am confident that efforts such as those collected and presented here offer productive opportunities to address the reflexive questions I outline briefly above.

In closing, let me reiterate my opening point: you never know when you may become aware of cycads—often when least expected. In May 2019, I was standing in Venice's Piazza San Marco, where I had been many times before, when I (serendipitously) looked up at the northern facade of the Basilica di

Figure E.4. A cycad representation on the northern facade of the Basilica di San Marco, Venice (*left* and *center*), compared with a young megasporophyll of *Cycas revoluta* (*right*) (photographs by the author).

San Marco and saw a piece of cycad art for the first time—what seemed to me to be an obvious representation of a *Cycas revoluta* megasporophyll (Figure E.4)—although the Basilica authorities insist it is not a cycad, but rather a "palm." With hindsight, I feel that this is redolent of the placement of cycad designs on church facades in Mexico documented extensively by Bonta et al. (2019:figs. 2–7), as well as other instances of the incorporation of cycads in syncretic religious practices in many contexts described in the introduction to this volume and elsewhere (e.g., Gamil 2014; Pérez-Farrera and Vovides 2006). While revising this epilogue with one of the editors, we both had the same questions: what would a representation of a cycad be doing on this paradigm of Venetian Renaissance architecture? Was this design element part of the original eleventh-century architectural program, or perhaps added in later additions following the Venetian-led conquest of Constantinople in the Fourth Crusade of 1204 and in the fourteenth and fifteenth centuries? Is it intended to contribute to what Howard (2004:28) calls a unique, Oriental feeling of exoticism combined with the traditional elements of the Venetian Renaissance artistic style—perhaps having been brought in as spolia from ancient or Byzantine buildings? (Although this possibility would beg the further question of what a cycad representation would have been doing on an ancient Venetian or medieval Byzantine construction . . .) In many ways, it simply defies explanation and must remain as yet another in a long list of potentially exciting topics for cycad research in the future.

In the end, however, what is clear is that more eyes can and should be opened to the fascinating world of cycads. This volume represents an important effort in this regard, and a significant advance. If we can heed its lessons and examples, we may continue to expand our understanding of these plants and their integral, millennial role in human cultures across the globe, and perhaps motivate the conservation and preservation of this unique biocultural patrimony. After all, these relics of the past are all around us, as they have been for the entirety of human history, and for eons before the rise of our own species, simply waiting to be discovered, and for their secrets to be unlocked.

References Cited

Abbad y Lasierra, Iñigo
1866   *Historia Geográfica, Civil y Natural de la Isla de San Juan Bautista de Puerto Rico.* Carlos Bailly-Bailliere, Madrid.
Akeroyd, John, Noel McGough, and Peter Wyse Jackson
1994   *A CITES Manual for Botanic Gardens.* Botanic Gardens Conservation International, London.
Akers, Hugh A.
1983   Isolation of the Siderophore Schizokinen from Soil of Rice Fields. *Applied and Environmental Microbiology* 45(5): 1704–1706.
Alcorn, Janice B.
1984   *Huastec Mayan Ethnobotany.* University of Texas Press, Austin.
Alcorn, Janice B., Barbara Edmondson, and Cándido Hernández Vidales
2006   Thipaak and the Origins of Maize in Northern Mesoamerica. In *Histories of Maize: Multidisciplinary Approaches to the Prehistory, Linguistics, Biogeography, Domestication, and Evolution of Maize,* edited by John E. Staller, Robert H. Tykot, and Bruce F. Benz, pp. 600–611. Academic, New York.
Alexander, David H., John Novembre, and Kenneth Lange
2009   Fast Model-Based Estimation of Ancestry in Unrelated Individuals. *Genome Research* 19(9): 1655–1664.
Amorim, Antonio
1999   Archaeogenetics. *Journal of Iberian Archaeology* 1: 15–25.
Anderson-Córdova, Karen F.
1990   Hispaniola and Puerto Rico: Indian Acculturation and Heterogeneity, 1492–1550. PhD dissertation, Yale University.
2017   *Surviving Spanish Conquest: Indian Fight, Flight, and Cultural Transformation in Hispaniola and Puerto Rico.* University of Alabama Press, Tuscaloosa.
Andrews, Henry N., Jr.
1961   *Studies in Paleobotany.* John Wiley and Sons, New York.
Aouizerat, Tzemach, Itai Gutman, Yitzhak Paz, Aren M. Maeir, Yuval Gadot, Daniel Gelman, Amir Szitenberg, Elyashiv Drori, Ania Pinkus, Miriam Schoemann, Rachel Kaplan, Tziona Ben-Gedalya, Shunit Coppenhagen-Glazer, Eli Reich, Amijai Saragovi, Oded Lipschits, Michael Klutstein, and Ronen Hazan
2019   Isolation and Characterization of Live Yeast Cells from Ancient Vessels as a Tool in Bio-Archaeology. *mBio* 10. doi: 10.1128/mBio.00388-19.

Ariel de Vidas, Anath
2004   *Thunder Doesn't Live Here Anymore: The Culture of Marginality among the Teeneks of Tantoyuca.* University Press of Colorado, Boulder.

Arnold, Philip J., III
2000   Sociopolitical Complexity and the Gulf Olmecs: A View from the Tuxtla Mountains. In *Olmec Art and Archaeology in Mesoamerica,* edited by John E. Clark and Mary E. Pye, pp. 117–135. Yale University Press, New Haven.
2005   The Shark–Monster in Olmec Iconography. *Mesoamerican Voices* 2: 1–38.

Atchison, Jennifer, and Richard Fullagar
1998   Starch Residues on Pounding Implements from Jinmium Rock-shelter. In *A Closer Look: Recent Australian Studies of Stone Tools,* edited by Richard Fullagar, pp. 109–125, Computing Laboratory, School of Archaeology, University of Sydney.

Ávila-Arcos, María, Kimberly F. Mcmanus, Karla Sandoval, Juan Esteban Rodríguez-Rodríguez, Viridiana Villa-Islas, Alicia R. Martin, Pierre Luisi, Rosenda I. Peñaloza-Espinosa, Celeste Eng, Scott Huntsman, Esteban G. Burchard, Christopher R. Gignoux, Carlos D. Bustamante, and Andrés Moreno-Estrada
2019   Population History and Gene Divergence in Native Mexicans Inferred from 76 Human Exomes. *Molecular Biology Evolution* 37(4): 994–1006.

Babot, Ma. del Pilar
2003   Starch Grain Damage as an Indicator of Food Processing. In *Phytolith and Starch Research in the Australian-Pacific-Asian Regions: The State of the Art.,* edited by Diane M. Hart and Lynley A. Wallis, pp. 69–81. Pandanus Books, Canberra.
2006   Box 4.4: Damage on Starch from Processing Andean Food Plants. In *Ancient Starch Research,* edited by Robin Torrence and Huw Barton, pp. 66–67. Left Coast Press, Walnut Creek, California.

Banks, Natalie Clare, Dean Ronald Paini, Kirsty Louise Bayliss, and Michael Hodda
2015   The Role of Global Trade and Transport Network Topology in the Human-Mediated Dispersal of Alien Species. *Ecology Letters* 18(2): 188–199.

Bano, Syeda A., and Sheikh M. Iqbal
2016   Biological Nitrogen Fixation to Improve Plant Growth and Productivity. *International Journal of Agricultural Innovations and Research* 4(4): 596–599.

Barone Lumaga, Maria Rosaria, Mario Coiro, Elisabeth Truernit, Boglárka Erdei, and Paolo De Luca
2015   Epidermal Micromorphology in Dioon: Did Volcanism Constrain Dioon Evolution? *Botanical Journal of the Linnean Society* 179(2): 236–254.

Bassie-Sweet, Karen
2002   Corn Deities and the Male/Female Principle. In *Ancient Maya Gender Identity and Relations,* edited by Lowell S. Gustafson and Amelia M. Trevelyan, pp. 169–190. Bergin and Garvey, Westport, Connecticut.
2008   *Maya Sacred Geography and the Creator Deities.* University of Oklahoma Press, Norman.

Bebrich, Carl, and Jack Wynn
1973   Mound B-V-6: A Late Formative Ceremonial Structure. In *The Pennsylvania*

State University Kaminaljuyu Project—1969, 1970 seasons (No. 9), edited by Joseph Michels and William Sanders. Pennsylvania State University, Department of Anthropology, University Park.

Beck, Wendy
1992   Aboriginal Preparation of Cycas Seeds in Australia. *Economic Botany* 46(2): 133–147.

Bello, Luis Arturo, and Octavio Paredes
1999   El Almidón: Lo Comemos pero no lo Conocemos. *Perspectivas* 50(3): 29–33.

Bennetzen, Jeff, Edward Buckler, Vicki Chandler, John Doebley, Jane Dorweiler, Brandon Gaut, Sarah Hake, Elizabeth Kellogg, R. Scott Poethig, Virginia Walbot, and Susan Wessler
2001   Genetic Evidence and the Origin of Maize. *Latin American Antiquity* 12(1): 84–86.

Benz, Bruce F.
1999   On the Origin, Evolution, and Dispersal of Maize. In *Pacific Latin America in Prehistory: The Evolution of Archaic and Formative Cultures*, edited by Michael Blake, pp. 25–38. Washington State University Press, Pullman.

Berdan, Frances F.
2014   *Aztec Archaeology and Ethnohistory.* Cambridge University Press, Cambridge.

Berman, Mary Jane, and Deborah M. Pearsall
2008   At the Crossroads: Starch Grain and Phytolith Analyses in Lucayan Prehistory. *Latin American Antiquity* 19(2): 182–204.

Bierhorst, David W.
1971   *Morphology of Vascular Plants.* Macmillan, New York.

Bishop, Stephanie L., and Susan J. Murch
2020   A Systematic Review of Analytical Methods for the Detection and Quantification of β-N-methylamino-L-alanine (BMAA). *Analyst* 145: 13–28.

Blackmore, Stephen, Mary Gibby, and David Rae
2011   Strengthening the Scientific Contribution of Botanic Gardens to the Second Phase of the Global Strategy for Plant Conservation. *Botanical Journal of the Linnean Society* 166(3): 267–281.

Blake, Michael
2015   *Maize for the Gods: Unearthing the 9,000-Year History of Corn.* University of California Press, Berkeley.

Blancas, José, Alejandro Casas, Diego Pérez-Salicrup, Javier Caballero, and Ernesto Vega
2013   Ecological and Socio-Cultural Factors Influencing Plant Management in Náhuatl Communities of the Tehuacán Valley, Mexico. *Journal of Ethnobiology and Ethnomedicine* 9(1): 39.

Blanco Rosas, Jose Luis
2006   Erosión de la agrodiversidad en la milpa de los Zoque Popoluca de Soteapan: Xutuchincon y Aktevet. Unpublished doctoral thesis, Universidad Iberoamericana, Puebla, Mexico.

Blom, Frans, and Oliver La Farge
1926   *Tribes and Temples*, 2 vols. Middle American Research Institute, Publication 1. Tulane University, New Orleans.
Bonomo, Mariano, Francisco J. Aceituno, Gustavo G. Politis, and Maria L. Pochettino
2011   Pre-Hispanic Horticulture in the Paraná Delta (Argentina): Archaeological and Historical Evidence. *World Archaeology* 43(4): 554–575.
Bonta, Mark A.
2007   Ethnobotany of Honduran Cycads. In *Proceedings of the Seventh International Conference on Cycad Biology (Xalapa, Mexico, 2005)*, edited by Andrew P. Vovides, Dennis W. Stevenson, and Roy Osborne, pp. 120–142. Memoirs of the New York Botanical Garden 97. New York Botanical Garden, New York.
2009   The Dilemma of Indigenous Identity Construction: The Case of the Newly-Recognized Nahoa of Olancho, Honduras. In *Temas de Geografía Latinoamericana, Reunión CLAG-Morelia*, edited by Pedro S. Urquijo Torres and Narciso Barrera-Bassols, pp. 49–86. UNAM Centro de Investigaciones en Geográfia Ambiental, Morelia.
2010a   Human Geography and Ethnobotany of Cycads in Xi'ui, Teenek, and Nahuatl Communities of Northeastern Mexico, Final Report. The Cycad Society.
2010b   Maize and Cycads: In Search of Sacred Ancestors. *Cycad Newsletter* 33: 4–7.
2012   Cycads and Human Life Cycles: Outline of a Symbology. In *Proceedings of Cycad 2008: The 8th International Congress on Cycad Biology, 13–15 January 2008, Panama City, Panama,* edited by Dennis W. Stevenson, Roy Osborne, and Alberto Sidney Taylor Blake, pp. 133–150. Memoirs of the New York Botanical Garden 106. New York Botanical Garden, New York.
Bonta, Mark A., and Samuel O. Bamigboye
2018   Use of Cycads as Ritual and Recreational Narcotics. *South African Journal of Botany* 115: 280–281.
Bonta, Mark A., Oscar Flores Pinot, Daniel Graham, Jody Haynes, and German Sandoval
2006   Ethnobotany and Conservation of Tiusinte (*Dioon mejiae* Standl. & L. O. Williams, Zamiaceae) in Northeastern Honduras. *Journal of Ethnobiology* 26(2): 228–257.
Bonta, Mark A., and Roy Osborne
2007   Cycads in the Vernacular—A Compendium of Local Names. In *Proceedings of the Seventh International Conference on Cycad Biology (Xalapa, Mexico, 2005)*, edited by Andrew P. Vovides, Dennis W. Stevenson, and Roy Osborne, pp. 143–175. Memoirs of the New York Botanical Garden 97. New York Botanical Garden, New York.
Bonta, Mark, María Teresa Pulido-Silva, Teresa Diego-Vargas, Aurelia Vite-Reyes, Andrew P. Vovides, and Angélica Cibrián-Jaramillo
2019   Ethnobotany of Mexican and Northern Central American Cycads (Zamiaceae). *Journal of Ethnobiology and Ethnomedicine* 15(4): 1–34.
Borstein, Joshua A.
2001   Tripping over Colossal Heads: Settlement Patterns and Population Develop-

ment in the Upland Olmec Heartland. Ph.D. dissertation, Department of Anthropology, Pennsylvania State University, State College.

Boserup, Ester
1965 *The Conditions of Agricultural Growth: The Economics of Agrarian Changes under Population Pressure.* George Allen & Unwin, London.

Braakhuis, Edwin
1990 The Bitter Flour: Birth-scenes of the Tonsured Maize God. In *Mesoamerican Dualism,* edited by Rudolf van Zantwijk, Rob de Ridder, and Edwin Braakhuis, pp. 125–147. ISOR Universiteit Utrecht, The Netherlands.
2009 The Tonsured Maize God and Chicome-Xochitl as Maize Bringers and Culture Heroes: A Gulf Coast Perspective. Wayeb Notes 32, electronic document, http://www.wayeb.org/notes/wayeb_notes0032.pdf, accessed June 27, 2020.
2014 Challenging the Lightnings: San Bartolo's West Wall Mural and the Maize Hero Myth. Wayeb Notes 46, electronic document, http://www.wayeb.org/notes/wayeb_notes0046.pdf, accessed June 27, 2020

Braakhuis, Edwin, and Kerry Hull
2014 Pluvial Aspects of the Mesoamerican Culture Hero. The "Kumix Angel" of the Ch'orti' Mayas and Other Rain-Bringing Heroes. *Anthropos* 109: 449–466. doi: 10.5771/0257-9774-2014-2-449.

Bradley, John J.
2005 "Same Time Poison, Same Time Good Tucker": The Cycad Palm in the Southwest Gulf of Carpentaria. *Journal of Australian Studies* 29: 119–133.

Brenner, Eric D., Nora Martínez-Barboza, Alexandra P. Clark, Quail S. Liang, Dennis W. Stevenson, and Gloria M. Coruzzi
2000 Arabidopsis Mutants Resistant to S(+)-β-Methyl-α, β-Diaminopropionic Acid, a Cycad-Derived Glutamate Receptor Agonist. *Plant Physiology* 124(4): 1615–1624.

Brenner, Eric D., Dennis W. Stevenson, Richard W. McCombie, Manpreet S. Katari, Stephen A. Rudd, Klaus F. X. Mayer, Peter M. Palenchar, Suzan J. Runko, Richard W. Twigg, Guangwei Dai, Rob A. Martienssen, Phillip N. Benfey, and Gloria M. Coruzzi
2003 Expressed Sequence Tag Analysis in *Cycas,* the Most Primitive Living Seed Plant. *Genome Biology* 4: R78.1-R78.11.

Brenner, Eric D., Dennis W. Stevenson, Joanna C. Chiu, Suzan J. Runko, Peter M. Palenchar, and Gloria M. Coruzzi
2007 Defining the Role of BMAA and its Potential Glutamate Receptor Targets in *Arabidopsis thaliana* and *Cycas rumphii.* In *Proceedings of the Seventh International Conference on Cycad Biology (Xalapa, Mexico, 2005),* edited by Andrew P. Vovides, Dennis W. Stevenson, and Roy Osborne, pp. 236–253. Memoirs of the New York Botanical Garden 97. New York Botanical Garden, New York.

Brenner, Eric D., Dennis W. Stevenson, Manpreet S. Katari, Richard W. McCombie, Stephen A. Rudd, Suzan J. Runko, Rob A. Martienssen, and Gloria M. Coruzzi
2007 Genomic Studies in *Cycas rumphii* Miquel. In *Proceedings of the Seventh International Conference on Cycad Biology (Xalapa, Mexico, 2005),* edited by

Andrew P. Vovides, Dennis W. Stevenson, and Roy Osborne, pp. 181–193. Memoirs of the New York Botanical Garden 97. New York Botanical Garden, New York.

Brenner, Eric D., Dennis W. Stevenson, and Richard W. Twigg
2003   Cycads: Evolutionary Innovations and the Role of Plant-derived Neurotoxins. *Trends in Plant Science* 8(9): 446–452.

Brent, Robert N., Hunter Wines, Joseph Luther, Nathan Irving, Joshua Collins, and Brandon Lee Drake
2017   Validation of Handheld X-Ray Fluorescence for In Situ Measurement of Mercury in Soils. *Journal of Environmental Chemical Engineering* 5: 768–776.

Brinkman, E. Pernilla, Wim H. Van der Putten, Evert-Jan Bakker, and Koen J. F. Verhoeven
2010   Plant-Soil Feedback: Experimental Approaches, Statistical Analyses and Ecological Interpretations. *Journal of Ecology* 98(5): 1063–1073.

Brongniart, Adolphe-Théodore
1846   Note sur un Nouveau Genre de Cycadées du Mexique. *Annales des Science Naturelles, Bot. Sér.* 3(5): 5–10.

Brown, Cecil
2010   Development of Agriculture in Prehistoric Mesoamerica: The Linguistic Evidence. In *Pre-Columbian Foodways: Interdisciplinary Approaches to Food, Culture, and Markets in Ancient Mesoamerica,* edited by John E. Staller and Michael D. Carrasco, pp. 71–107. Springer, New York.

Brummitt, Neil A., Steven P. Bachman, Janine Griffiths-Lee, Maiko Lutz, Justin F. Moat, Aljos Farjon, John S. Donaldson, Craig Hilton-Taylor, Thomas R. Meagher, Sara Albuquerque, Elina Aletrari, A. Kei Andrews, Guy Atchison, Elisabeth Baloch, Barbara Barlozzini, Alice Brunazzi, Julia Carretero, Marco Celesti, Helen Chadburn, Eduardo Cianfoni, Chris Cockel, Vanessa Coldwell, Benedetta Concetti, Sara Contu, Vicki Crook, Philippa Dyson, Lauren Gardiner, Nadia Ghanim, Hannah Greene, Alice Groom, Ruth Harker, Della Hopkins, Sonia Khela, Poppy Lakeman-Fraser, Heather Lindon, Helen Lockwood, Christine Loftus, Debora Lombrici, Lucia Lopez-Poveda, James Lyon, Patricia Malcolm-Tompkins, Kirsty McGregor, Laura Moreno, Linda Murray, Keara Nazar, Emily Power, Mireya Quiton Tuijtelaars, Ruth Salter, Robert Segrott, Hannah Thacker, Leighton J. Thomas, Sarah Tingvoll, Gemma Watkinson, Katerina Wojtaszekova, and Eimear M. Nic Lughadha
2015   Green Plants in the Red: A Baseline Global Assessment for the IUCN Sampled Red List Index for Plants. *PLOS ONE* 10(8): e0135152.

Brummitt, Richard K.
1987   Report of the Committee for Spermatophyta, 31. *Taxon* 36(1): 72–78.

Brüssow, Harald
2020   Bioarchaeology: A Profitable Dialogue between Microbiology and Archaeology. *Microbial Biotechnology* 13(2): 406–409. doi: 10.1111/1751-7915.13527.

Bryant, Vaughn M.
2007   Microscopic Evidence for the Domestication and Spread of Maize. *PNAS* 104(50): 19659-19660.

Buléon, Alain, Paul Colona, Véronique Planchot, and Steven G. Ball
1998 Starch Granules: Structure and Biosynthesis. *International Journal of Biological Macromolecules* 23(2): 85–112.

Bull, William
1881 A Retail List of New, Beautiful, and Rare Plants Offered by William Bull. Electronic document, https://www.biodiversitylibrary.org/page/49019680, accessed April 8, 2022.

Burkhardt, Mrs. Henry J.
1952 Starch Making: A Pioneer Florida Industry. *Tequesta* 12: 47–51.

Bustos-Díaz, Edder D., Francisco Barona-Gómez, and Angélica Cibrián-Jaramillo
2019 Cyanobacteria in Nitrogen-Fixing Symbioses. In *Cyanobacteria: From Basic Science to Applications*, edited by Arun K. Mishra, D. N. Tiwari, and Amit N. Rai, pp. 29–42. Academic, New York.

Buyer, Jeffrey S., and Lawrence J. Sikora
1990 Rhizosphere Interactions and Siderophores. *Plant and Soil* 129(1): 101–107.

Byrne, Margaret, and Sidney H. James
1991 Genetic Diversity in the Cycad *Macrozamia riedlei*. *Heredity* 67(1): 35–39.

Caballero Zamora, Carlos
2009 Tutu-Nakú: Testimonios II. Electronic document, http://www.tutu-naku.tlax.com/2009/05/testimonios-ii/, accessed July 9, 2021.

Cabrera-Toledo, Dánae, Jorge González-Astorga, and Juan Carlos Flores-Vázquez
2012 Fine-Scale Spatial Genetic Structure in Two Mexican Cycad Species *Dioon caputoi* and *Dioon merolae* (Zamiaceae, Cycadales): Implications for Conservation. *Biochemical Systematics and Ecology* 40: 43–48.

Cabrera-Toledo, Dánae, Jorge González-Astorga, Fernando Nicolalde-Morejón, Francisco Vergara-Silva, and Andrew P. Vovides
2010 Allozyme Diversity Levels in Two Congeneric *Dioon* spp. (Zamiaceae, Cycadales) with Contrasting Rarities. *Plant Systematics and Evolution* 290(1): 115–125.

Cabrera-Toledo, Dánae, Jorge González-Astorga, and Andrew P. Vovides
2008 Heterozygote Excess in Ancient Populations of the Critically Endangered *Dioon caputoi* (Zamiaceae, Cycadales) from Central Mexico. *Botanical Journal of the Linnean Society* 158(3): 436–447.

Cabrera-Toledo, Dánae, Jorge González-Astorga, Andrew P. Vovides, Alejandro Casas, Ofelia Vargas-Ponce, Pablo Carrillo-Reyes, Janet Nolasco-Soto, and Ernesto Vega
2019 Surviving Background Extinction: Inferences from Historic and Current Dynamics in the Contrasting Population Structures of Two Endemic Mexican Cycads. *Population Ecology* 61(1): 62–73.

Calonje, Michael
2009 A New Cliff-Dwelling Species of *Zamia* (Zamiaceae) from Belize. *Journal of the Botanical Research Institute of Texas* 3(1): 23–29.

Calonje, Michael, C. Calonje, and S. Cuestas
2016 In Search of the World's Most Massive Seed Cone. *Encephalartos* 124: 12–14.

Calonje, Michael, and Jan Meerman
2009 What is *Zamia prasina* (Zamiaceae: Cycadales)? *Journal of the Botanical Research Institute of Texas* 3(1): 43–49.

Calonje, Michael, Jan Meerman, M. Patrick Griffith, and Goeffrey B. Hoese
2009 A New Species of *Zamia* (Zamiaceae) from the Maya Mountains of Belize. *Journal of the Botanical Research Institute of Texas* 3(1): 31–41.

Calonje, Michael, Alan W. Meerow, M. Patrick Griffith, Dayana Salas-Leiva, Andrew P. Vovides, Mario Coiro, and Javier Francisco-Ortega
2019 A Time-Calibrated Species Tree Phylogeny of the New World Cycad Genus *Zamia* L. (Zamiaceae, Cycadales). *International Journal of Plant Sciences* 180(4): 286–314.

Calonje, Michael, Alan W. Meerow, Lindy Knowles, David Knowles, M. Patrick Griffith, Kyoko Nakamura, and Javier Francisco-Ortega
2013 Cycad Biodiversity in the Bahamas Archipelago and Conservation Genetics of the Threatened *Zamia lucayana* (Zamiaceae). *ORYX* 47(2): 190–198.

Calonje, Michael, and Alexander N. Sennikov
2017 In the Process of Saving Plant Names from Oblivion: The Revised Nomenclature of *Ceratozamia fuscoviridis* (Zamiaceae). *Taxon* 66(1): 158–164.

Calonje, Michael, Dennis William Stevenson, and Roy Osborne
2022 The World List of Cycads, online edition. http://cycadlist.org, accessed April 8, 2022.

Campbell, Lyle
1997 *American Indian Languages: The Historical Linguistics of Native America*. Oxford Studies in Anthropological Linguistics, 4. Oxford University Press, Oxford.

Cappellini, Enrico, M. Thomas P. Gilbert, Filippo Geuna, Girolamo Fiorentino, Allan Hall, Jane Thomas-Oates, Peter D. Ashton, David A. Ashford, Paul Arthur, Paula F. Campos, Johan Kool, Eske Willerslev, and Matthew J. Collins
2010 A Multidisciplinary Study of Archaeological Grape Seeds. *Naturwissenschaften* 97(2): 205–217.

Caputo, Paolo, Salvatore Cozzolino, Paolo De Luca, Aldo Moretti, and Dennis W. Stevenson
2004 Molecular Phylogeny of *Zamia* (Zamiaceae). In *Cycad Classification: Concepts and Recommendations*, edited by Terrence Walters and Roy Osborne, pp. 149–158. CABI, Oxford.

Caputo, Paolo, C. Marquis, T. Wurtzel, Dennis W. Stevenson, and E. T. Wurtzel
1990 Molecular Biology in Cycad Systematics. In *Second International Conference on Cycad Biology*, edited by Dennis W. Stevenson and Knut J. Norstog. Palm and Cycad Societies of Australia, Townsville.

Carbajal-Esquivel, Haydeé, Javier Fortanelli Martínez, José García-Pérez, Juan A. Reyes-Agüero, Laura Yáñez-Espinosa, and Mark A. Bonta
2012 Use Value of Food Plants in the Xi'iuy Indigenous Community of Las Guapas, Rayon, San Luis Potosi, Mexico. *Ethnobiology Letters* 3: 39–55.

Carrasco, Michael D.
2010 From Field to Hearth: An Earthly Interpretation of Maya and Other Me-

soamerican Creation Myths. In *Pre-Columbian Foodways: Interdisciplinary Approaches to Food, Culture, and Markets in Mesoamerica*, edited by John Staller and Michael D. Carrasco, pp. 601-634. Springer, New York.

2012   Mǎyǎ nóngyè shénhuà: Yùmǐ hé rén de shēngmìng lúnhuí [Mayan Agrarian Mythologies: The Lifecycle of Maize and Humans]. *Shìjiè zōngjiào wénhuà* [*The Religious Cultures in the World*] 76: 18–24.

2015   Cycads, Maize, and Garfish: The Representation of Ethnoecological Systems in Olmec Iconography. Paper presented at the 10th International Conference on Cycad Biology, Medellín, Colombia, August 16th–21st.

2020   Cycads, Maize, and Garfish: The Representation of Ethnoecological Systems in Olmec Iconography. In *Proceedings of Cycad 2015: The 10th International Conference on Cycad Biology*, edited by Cristina López-Gallego, Michael Calonje, M. Patrick Griffith, and Jibankumar S. Khuraijam, pp. 1–12. IUCNS/Species Survival Commission Cycad Specialist Group.

Casas, Alejandro, Adriana Otero-Arnaiz, Edgar Pérez-Negrón, and Alfonso Valiente-Banuet

2007   In Situ Management and Domestication of Plants in Mesoamerica. *Annals of Botany* 100(5): 1101–1115

Caso, Alfonso

1953   *El Pueblo del Sol*. Fondo de Cultura Económica, Mexico City.

Castilla-Beltrán, Alvaro, Henry Hooghiemstra, Menno L. P. Hoogland, Jaime R. Pagán-Jiménez, Bas van Geel, Michael H. Field, Maarten Prins, Timme Donders, Eduardo Herrera Malatesta, Jorge Ulloa Hung, Crystal H. McMichael, William D. Gosling, and Corinne L. Hofman

2018   Columbus' Footprint in Hispaniola: A Paleoenvironmental Record of Indigenous and Colonial Impacts on the Landscape of the Central Cibao Valley, Northern Dominican Republic. *Anthropocene* 22: 66–80.

Castillo-Lara, Pedro, Pablo Octavio-Aguilar, and José Arturo De-Nova

2018   *Ceratozamia Zaragozae* Medellín-Leal (Zamiaceae), an Endangered Mexican Cycad: New Information on Population Structure and Spatial Distribution. *Brittonia* 70(2): 155–165.

Ceballos, Gerardo, Paul R. Ehrlich, Anthony D. Barnosky, Andrés García, Robert M. Pringle, and Todd M. Palmer

2015   Accelerated Modern Human-Induced Species Losses: Entering the Sixth Mass Extinction. *Science Advances* 1(5): e1400253.

Chacón S., Maria I., Barbara Pickersgill, and Daniel G. Debouck

2005   Domestication Patterns in Common Bean (*Phaseolus vulgaris* L.) and the Origin of the Mesoamerican and Andean Cultivated Races. *Theoretical and Applied Genetics* 110(3): 432–444.

Chakra, Prashantkumar S., P. G. Vinay Kumar, and C. T. Swamy

2019   Isolation and Biochemical Characterization of Plant Growth Promoting Bacteria from a Maize Crop Field. *International Journal of Current Microbiology and Applied Sciences* 8(4): 1415–1422.

Chamberlain, Charles J.
1906   The Ovule and Female Gametophyte of Dioon. *Botanical Gazette* 42: 321–358.
1911   The Adult Cycad Trunk. *Botanical Gazette* 52: 81–104.
1919   *The Living Cycads*. University of Chicago Press, Chicago.
1935   Cycadophytes—Cycadales. In *Gymnosperms. Structure and Evolution,* edited by Charles J. Chamberlain, pp. 60–164. University of Chicago Press, Chicago.
n.d.   "A Taxonomic Monograph of the Cycads" (unpublished manuscript).

Chang, Jui-Tse, Bing-Hong Huang, and Pei-Chun Liao
2019   Genetic Evidence of the Southward Founder Speciation of *Cycas taitungensis* from Ancestral *C. revoluta* along the Ryukyu Archipelagos. *Conservation Genetics* 20: 1045–1056.

Chaves, Ramiro, and Yuriet Ferrer
2007   Some News of Microcycas in Cuba. *Cycad Newsletter* 30(4): 10–14.

Chemín-Bassler, Heidi
1984   *Los pames septentrionales de San Luís Potosí*. Instituto Nacional Indigenista, Mexico City.
2000   *Recetario Pame de San Luís Potosí y Querétaro*. Cocina Indígena y Popular No. 26. CONACULTA, Mexico City.
2012   El chamal, alimento divino de los Pames-Xi'iui de San Luis Potosí y Querétaro. In *Los pueblos indígenas de la Huasteca y el semidesierto queretano (Atlas etnográfico),* edited by Julieta Valle Esquivel, Diego Prieto Hernández, and Beatriz Utrilla Sarmiento, pp. 275–279. Instituto Nacional de Antropología e Historia, Instituto Nacional de Lenguas Indígenas, Universidad Autónoma de Querétaro, and Instituto Queretano de Cultura y las Artes, Querétaro, Mexico.

Chevalier, Jacques M., and Andrés Sánchez Bain
2003   *The Hot and the Cold: Ills of Humans and Maize in Native Mexico*. University of Toronto Press, Toronto.

Chinchilla, Oswaldo
2011   *Imágenes de la mitología Maya*. Museo Popol Vuh, Universidad Francisco Marroquín, Guatemala.

Chinique de Armas, Yadira C., William M. Buhay, Roberto Rodríguez Suárez, Sheahan Bestel, David G. Smith, Stephanie D. Mowat, and Mirjana Roksandic
2015   Starch Analysis and Isotopic Evidence of Consumption of Cultigens among Fisher-Gatherers in Cuba: The Archaeological Site of Canímar Abajo, Matanzas. *Journal of Archaeological Science* 58: 121–132.

Christenhusz, Maarten J. M., Mark W. Chase, and Michael F. Fay
2011   Linear Sequence, Classification, Synonymy, and Bibliography of Vascular Plants: Lycophytes, Ferns, Gymnosperms and Angiosperms. *Phytotaxa* 19: 1–134.

Christenhusz, Maarten J. M., James L. Reveal, Aljos Farjon, Martin F. Gardner, Robert R. Mill, and Mark W. Chase
2011   A New Classification and Linear Sequence of Extant Gymnosperms. *Phytotaxa* 19: 55–70.

Christenson, Allen J.
2003   *Popol Vuh: The Sacred Book of the Maya.* University of Oklahoma Press, Norman.

Cibrián-Jaramillo, Angélica, Aidan Daly, Eric Brenner, Robert Desalle, and Thomas Marler
2010   When North and South Don't Mix: Genetic Connectivity of a Recently Endangered Oceanic Cycad, *Cycas micronesica,* in Guam Using EST-Microsatellites. *Molecular Ecology* 19(12): 2364–2379.

Cibrián-Jaramillo, Angélica, Jose Eduardo de la Torre-Barcena, Ernest K. Lee, Manpreet S. Katari, Damon P. Little, Dennis W. Stevenson, Rob A. Martienssen, Gloria M. Coruzzi, and Robert Desalle
2010   Using Phylogenomic Patterns and Gene Ontology to Identify Proteins of Importance in Plant Evolution. *Genome Biology and Evolution* 2(1): 225–239.

Ciofalo, Andy J., William F. Keegan, Michael P. Pateman, Jaime R. Pagán-Jiménez, and Corinne L. Hofman
2018   Determining Precolonial Botanical Foodways: Starch Recovery and Analysis, Long Island, The Bahamas. *Journal of Archaeological Science Reports* 21: 305–317.

Ciofalo, Andy J., Peter T. Sinelli, and Corinne L. Hofman
2019   Late Precolonial Culinary Practices: Starch Analysis on Griddles from the Northern Caribbean. *Journal of Archaeological Method and Theory.* doi: 10.1007/s10816-019-09421-1.

Clarkson, Chris, Mike Smith, Ben Marwick, Richard Fullagar, Lynley A. Wallis, Patrick Faulkner, Tiina Manne, Elspeth Hayes, Richard G. Roberts, Zenobia Jacobs, Xavier Carah, Kelsey M. Lowe, Jacqueline Matthews, and S. Anna Florin
2015   The Archaeology, Chronology and Stratigraphy of Madjedbebe (Malakunanja II): A Site in Northern Australia with Early Occupation. *Journal of Human Evolution* 83: 46–64. doi: 10.1016/j.jhevol.2015.03.014.

Cody, Robert B., James A. Laramée, and H. Dupont Durst
2005   Versatile New Ion Source for the Analysis of Materials in Open Air under Ambient Conditions. *Analytical Chemistry* 77(8): 2297–2302.

Coe, Michael D.
1968   *America's First Civilization: Discovering the Olmec.* American Heritage Press, New York.
1973   *The Maya Scribe and His World.* Grolier Club, New York.
1976   Early Steps in the Evolution of Mayan Writing. In *Origins of Religious Art and Iconography in Preclassic Mesoamerica,* edited by Henry B. Nicholson, pp. 107–122. UCLA Latin American Studies Series 31. UCLA, Los Angeles.

Coiro, Mario, and Christian Pott
2017   *Eobowenia* gen. nov. from the Early Cretaceous of Patagonia: Indication for an Early Divergence of *Bowenia*? *BMC Evolutionary Biology* 17(1): 97. doi: 10.1186/s12862-017-0943-x.

Collins, Guy N.
1921   Teosinte in Mexico. *Journal of Heredity* 12: 339–350.

Comisión Nacional para el Conocimiento y Uso de la Biodiversidad (CONABIO), Instituto Nacional Electoral (INE), and Instituto Nacional de Investigaciones Forestales, Agrícolas y Pecuarias (INIFAP)
2008　　Componente 2: Distribución geográfica del teocintle (*Zea* spp.) en México y situación actual de las poblaciones. Report to the Comisión Nacional para el Conocimiento y Uso de la Biodiversidad. Guadalajara, Jalisco, Mexico.

Community Science
2019　　Building Community Capacity in Puerto Rico through Food Sovereignty Advocacy. Electronic document, http://www.communityscience.com/news-detail.php?news=307, accessed November 15, 2019.

Condamine, Fabien L., Nathalie S. Nagalingum, Charles R. Marshall, and Hélène Morlon
2015　　Origin and Diversification of Living Cycads: A Cautionary Tale on the Impact of the Branching Process Prior in Bayesian Molecular Dating. *BMC Evolutionary Biology* 15. doi:10.1186/s12862-015-0347-8.

Contreras-Medina, Raúl, Carlos A. Ruiz-Jiménez, and Isolda Luna Vega
2003　　Caterpillars of *Eumaeus childrenae* (Lepidoptera: Lycaenidae) Feeding on Two Species of Cycads (Zamiaceae) in the Huasteca Region, Mexico. *Revista de Biologia Tropical* 51(1): 201–204.

Cooper, Andrew M., Didra Felix, Fatima Alcantara, Ilya Zaslavsky, Amy Work, Paul L. Watson, Keith Pezzol, Qi Yu, Dan Zhu, Alexander J. Scavo, Yasman Zarabi, and Julian I. Schroeder
2020　　Monitoring and Mitigation of Toxic Heavy Metals and Arsenic Accumulation in Food Crops: A Case Study of an Urban Community Garden. *Plant Direct* 4: 1–12.

Cortés, Hernan
1963[1526]　　Quinta carta relación de Hernán Cortés al Emperador Carlos V, Tenuxtitan, 3 de septiembre de 1526. *Cartas y Documentos 242–328*. Editorial Porrúa, Mexico City.

Cousins, Stephen R., Vivienne L. Williams, and Ed T. F. Witkowski
2012　　Uncovering the Cycad Taxa (*Encephalartos* species) Traded for Traditional Medicine in Johannesburg and Durban, South Africa. *South African Journal of Botany* 78: 129–138.

Cox, Paul A., and Oliver W. Sacks
2002　　Cycad Neurotoxins, Consumption of Flying Foxes, and ALS-PDC Disease in Guam. *Neurology* 58: 956–959.

Crane, P. R.
1985　　Seed Plants and Progymnosperms, Except Angiosperms and Ingroup Phylogeny of Conifers. *Annals of the Missouri Botanical Garden* 72: 716–793.
1988　　Major Clades and Relationships in the "Higher" Gymnosperms. In *Origin and Evolution of Gymnosperms,* edited by Charles B. Beck, pp. 218–272. Columbia University Press, New York.

Crisp, Michael D., and Lyn G. Cook
2011　　Cenozoic Extinctions Account for the Low Diversity of Extant Gymnosperms Compared with Angiosperms. *New Phytologist* 192(4): 997–1009.

Crowther, Alison
2012   The Differential Survival of Native Starch during Cooking and Implications for Archaeological Analyses: A Review. *Archaeological and Anthropological Sciences* 4(3): 221–235.

Cruz-Morales, Pablo, Antonio Corona-Gómez, Nelly Selem-Mójica, Miguel A. Pérez-Farrera, Francisco Barona-Gómez, and Angélica Cibrián-Jaramillo
2017   The Cycad Coralloid Root Contains a Diverse Endophytic Bacterial Community with Novel Biosynthetic Gene Clusters Unique to its Microbiome. *bioRxiv*: 121160.

David, Bruno, Bryce Barker, and Ian J. McNiven (eds.)
2006   *The Social Archaeology of Australian Indigenous Societies*. Aboriginal Studies Press, Canberra.

Davidson, William V.
1991   Geographical Perspectives on Spanish-Pech (Paya) Indian Relationships, Northeast Honduras, Sixteenth-Century. In *Columbian Consequences,* edited by David Hurst Thomas, Volume 3, pp. 205–226. Smithsonian Institution Press, Washington, DC.

Deance Bravo y Troncoso, Ivan Gerardo
2012   Mas allá del alimento: El papel cultural del maíz entre los pueblos totonacos. *Alter, Enfoques críticos* 3: 57–68.

Debaene, Vincent
2013   Lévi-Strauss: What Legacy? (Translated by Caroline Vial.) *Yale French Studies* 123: 14–40.

de Boyrie Moya, Emile, Marguerita K. Krestensen, and John M. Goggin
1957   Zamia Starch in Santo Domingo: A Contribution to the Ethnobotany of the Dominican Republic. *Florida Anthropologist* 10(3–4): 17–40.

DeGiorgio, Michael, Mattias Jakobsson, and Noah A. Rosenberg
2009   Explaining Worldwide Patterns of Human Genetic Variation Using a Coalescent-Based Serial Founder Model of Migration Outward from Africa. *PNAS* 106(38): 16057–16062.

Dehgan, Bijan, and Nancy B. Dehgan
1988   Comparative Pollen Morphology and Taxonomic Affinities in Cycadales. *American Journal of Botany* 75(10): 1501–1516.

de las Casas, Fray Bartolomé
1876   *Historia de Las Indias*. Tomo V. Imprenta de Miguel Ginesta, Madrid.

del Paso y Troncoso, Francisco
1903[1558]   *Leyenda de los Soles*. Tipograffa de Salvador Landi, Florence, Italy.

De Luca, Paolo
1990   A Historical Perspective on Cycads from Antiquity to the Present. *Memoirs of the New York Botanical Garden* 57: 1–7.

De Luca, Paolo, Aldo Moretti, Sergio Sabato, and Mario Vázquez-Torres
1980   *Dioon rzedowskii* (Zamiaceae), a New Species from Mexico. *Brittonia* 32(2): 225–229.

De Luca, Paolo, and Sergio Sabato
1979   *Dioon califanoi* (Zamiaceae), a New Species from Mexico. *Brittonia* 31(1): 170–173.

De Luca, Paolo, Sergio Sabato, and Mario Vázquez-Torres
1980   *Dioon caputoi* (Zamiaceae), a New Species from Mexico. *Brittonia* 32(1): 43–46.
1981a   *Dioon holmgrenii* (Zamiaceae), a New Species from Mexico. *Brittonia* 33(4): 552–555.
1981b   *Dioon merolae* (Zamiaceae), a New Species from Mexico. *Brittonia* 33(2): 179–185.
1982   Distribution and Variation of *Dion edule* (Zamiaceae). *Brittonia* 34(3): 355–362.
1984   *Dion tomasellii* (Zamiaceae), a New Species with Two Varieties from Western Mexico. *Brittonia* 36(3): 223–227.

De Luca, Paolo, Sergio Sabato, and Dennis W. Stevenson
1984   Comments Opposed to (708) Proposal to Conserve the Spelling *Dioon* against *Dion* (Zamiaceae). *Taxon* 33(4): 728–730.

Del Valle, Liz Yanira
2017   Desde Holanda tras el marunguey (Spanish, interview, *El Nuevo Día*, Puerto Rico, 2017). https://www.researchgate.net/publication/333985279_Desde_Holanda_tras_el_marunguey_Spanish_interview_El_Nuevo_Dia_Puerto_Rico_2017, accessed September 26, 2019.

de Molina, Alonso, and Antonio de Spinosa
1966[1500s]   *Vocabulario nahuatl-castellano, castellano-nahuatl*. Ediciones Colofón, Mexico City.

Descola, Philippe
2013   *Beyond Nature and Culture*. (Translated by Janet Lloyd.) University of Chicago Press, Chicago.

Dickau, Ruth
2005   *Resource Use, Crop Dispersals, and the Transition to Agriculture in Prehistoric Panama: Evidence from Starch Grains and Macroremains*. Temple University, Philadelphia.
2010   Microbotanical and Macrobotanical Evidence of Plant Use and the Transition to Agriculture in Panama. In *Integrating Zooarchaeology and Paleoethnobotany: A Consideration of Issues, Methods, and Cases,* edited by Amber M. VanDerwarker and Tanya M. Peres, pp. 99–134. Springer, New York.

Dickau, Ruth, Anthony J. Ranere, and Richard G. Cooke
2007   Starch Grain Evidence for the Preceramic Dispersals of Maize and Root Crops into Tropical and Humid Forests of Panama. *PNAS* 104(9): 3651–3656.

Diego-Vargas, Teresa
2017   Relaciones culturales entre las cícadas y el maíz en localidades nahuas y teenek de la Huasteca. Master's thesis, Universidad Autónoma del Estado de Hidalgo, Mexico.

Dieseldorff, Erwin P.
1922   Welchen Gott Stellen die Steindollen der Mayavölker dar? In *Festschrift Edu-*

ard Seler, edited by Walter Lehmann, pp. 47-58. Strecker und Schröder, Stuttgart.

Donaldson, John S.
2003 Introduction. In *Cycads: Status, Survey, and Conservation Action Plan,* edited by John S. Donaldson, pp. 1–2. IUCN/SSC Cycad Specialist Group, Gland, Switzerland.

Donaldson, John S., K. D. Hill, and Dennis W. Stevenson
2003 Cycads of the World: An Overview. In *Cycads: Status, Survey, and Conservation Action Plan,* edited by John S. Donaldson, pp. 3–8. IUCN/SSC Cycad Specialist Group, Gland, Switzerland.

Donnegan, Joseph A., Sarah L. Butler, Walter Grabowiecki, Bruce A. Hiserote, and David Limtiaco
2004 Guam's Forest Resources, 2002, edited by U.S. Forest Service, U.S. Department of Agriculture, Pacific Northwest Research Station, p. 32, Portland, Oregon.

Doolittle, William E.
1992 House-lot Gardens in the Gran Chichimeca: Ethnographic Cause for Archaeological Concern. In *Gardens of Prehistory: The Archaeology of Settlement Agriculture in Greater Mesoamerica,* edited by Thomas W. Killion, pp. 69–91. University of Alabama Press, Tuscaloosa.

Dorsey, Brian L., Timothy J. Gregory, Chodon Sass, and Chelsea D. Specht
2018 Pleistocene Diversification in an Ancient Lineage: A Role for Glacial Cycles in the Evolutionary History of *Dioon Lindl.* (Zamiaceae). *American Journal of Botany* 105(9): 1512–1530.

Duminil, Jérôme, Silvia Fineschi, Arndt Hampe, Pedro Jordano, Daniela Salvini, Giovanni G. Vendramin, and Rémy Petit
2007 Can Population Genetic Structure be Predicted from Life-History Traits? *American Naturalist* 169: 662–672.

Dyer, Rodney J., and John D. Nason
2004 Population Graphs: The Graph Theoretic Shape of Genetic Structure. *Molecular Ecology* 13: 1713–1727.

Eckenwalder, James E.
1980 Taxonomy of the West Indian Cycads. *Journal of the Arnold Arboretum* 61: 701–722.

Edmonson, Munroe S. (translator and editor)
1986 *Heaven Born Merida and Its Destiny: The Book of Chilam Balam of Chumayel.* University of Texas Press, Austin.

Eguiarte, Luis, Jonas Aguirre-Liguori, Lev Jardón-Barbolla, Erika Aguirre-Planter, and Valeria Souza
2013 Genómica de Poblaciones: Nada en evolución va a tener sentido si no es a la luz de la genómica, y nada en genómica tendrá sentido si no es a la luz de la evolución. *TIP Revista Especializada En Ciencias Químico-Biológicas* 16: 42–56.

Ehrenfeld, Joan G., Beth Ravit, and Kenneth Elgersma
2005   Feedback in the Plant-Soil System. *Annual Review of Environment and Resources* 30: 75–115.

Ekué, Marius, Oliver Gailing, Dirk Hölscher, Brice Sinsin, and Reiner Finkeldey
2008   Population Genetics of the Cycad *Encephalartos barteri* ssp. *barteri* (Zamiaceae) in Benin, with Notes on Leaflet Morphology and Implications for Conservation. *Belgian Journal of Botany* 141: 78–94.

Ellstrand, Norman C., Robert Ornduff, and Janet M. Clegg
1990   Genetic Structure of the Australian Cycad, *Macrozamia communis* (Zamiaceae). *American Journal of Botany* 77(5): 677.

Elson, Ben
1947   The Homshuk: A Sierra Popoluca Text. *Tlalocan* 2: 193–214.

Englehardt, Joshua D., and Michael D. Carrasco
2020   Representación Densa: Topofilia y Traslape Conceptual entre Imágenes de la Vida Acuática en el Arte Olmeca. In *Uso y Representación del Agua en la Costa del Golfo,* edited by Lourdes Budar and Sara Ladrón de Guevara, pp. 345–362. Universidad Veracruzana, Xalapa Mexico.

2022   From "Cycad Hell" to Sacred Landscapes: Tracing the Cultural Significance of Cycads in the Ryukyu Islands and Japan. *Journal of Ethnobiology,* in press.

Englehardt, Joshua D., Angélica Cibrián-Jaramillo, and Michael D. Carrasco
2020   Elucidating Ancient Mesoamerican Human-Plant Interactions. In 文化遗产新探: 跨界话语研究 [*New Approach to Cultural Heritage: Profiling Discourse Across Borders*], edited by Le Cheng, Jianpin Yang, and Jianming Cai, pp. 48–76. ACHS and Zhejiang University Press, Hangzhou, China.

Epperson, Bryan K., and Elena R. Alvarez-Buylla
1997   Limited Seed Dispersal and Genetic Structure in Life Stages of *Cecropia obtusifolia*. *Evolution* 51(1): 275.

Escobar Fuentes, Tania
2016   Importancia cultural y prácticas de manejo de las plantas asociadas a dos rituales totonacos en la Sierra Norte de Puebla. Master's thesis, Universidad Veracruzana, Xalapa, Mexico.

Evans, Susan T.
1992   The Productivity of Maguey Terrace Agriculture in Central Mexico during the Aztec Period. In *Gardens of Prehistory: The Archaeology of Settlement Agriculture in Greater Mesoamerica,* edited by Thomas W. Killion, pp. 92–115. University of Alabama Press, Tuscaloosa.

Excoffier, Laurent, Peter E. Smouse, and Joseph M. Quattro
1992   Analysis of Molecular Variance Inferred from Metric Distances Among DNA Haplotypes: Application to Human Mitochondrial DNA Restriction. *Genetics* 131: 479–491.

Fankhauser, Barry
1997   Amino Acid Analysis of Food Residues in Pottery: A Field and Laboratory Study. *Archaeology in Oceania* 32(1): 131–140.

Feider, Clara L., Anna Krieger, Rachel J. DeHoog, and Livia S. Eberlin
2019   Ambient Ionization Mass Spectrometry: Recent Developments and Applications. *Analytical Chemistry* 91: 4266–4290.

Feinman, Gary M.
2006   The Economic Underpinnings of Prehispanic Zapotec Civilization: Small-scale Production, Economic Interdependence, and Market Exchange. In *Agricultural Strategies*, edited by Joyce Marcus and Chip Stanish, pp. 255–280. Cotsen Institute of Archaeology, University of California, Los Angeles.

Feng, Xiuyan, Yuehua Wang, and Xun Gong
2014   Genetic Diversity, Genetic Structure and Demographic History of *Cycas simplicipinna* (Cycadaceae) Assessed by DNA Sequences and SSR Markers. *BMC Plant Biology* 14(1): 187.

Fernández, Juan, and Victoria Sork
2007   Genetic Variation in Fragmented Forest Stands of the Andean Oak *Quercus humboldtii* Bonpl. (Fagaceae). *Biotropica* 39: 72–78.

Fields, Virginia M.
1991   The Iconographic Heritage of the Maya Jester God. In *Sixth Palenque Roundtable, 1986*, edited by Merle Green Robertson and Virginia M. Fields, pp. 167–174. University of Oklahoma Press, Norman.

Fischer, Martin C., Christian Rellstab, Marianne Leuzinger, Marie Roumet, Felix Gugerli, Kentaro K. Shimizu, Rolf Holderegger, and Alex Widmer
2017   Estimating Genomic Diversity and Population Differentiation: An Empirical Comparison of Microsatellite and SNP Variation in *Arabidopsis halleri*. *BMC Genomics* 18(1).

Flannery, Kent V.
1986   *Guilá Naquitz: Archaic Foraging and Early Agriculture in Oaxaca, Mexico*. Academic, New York.

Flores, José S., Hector Narave, and Andrew Vovides
1992   Etnoflora yucatanense: fasciculo 5. *Gymnospermae: taxonomia y etnobotanica*. Universidad Autónoma de Yucatán, Merida, Mexico.

Flores Martínez, Jesus Alberto
2010   Ma tiyolchikawakah tomasewalnemilis wan tomasewaltlahtol. Propuestas de intervención para el mantenimiento y desarrollo de la lengua náhuatl en la comunidad de La Reforma, Benito Juárez, Veracruz. Unpublished undergraduate thesis, Universidad Veracruzana Intercultural, Ixhuatlan de Madero, Mexico.

Florescano, Enrique
1997   El mito nahua de la creación del cosmos y el principio de los reinos. Master's thesis, Universidad Veracruzana, Xalapa, Mexico.

Florin, C. R.
1931   Untersuchungen zur Stammesgeschichte der Coniferales und Cordiatales. I. Morphologie und Epidermisstruktur der Assimilationsorgane beiden rezenten Koniferen. *Kongliga Svenska Vetenskaps Academiens Handlingar* 10: 1–588.

Ford, Richard I.
1985 The Processes of Plant Production in Prehistoric North America. In *Prehistoric Food Production in North America,* edited by Richard I. Ford, pp. 1–18. Anthropological Papers 75. Museum of Anthropology, University of Michigan, Ann Arbor.

Forno, Mario
1967 I Piaroa, attivita' economiche. *Journal de la Société des Américanistes* 56(2): 388–409.

Forsyth, Miranda
2009 *A Bird that Flies with Two Wings: The Kastom and State Justice Systems in Vanuatu.* ANU Press, Canberra.

Foster, Adriance S., and Ernest M. Gifford, Jr.
1974 *Comparative Morphology of Vascular Plants.* San Francisco, Freeman.

Foster, George M.
1945 Sierra Popoluca Folklore and Beliefs. *American Archaeology and Ethnology* 42(2): 177–250.

Fragnière, Yann, Sébastien Bétrisey, Léonard Cardinaux, Markus Stoffel, and Gregor Kozlowski
2015 Fighting their Last Stand? A Global Analysis of the Distribution and Conservation Status of Gymnosperms. *Journal of Biogeography* 42(5): 809–820.

Frankham, Richard, and Jonathan Briscoe
2004 *A Primer of Conservation Genetics.* Cambridge University Press, Cambridge.

Frederiksen, Norman O.
1978 Preservation of Cycad and Ginkgo Pollen. *Review of Palaeobotany and Palynology* 25(2): 163–179.

Freeman, Jacob, Matthew Peeples, and John M. Anderies
2015 Toward a Theory of Non-linear Transitions from Foraging to Farming. *Journal of Anthropological Archaeology* 40: 109–122.

Fuller, Dorian, Cristina Castillo, and Charlene Murphy
2016 How Rice Failed to Unify Asia: Globalization and Regionalism of Early Farming Traditions in the Monsoon World. In *The Routledge Handbook of Archaeology and Globalization,* edited by Tamar Hodos, pp. 711–729. Routledge, New York.

Gamil, Alma P.
2014 Palm Sunday 2014 Images in Bulusan. Electronic document, https://bulusanruralvagabond.wordpress.com/2014/04/14/palm-sunday-2014-images-in-bulusan/, accessed July 9, 2021.

Ganal, Martin W., Thomas Altmann, and Marion S. Röder
2009 SNP Identification in Crop Plants. *Current Opinion in Plant Biology* 12(2): 211–217.

Gao, Zhifeng, and Barry A. Thomas
1989 A Review of Fossil Cycad Megasporophylls, with New Evidence of *Crossozamia* Pomel and its Associated Leaves from the Lower Permian of Taiyuan, China. *Review of Palaeobotany and Palynology* 60: 205–223.

García de León, Antonio
1968   El dueño del maíz y otros relatos nahuas del sur de Veracruz. *Tlalocan* 5: 349–352.

García-Montes, Mario Adolfo, Cuauhtémoc Alain Rubio-Tobón, Yanin Islas-Barrios, Alejandra Serrato-Díaz, Carmen Julia Figueredo-Urbina, Dulce María Galván-Hernández, and Pablo Octavio-Aguilar
2020   The Influence of Anthropogenic Disturbance on the Genetic Diversity of *Ceratozamia fuscoviridis* (Zamiaceae). *International Journal of Plant Sciences.* doi: 10.1086/707108.

Gates, William
1939[1552]   *The de la Cruz-Badiano Aztec Herbal of 1552.* Maya Society, Baltimore.

Gepts, Paul, and Daniel Debouck
1991   Origin, Domestication, and Evolution of the Common Bean (*Phaseolus vulgaris* L.). In *Common Beans: Research for Crop Improvement,* edited by A. van Schoonhoven and O. Voyest, pp. 7–53. CAB International, Wallingford, UK.

Gibson, Lorna F., Donald Olson, and Anne Olson
1963   Four Pame Texts. *Tlalocan* 4: 125–143.

Girard, Rafael
1952   *El Popol-Vuh, Fuente Historica-El Popol-Vuh como Fundamento de la Historia Maya-Quiche.* Ministerio de Educación, Guatemala.

Goel, Anil K., and Jibankumar S. Khuraijam
2015   Cycads: An Overview. In *Plant Biology and Biotechnology: Volume I: Plant Diversity, Organization, Function, and Improvement,* edited by B. Bahadur, M. Venkat Rajam, L. Sahijram, and K. V. Krishnamurthy, pp. 349–360. Springer, New Delhi.

Gong, Yi-Qing, Qing-Qing Zhan, Khang Sinh Nguyen, Hiep Tien Nguyen, Yue-Hua Wang, and Xun Gong
2015   The Historical Demography and Genetic Variation of the Endangered *Cycas multipinnata* (Cycadaceae) in the Red River Region, Examined by Chloroplast DNA Sequences and Microsatellite Markers. *PLOS ONE* 10(2): e0117719.

Gong, Yi-Qing, and Xun Gong
2016   Pollen-Mediated Gene Flow Promotes Low Nuclear Genetic Differentiation among Populations of *Cycas debaoensis* (Cycadaceae). *Tree Genetics and Genomes* 12(5): 1–15.

González, Dolores, and Andrew P. Vovides
2002   Low Intralineage Divergence in *Ceratozamia* (Zamiaceae) Detected with Nuclear Ribosornal DNA ITS and Chloroplast DNA trnL-F Non-coding Region. *Systematic Botany* 27(4): 654–661.
2012   A Modification to the SCAR (Sequence Characterized Amplified Region) method provides phylogenetic insights within *Ceratozamia* (Zamiaceae). *Revista Mexicana de Biodiversidad* 83: 929–938.

González, Dolores, Andrew P. Vovides, and Cristina Bárcenas
2008   Phylogenetic Relationships of the Neotropical Genus *Dioon* (Cycadales, Zamiaceae) Based on Nuclear and Chloroplast DNA Sequence Data. *Systematic Botany* 33(2): 229–236.

González-Astorga, Jorge, Andrew P. Vovides, Andrea Cruz-Angón, Pablo Octavio-Aguilar, and Carlos Iglesias
2005   Allozyme Variation in the Three Extant Populations of the Narrowly Endemic Cycad *Dioon angustifolium* Miq. (Zamiaceae) from North-Eastern Mexico. *Annals of Botany* 95(6): 999–1007.

González-Astorga, Jorge, Andrew P. Vovides, Miriam M. Ferrer, and Carlos Iglesias
2003   Population Genetics of *Dioon edule* Lindl. (Zamiaceae, Cycadales): Biogeographical and Evolutionary Implications. *Biological Journal of the Linnean Society* 80(3): 457–467.

González-Astorga, Jorge, Andrew P. Vovides, Pablo Octavio-Aguilar, Daniel Aguirre-Fey, Fernando Nicolalde-Morejón, and Carlos Iglesias
2006   Genetic Diversity and Structure of the Cycad *Zamia loddigesii* Miq. (Zamiaceae): Implications for Evolution and Conservation. *Botanical Journal of the Linnean Society* 152(4): 533–544.

González-Astorga, Jorge, Francisco Vergara-Silva, Andrew P. Vovides, Fernando Nicolalde-Morejón, Dánae Cabrera-Toledo, and Miguel Angel Pérez-Farrera
2008   Diversity and Genetic Structure of Three Species of *Dioon* Lindl. (Zamiaceae, Cycadales) from the Pacific Seaboard of Mexico. *Biological Journal of the Linnean Society* 94(4): 765–776.

González Christen, Alvar
1990   Algunas Interacciones entre *Dioon edule* (Zamiaceae) y *Peromyscus mexicanus* (Rodentia: Cricetidae). *La Ciencia y el Hombre* 5: 77–92.

González-Géigel, Lutgarda
2003   Zamiaceae. In *Flora de la República de Cuba*. A.R. Ganter Verlag, Liechtenstein.

González Torres, Yolotl
2007   Etnografía del maíz: variedades, tipos de suelo y rituales en treinta monografías. In *Etnografía de los confines. Andanzas de Anne Chapman,* edited by A. M. Juárez Becerril, Andrés Medina, and Ángela Ochoa, pp. 179–219. Anales de Antropología 41, Mexico City.

Goman, Michelle, and Roger Byrne
1998   A 5000-Year Record of Agriculture and Tropical Forest Clearance in the Tuxtlas, Veracruz, Mexico. *Holocene* 8(1): 83–89.

Goodman, J. T.
1897   The Archaic Maya Inscriptions. Appendix to A. P. Maudslay, *Biologia-Centrali Americana; Archaeology*. R. H. Porter and Dulau and Company, London.

Gott, Beth, Huw Barton, Delwen Samuel, and Robin Torrence
2006   Biology of Starch. In *Ancient Starch Research,* edited by Robin Torrence and Huw Barton, pp. 35–45. Left Coast Press, Walnut Creek, California.

Greguss, Pál
1968   *Xylotomy of the Living Cycads*. Akademica Kiado, Budapest.

Gremillion, Kristen J., Loukas Barton, and Dolores R. Piperno
2014   Particularism and the Retreat from Theory in the Archaeology of Agricultural Origins. *PNAS* 111(17): 6171–6177.

Griffith, M. Patrick, Alan W. Meerow, Michael Calonje, Eliza Gonzalez, Kyoko Nakamura, and Javier Francisco-Ortega
2022 Genetic Patterns of Zamia in Florida Are Consistent with Ancient Human Influence and Recent Near Extirpation. *International Journal of Plant Sciences* 183(3):169–185.

Gross, Briana, and Zhijun Zhao
2014 Archaeological and Genetic Insights into the Origins of Domesticated Rice. *PNAS* 111(17): 6190–6197.

Grove, T. S., A. M. O'Connell, and N. Malajczuk
1980 Effects of Fire on the Growth, Nutrient Content and Rate of Nitrogen Fixation of the Cycad *Macrozamia riedlei*. *Australian Journal of Botany* 28: 271–281.

Grube, Nikolai, and Maria Gaida
2006 *Die Maya: Schrift und Kunst*. SMB DuMont, Berlin.

Gutaker, Rafal M., Simon C. Groen, Emily S. Bellis, Jae Y. Choi, Inês S. Pires, R. Kyle Bocinsky, Emma R. Slayton, Olivia Wilkins, Cristina C. Castillo, Sónia Negrão, M. Margarida Oliveira, Dorian Q. Fuller, Jade A. d'Alpoim Guedes, Jesse R. Lasky, and Michael D. Purugganan
2019 Genomic History and Ecology of the Geographic Spread of Rice. *BioRxiv* 748178.

Gutiérrez, Gerardo, and Lorenzo Ochoa
2009 Los límites culturales de la región Huasteca. In *Memoria del Taller de Arqueología de la Huasteca. Homenaje a Leonor Merino Carrión*, edited by Diana Zaragoza Ocaña, pp. 77–92. Instituto Nacional de Antropología e Historia, Mexico City.

Gutiérrez-Arroyo, Naishla Miroslava, Pablo Octavio-Aguilar, and José Arturo De-Nova
2018 Variabilidad y Estructura Genética de *Ceratozamia zaragozae* Medellín-Leal (Cycadales, Zamiaceae) en San Luis Potosí, México. Undergraduate thesis, Universidad Autónoma del Estado de Hidalgo, Pachuca, Mexico.

Gutiérrez-García, Karina, Edder D. Bustos-Díaz, José Antonio Corona-Gómez, Hilda E. Ramos-Aboites, Nelly Sélem-Mojica, Pablo Cruz-Morales, Miguel Angel Pérez-Farrera, Francisco Barona-Gómez, and Angélica Cibrián-Jaramillo
2019 Cycad Coralloid Roots Contain Bacterial Communities Including Cyanobacteria and *Caulobacter* spp. that Encode Niche-Specific Biosynthetic Gene Clusters. *Genome Biology and Evolution* 11(1): 319–334.

Gutiérrez-Ortega, José Said, Francisco Molina-Freaner, José F. Martínez, Miguel Angel Pérez-Farrera, Andrew P. Vovides, Antonio Hernández-López, Ayumi Tezuka, Atsushi J. Nagano, Yasuyuki Watano, Yuma Takahashi, Masashi Murakami, and Tadashi Kajita
2021 Speciation along a Latitudinal Gradient: The Origin of the Neotropical Cycad Sister Pair *Dioon sonorense–D. vovidesii* (Zamiaceae). *Ecology and Evolution* 11(11): 6962–6976. doi: 10.1002/ece3.7545.

Gutiérrez-Ortega, José Said, María Magdalena Salinas-Rodríguez, Takuro Ito, Miguel Angel Pérez-Farrera, Andrew P. Vovides, José F. Martínez, Francisco Molina-Freaner, Antonio Hernández-López, Lina Kawaguchi, Atsushi J. Nagano, Tadashi Kajita, Yasuyuki Watano, Takashi Tsuchimatsu, Yuma Takahashi, and Masashi Murakami
2020   Niche Conservatism Promotes Speciation in Cycads: The Case of *Dioon merolae* (Zamiaceae) in Mexico. *New Phytologist* doi: 10.1111/nph.16647.

Gutiérrez-Ortega, José Said, María Magdalena Salinas-Rodríguez, José F. Martínez, Francisco Molina-Freaner, Miguel Angel Pérez-Farrera, Andrew P. Vovides, Yu Matsuki, Yoshihisa Suyama, Takeshi A. Ohsawa, Yasuyuki Watano, and Tadashi Kajita
2018   The Phylogeography of the Cycad Genus *Dioon* (Zamiaceae) Clarifies its Cenozoic Expansion and Diversification in the Mexican Transition Zone. *Annals of Botany* 121(3): 535–548.

Gutiérrez-Ortega, José Said, Takashi Yamamoto, Andrew P. Vovides, Miguel Angel Pérez-Farrera, José F. Martínez, Francisco Molina-Freaner, Yasuyuki Watano, and Tadashi Kajita
2018a   Aridification as a Driver of Biodiversity: A Case Study for the Cycad Genus *Dioon* (Zamiaceae). *Annals of Botany* 121(1): 47–60.
2018b   Species Definition of *Dioon sonorense* (Zamiaceae, Cycadales), and Description of *D. vovidesii*, a New Cycad Species from Northwestern Mexico. *Phytotaxa* 369: 107–114.

Hall, John A., Gimme H. Walter, Dana M. Bergstrom, and Peter Machin
2004   Pollination Ecology of the Australian Cycad *Lepidozamia Peroffskyana* (Zamiaceae). *Australian Journal of Botany* 52(3): 333–343.

Halliday, Jake, and John S. Pate
1976   Symbiotic Nitrogen Fixation by Coralloid Roots of the Cycad *Macrozamia riedlei*: Physiological Characteristics and Ecological Significance. *Australian Journal of Plant Physiology* 3: 349–358.

Hajslova, Jana, Tomas Cajka, and Lukas Vaclavik
2011   Challenging Applications Offered by Direct Analysis in Real Time (DART). *Trends in Analytical Chemistry* 30(2): 204–218.

Hamada, Trenton, Irene Terry, Robert Roemer, and Thomas Marler
2015   Potential Drift of Pollen of *Cycas micronesia* on the Island of Guam: A Comparative Study. *HortScience* 50(7): 1106–1117.

Hamilton, Matthew
2009   *Population Genetics*. Wiley-Blackwell, Oxford.

Hamrick, J. L., and M.J.W. Godt
1996   Effects of Life History Traits on Genetic Diversity in Plant Species. *Philosophical Transactions of the Royal Society* B 351: 1291–1298.

Handley, L. L., and John A. Raven
1992   The Use of Natural Abundance of Nitrogen Isotopes in Plant Physiology and Ecology. *Plant, Cell, and Environment* 15: 965–985.

Hanselka, J. Kevin
2011   Prehistoric Plant Procurement, Food Production, and Land Use in Southwestern Tamaulipas, Mexico. PhD dissertation, Department of Anthropology, Washington University in St. Louis. University Microfilms, Ann Arbor.

2017 Revisiting the Archaeobotanical Record of Romero's Cave in the Ocampo Region of Tamaulipas, Mexico. *Journal of Ethnobiology* 37(1): 37–59.

Hardy, Karen, Tony Blakeney, Les Copeland, Jennifer Kirkham, Richard Wrangham, and Matthew J. Collins

2009 Starch Granules, Dental Calculus and New Perspectives on Ancient Diet. *Journal of Archaeological Science* 36(2): 248–255.

Hardy, Karen, Stephen Buckley, Matthew J. Collins, Almudena Estalrrich, Don Brothwell, Les Copeland, Antonio García-Tabernero, Samuel García-Vargas, Marco Rasilla, Carles Lalueza-Fox, Rosa Huguet, Markus Bastir, David Santamaría, Marco Madella, Julie Wilson, Ángel Fernández Cortés, and Antonio Rosas

2012 Neanderthal Medics? Evidence for Food, Cooking, and Medicinal Plants Entrapped in Dental Calculus. *Naturwissenschaften* 99(8): 617–626.

Harris, Tom M.

1961 The Fossil Cycads. *Palaeontology* 4(3): 313–323.

Hart, John P., David L. Asch, C. Margaret Scarry, and Gary W. Crawford

2002 The Age of the Common Bean (*Phaseolus vulgaris* L.) in the Northern Eastern Woodlands of North America. *Antiquity* 76: 377–385.

Hastorf, Christine A. and Sissel Johannessen (eds.)

1994 *Corn and Culture in the Prehistoric New World*. University of Minnesota Publications in Anthropology No. 5. Boulder: Westview Press.

Haynes, Jody L., and Mark A. Bonta

2003 Montgomery Botanical Center Honduras 2003 Cycad Expedition Final Report. Submitted to AFE-COHDEFOR, Tegucigalpa, Honduras.

2007 An Emended Description of *Dioon mejiae* Standl. & LO Williams (Zamiaceae). In *Proceedings of the Seventh International Conference on Cycad Biology (Xalapa, Mexico, 2005)*, edited by Andrew P. Vovides, Dennis W. Stevenson, and Roy Osborne, pp. 418–443. Memoirs of the New York Botanical Garden 97. New York Botanical Garden, New York.

Hayward, Philip, and Sueo Kuwahara

2012 Sotetsu Heritage: Cycads, Sustenance and Cultural Landscapes in the Amami Islands. *Locale: The Australasian-Pacific Journal of Regional Food Studies* 2: 26–46.

Hedges, S. Blair, and Sudhir Kumar (eds.)

2009 *The Timetree of Life*. Oxford University Press, Oxford.

Henry, Amanda G., Holly F. Hudson, and Dolores R. Piperno

2009 Changes in Starch Grain Morphologies from Cooking. *Journal of Archaeological Science* 36(3): 915–922.

Hepp, Guy David

2019 *La Consentida: Settlement, Subsistence, and Social Organization in an Early Formative Mesoamerican Community*. University of Colorado Press, Louisville.

Hernández, Esther

2008 Indigenismos en el vocabulario de la lengua Cakchiquel atribuido a fray Domingo de Vico, Ms. BNF R. 7507. *Revista de Filología Española* 88: 67–88.

Hernández, Francisco
1946[mid-1500s]   *Historia de las plantas de la Nueva España.* Imprenta Universitaria. UNAM, Mexico City.
Hernández Ferrer, Marcela
2004   Idhidh Kwitot. Niño maíz. Los niños en los rituales agrícolas de los Teenek de la Huasteca Potosina. In *Historia y vida ceremonial en las comunidades mesoamericanas: los ritos agrícolas,* edited by Johanna Broda and Catharine Good Eshelman, pp. 215–233. UNAM, INAH, Mexico City.
Hernández-Tapia, Jessica E., Jaime Jiménez-Ramirez, and Andrew P. Vovides
2020   Taxonomic Revision of the Genus *Dioon* (Zamiaceae). *Phytotaxa* 442: 267–290.
Hernández Vaca, Victor
2010   Son huasteco, son de costumbre. Etnolaudería del son a lo humano y a lo divino en Texquitote, San Luis Potosí. *Revista de Literaturas Populares* 10: 238–269.
Hewitt, Godfrey M.
1996   Some Genetic Consequences of Ice Ages, and their Role in Divergence and Speciation. *Biological Journal of the Linnean Society* 58(3): 247–276.
Historia de los Mexicanos por sus Pinturas
1941   In Joaquín García Icazbalceta, ed., *Nueva Colección de documentos para la historia de México,* vol. 3, pp. 209–240. Salvador Chavez Hayhoe, Mexico City.
Hoddle, M. S., Christina D. Stosic, J. R. Faleiro, Hamadttu A. F. El-Shafie, Daniel R. Jeske, and Abdel-Kader Sallam
2015   How Far Can the Red Palm Weevil (Coleoptera: Curculionidae) Fly?: Computerized Flight Mill Studies With Field-Captured Weevils. *Journal of Economic Entomology* 108(6): 2599–2609.
Holderegger, Rolf, and Helene H. Wagner
2006   A Brief Guide to Landscape Genetics. *Landscape Ecology* 21(6): 793–796.
Holst, Irene, J. Enrique Moreno, and Dolores R. Piperno
2007   Identification of Teosinte, Maize, and Tripsacum in Mesoamerica by using Pollen, Starch Grains, and Phytoliths. *PNAS* 104(45): 17608-17613.
Howard, Deborah
2004   *The Architectural History of Venice* (2nd ed.). Yale University Press, New Haven.
Huang, Shong, Hui-Ting Hsieh, Kang Fang, and Yu-Chung Chiang
2004   Patterns of Genetic Variation and Demography of *Cycas taitungensis* in Taiwan. *Botanical Review* 70(1): 86–92.
Hull, Kerry M.
2009   The Grand Ch'orti' Epic: The Story of the Kumix Angel. In *The Maya and Their Sacred Narratives: Text and Context in Maya Mythologies,* edited by Geneviève Le Fort, Raphaël Gardiol, Sebastian Matteo, and Christophe Helmke. Proceedings of the 12th European Maya Conference Geneva, December 7–8, 2007. Verlag Anton Saurwein, Markt Schwaben, Germany.

Hunt, Alice M. W., and Robert J. Speakman
2015   Portable XRF Analysis of Archaeological Sediments and Ceramics. *Journal of Archaeological Science* 53: 626–638.

Hunt, Eva
1977   *The Transformation of the Hummingbird: Cultural Roots of a Zinacantecan Mythical Poem.* Cornell University Press, Ithaca, New York.

Hurley, James D., Linda J. Engle, Jesse T. Davis, Adam M. Welsh, and John E. Landers
2004   A Simple, Bead-Based Approach for Multi-SNP Molecular Haplotyping. *Nucleic Acids Research* 32(22): e186. doi: 10.1093/nar/gnh187.

Hurst, W. Jeffrey
2006   The Determination of Cacao in Samples of Archaeological Interest. In *Chocolate in Mesoamerica: A Cultural History of Cacao,* edited by Cameron L. McNeil, pp. 105–113. University Press of Florida, Gainesville.

Husemann, Martin, Frank E. Zachos, Robert J. Paxton, and Jan Christian Habel
2016   Effective Population Size in Ecology and Evolution. *Heredity* 117(4): 191–192.

Iglesias-Andreu, Lourdes G., Pablo Octavio-Aguilar, Andrew P. Vovides, Alan W. Meerow, Francisco N. De Cáceres-González, and Dulce María Galván-Hernández
2017   Extinction Risk of *Zamia inermis* (Zamiaceae): A Genetic Approach for the Conservation of Its Single Natural Population. *International Journal of Plant Sciences* 178(9): 715–723.

Iltis, Hugh H.
2000   Homeotic Sexual Translocations and the Origins of Maize (*Zea mays,* Poacaea): A New Look at an Old Problem. *Economic Botany* 54(1): 7–42.
2006   Origin of Polystichy in Maize. In *Histories of Maize–Multidisciplinary Approaches to Prehistory, Linguistics, Biogeography, Domestication, and Evolution of Maize,* edited by John J. Staller, Robert H. Tykot, and Bruce F. Benz. Academic, Burlington, Massachusetts.

Iriarte, José
2007   New Perspectives on Plant Domestication and the Development of Agriculture in the New World. In *Rethinking Agriculture: Archaeological and Ethnoarchaeological Perspectives,* edited by Timothy P. Denham, José Iriarte, and Luc Vrydaghs, pp. 167–188. Left Coast Press, Walnut Creek, California.

Islam, M. Anowarul, and Albert T. Adjesiwor
2018   Nitrogen Fixation and Transfer in Agricultural Production Systems. In *Nitrogen in Agriculture Updates,* edited by A. Fahad and S. Fahad, pp. 95–110. InTech Publishers, Rijeka, Croatia.

Jiarui, Chen, and Dennis W. Stevenson
1999   Cycadaceae. *Flora of China* 4: 1–7.

Joralemon, Peter D.
1971   *A Study of Olmec Iconography.* Studies in Pre-Columbian Art and Archaeology 7. Dumbarton Oaks, Washington, DC.

Kantar, Michael, Michael Bruford, and Loren Rieseberg
2019   The Genomics of Domestication Special Issue Editorial. *Evolutionary Applications* 12(1): 3–5.

Kaplan, Lawrence
1981 What Is the Origin of the Common Bean? *Economic Botany* 35(2): 241–254.
Kaplan, Lawrence, and Thomas F. Lynch
1999 *Phaseolus* (Fabaceae) in Archaeology: AMS Radiocarbon Dates and their Significance for Pre-Columbian Agriculture. *Economic Botany* 53(3): 261–272.
Kardol, Paul, Gerlinde B. De Deyn, Etienne Laliberté, Pierre Mariotte, and Christine V. Hawkes
2013 Biotic Plant–Soil Feedbacks across Temporal Scales. *Journal of Ecology* 101(2): 309–315.
Karttunen, Frances
1985 Nahuatl and Maya in Contact with Spanish. *Texas Linguistic Forum* 26: 1–135.
1992 *An Analytical Dictionary of Nahuatl.* University of Oklahoma Press, Norman.
Kaufman, Terrence
1976 Archaeological and Linguistic Correlations in Mayaland and Associated Areas of Mesoamerica. *World Archaeology* 8(1): 101–118.
2001 The History of the Nawa Language Group from the Earliest Times to the Sixteenth Century: Some Initial Results. Electronic document, Mesoamerican Languages Documentation Project, https://www.albany.edu/pdlma/Nawa.pdf, accessed May 23, 2020.
2016a *Soteapan Gulf Sokean Ethnobotanical Terminology.* Project for the Documentation of the Languages of Mesoamerica. Institute for Mesoamerican Studies, Unviersity at Albany, SUNY, digital publication accessed January 31, 2021: https://www.albany.edu/ims/pdlma/2016%20Additions/Kaufman-1995-2016ms-Soteapan%20Gulf%20Sokean%20ethnobotany.pdf.
2016b *SAYula Ethnobotany Plant Names and Their Taxonomy.* Project for the Documentation of the Languages of Mesoamerica. Institute for Mesoamerican Studies, Unviersity at Albany, SUNY, digital publication accessed January 31, 2021: https://www.albany.edu/ims/pdlma/2016%20Additions/Kaufman-2005-2016ms-Sayula%20Mijean%20ethnobotany.pdf.
Kaufman, Terrence, and John S. Justeson
2009 Historical Linguistics and Pre-Columbian Mesoamerica. *Ancient Mesoamerica* 20(2): 221–231.
Kelley, David H.
1954a Valenzuela's Cave (Tmc 248): Report. Peabody Museum of Archaeology, Phillips Academy, Andover, Massachusetts.
1954b Catalog of the Artifacts from Cave 248 of the Survey of Tamaulipas. Catalog on file. Peabody Museum of Archaeology, Phillips Academy, Andover, Massachusetts.
1954c Description of squares dug in Tmc 248, March 1954. Notes on file. Peabody Museum of Archaeology, Phillips Academy, Andover, Massachusetts.
1965 The Birth of the Gods at Palenque. *Estudios de Cultura Maya* 5: 93–134.
Kempton, James Howard, and Frederick Wilson Popenoe
1937 *Teosinte in Guatemala: Report of an Expedition to Guatemala, El Salvador, and Chiapas, Mexico.* Carnegie Institution, Washington, DC.

Kennett, Douglas J., Keith M. Prufer, Brendan J. Culleton, Richard J. George, Mark Robinson, Willa R. Trask, Gina M. Buckley, Emily Moes, Emily J. Kate, Thomas K. Harper, Lexi O'Donnell, Erin E. Ray, Eythan C. Hill, Asia Alsgaard, Christopher Merriman, Clayton Meredith, Heather J. H. Edgar, Jaime J. Awe, and Said M. Gutíerrez
2020 Early Isotopic Evidence for Maize as a Staple Grain in the Americas. *Science Advances* 6(23): eaba3245. doi: 10.1126/sciadv.aba3245

Kennett, Douglas J., Dolores Piperno, John Jones, Hector Neff, Barbara Voorhies, Megan Walsh, and Brendan Culleton
2010 Pre-pottery Farmers on the Pacific Coast of Southern Mexico. *Journal of Archaeological Science* 37: 3401–3411.

Kennett, Douglas J., Heather B. Thakar, Amber M. VanDerwarker, David L. Webster, Brendan J. Culleton, T. K. Harper, Logan Kistler, Timothy E. Scheffler, and Kenneth Hirth
2017 High-Precision Chronology for Central American Maize Diversification from El Gigante Rockshelter, Honduras. *PNAS* 114(34): 9026–9031.

Keppel, Gunnar
2002 Low Genetic Variation in a Pacific Cycad: Conservation Concerns for *Cycas seemannii* (Cycadaceae). *ORYX* 36(1): 41–49.

Kerrin, Elliott S., Robert L. White, and Michael A. Quilliam
2017 Quantitative Determination of the Neurotoxin β-N-methylamino-L-alanine (BMAA) by Capillary Electrophoresis–tandem Mass Spectrometry. *Analytical and Bioanalytical Chemistry* 409: 1481–1491.

Khuraijam, Jibankumar S., and Rita Singh
2012 The Ethnobotany of Cycas in the States of Assam and Meghalaya, India. In *Proceedings of Cycad 2008: The 8th International Congress on Cycad Biology, 13–15 January 2008, Panama City, Panama*, edited by Dennis W. Stevenson, Roy Osborne, and Alberto Sidney Taylor Blake, pp. 151–164. Memoirs of the New York Botanical Garden 106. New York Botanical Garden, New York.

Killion, Thomas
1992 Agriculture and Residential Site Structure among Campesinos in Southern Veracruz, Mexico. PhD dissertation, Department of Anthropology, University of New Mexico.

Kingman, John F. C.
1982 The Coalescent. *Stochastic Processes and Their Applications* 13(3): 235–248.

Kinniburgh, Alan, Janet E. Mertz, and Jeffrey Ross
1978 The Precursor of Mouse β-globin Messenger RNA Contains Two Intervening RNA Sequences. *Cell* 14(3): 681–693.

Kipp, Michael A., Eva E. Stüeken, Michelle M. Gehringer, Kim Sterelny, John K. Scott, Paul I. Forster, Caroline A. E. Strömberg, and Roger Buick
2019 Exploring Cycad Foliage as an Archive of the Isotopic Composition of Atmospheric Nitrogen. *Geobiology* 18(2): 152–166.

Kira, Kesayoshi, and Aki Miyoshi
2000 Utilization of SOTETSU (*Cycas revoluta*) in the Amami Islands. *Kagoshima University Forests Research Bulletin* 22: 31–37.

Kopittke, Peter M., Enzo Lombi, Antony van der Ent, Peng Wang, Jamie Laird, Katie Moore, Daniel Pergament Persson, and Søren Husted
2020 Methods to Visualize Elements in Plants. *Plant Physiology* 1852(3). doi: 10.1104/pp.19.01306.

Krige, Eileen Jensen, and Jacob D. Krige
1943 *The Realm of a Rain-Queen. A Study of the Pattern of Lovedu Society.* Oxford University Press for the International African Institute, London.

Krishnamurthy, Vandana
2014 Ethnobotany, Trade and Population Dynamics of *Cycas Circinalis* L. and *Cycas swamyi* Singh & Radha in the Western Ghats of Southern India. PhD dissertation, University of Hawai'i at Manoa.

Kuwahara, Sueo
2013 Research Issues in the Culture and Society of the Amami Islands. In *The Islands of Kagoshima,* pp. 5–13. Kagoshima University Research Center for the Pacific Islands, Kagoshima.

Kwak, Myounghai, James A. Kami, and Paul Gepts
2009 The Putative Mesoamerican Domestication Center of *Phaseolus vulgaris* Is Located in the Lerma-Santiago Basin of Mexico. *Crop Science* 49: 554–563.

Kyoda, Shogo, and Hiroaki Setoguchi
2010 Phylogeography of *Cycas revoluta* Thunb. (Cycadaceae) on the Ryukyu Islands: Very Low Genetic Diversity and Geographical Structure. *Plant Systematics and Evolution* 288(3–4): 177–189.

Lamb, M. Alice
1923 Leaflets of the Cycadaceae. *Botanical Gazette* 76: 185–202.

Lane, Chad S., Sally P. Horn, Claudia I. Mora, and Kenneth H. Orvis
2009 Late-Holocene Paleoenvironmental Change at Mid-Elevation on the Caribbean Slope of the Cordillera Central, Dominican Republic: A Multi-Site, Multi-Proxy Analysis. *Quaternary Science Reviews* 28: 2239–2260.

Law, Howard W.
1957 Tamakasti: A Gulf Nahuat Text. *Tlalocan* 3: 344–360.

Layard, John
1942 *Stone Men of Malekula.* Chatto & Windus, London.

Lazcano-Lara, Julio C.
2015 *The Reproductive Biology of Zamia (Cycadales: Zamiaceae) in Puerto Rico: Implications for Patterns of Genetic Structure and Species Conservation.* PhD dissertation. Department of Biology, Faculty of Natural Sciences, Universidad de Puerto Rico, Río Piedras.

Lee, Shin Ae, Yiseul Kim, Jeong Myeong Kim, Bora Chu, Jae-Ho Joa, Mee Kyung Sang, Jaekyeong Song, and Hang-Yeon Weon
2019 A Preliminary Examination of Bacterial, Archaeal, and Fungal Communities Inhabiting Different Rhizocompartments of Tomato Plants under Real-World Environments. *Scientific Reports* 9, 9300. doi: 10.1038/s41598-019-45660-8.

Lemaire, Charles, and Ambroise Verschaffelt
1869   Enumeration Synonymique des Genres et des Espèces de la Famille des Cycadées. *L'illustration Horticole* 16: 97–101.

León-Portilla, Miguel
1966   *La Filosofía Náhuatl estudiada en sus fuentes*. UNAM-IIH, Mexico City.

Lesure, Richard G.
2008   The Neolithic Demographic Transition in Mesoamerica? Larger Implications of the Strategy of Relative Chronology. In *The Neolithic Demographic Transition and Its Consequences*, edited by Jean-Pierre Bocquet-Appel and Ofer Bar-Yosef, pp. 107–138. Springer, Dordrecht, The Netherlands.

Lévi-Strauss, Claude
1962   *La Pensée Sauvage*. Librairie Plon, Paris.
1969–1981[1964–1971]   *Mythologiques I–IV*. (Translated by John Weightman and Doreen Weightman.) Harper and Row, New York.

Libby, Willard F.
1952   *Radiocarbon Dating*. University of Chicago Press, Chicago.

Lin, Tsan Piao, Yu Ching Sun, Hann Chung Lo, and Yu Pin Cheng
2000   Low Genetic Diversity of *Cycas taitungensis* (Cycadaceae), an Endemic Species in Taiwan, Revealed by Allozyme Analysis. *Taiwan Journal of Forest Science* 15(1): 13–19.

Lindblad, Peter
2008   Cyanobacteria in Symbiosis with Cycads. In *Prokaryotic Symbionts in Plants*, edited by Katharina Pawlowski, pp. 225–233. Microbiology Monographs Vol. 8. Springer, Berlin.

Linnaeus, Filus C.
1789   Zamia Furfuracea. In *Hortus Kewensis* Vol 3, edited by William Aiton, pp. 477. George Nicol, London.

Linton, Charlotte
2020   "Making It for Our Country": An Ethnography of Mud-Dyeing on Amami Ōshima Island. *Textile* 18(3): 250–277. doi: 10.1080/14759756.2019.1690837.

Liu, Sian, Anding Li, Caihui Chen, Guojun Cai, Limin Zhang, Chunyan Guo, and Meng Xu
2017   De Novo Transcriptome Sequencing in *Passiflora edulis* Sims to Identify Genes and Signaling Pathways Involved in Cold Tolerance. *Forests* 8(11): 435.

Looper, Matthew G.
1993   Observations on the Morphology of Sprouts in Olmec Art. *Texas Notes on Precolumbian Art, Writing, and Culture* 58(1).

López Austin, Alfredo
1992   Homshuk. Análisis temático del relato. *Anales de Antropología* 29: 261–283.

López-Gallego, Cristina, and Pamela O'Neil
2010   Life-History Variation Following Habitat Degradation Associated with Differing Fine-Scale Spatial Genetic Structure in a Rainforest Cycad. *Population Ecology* 52(1): 191–201.

López-Montalvo, Esther, Clodoaldo Roldán, Ernestina Badal, Sonia Murcia-Mascarós, and Valentin Villaverde
2017 Identification of Plant Cells in Black Pigments of Prehistoric Spanish Levantine Rock Art by Means of a Multi-Analytical Approach: A New Method for Social Identity Materialization Using *Chaîne Opératoire*. PLOS ONE 12(2): e0172225. doi: 10.1371/journal.pone.0172225.

Loy, Thomas H., Matthew Spriggs, and Stephen Wickler
1992 Direct Evidence for Human Use of Plants 28,000 Years Ago: Starch Residues on Stone Artifacts from the Northern Solomon Islands. *Antiquity* 66(253): 898–912.

Lucero, Lisa J.
2006 Agricultural Intensification, Water, and Political Power in the Southern Maya Lowlands. In *Agricultural Strategies,* edited by Joyce Marcus and Chip Stanish, pp. 281–305. Cotsen Institute of Archaeology, University of California, Los Angeles.

Luna, Ruth, Bryan K. Epperson, and Ken Oyama
2005 Spatial Genetic Structure of Two Sympatric Neotropical Palms with Contrasting Life Histories. *Heredity* 95(4): 298–305.

MacLeod, Murdo J.
2010 *Spanish Central America: A Socioeconomic History, 1520–1720.* University of Texas Press, Austin.

MacNeish, Richard S.
1954a Ojo de Agua Cave—Tmc 174. Report and catalog of excavated material. Peabody Museum of Archaeology, Phillips Academy, Andover, Massachusetts.
1954b Romero's Cave: Report. Peabody Museum of Archaeology, Phillips Academy, Andover, Massachusetts.
1954c Catalog of Excavated Material from Romero's Cave (Tmc 247) Municipio de Ocampo, Tamaulipas, Mexico. Catalog on file. Peabody Museum of Archaeology, Phillips Academy, Andover, Massachusetts.
1958 Preliminary Archaeological Investigations in the Sierra de Tamaulipas, Mexico. *Transactions of the American Philosophical Society* 48(6): 1–210.
1964 The Food-Gathering and Incipient Agriculture Stage of Prehistoric Middle America. In *Natural Environment and Early Cultures,* edited by Robert C. West, pp. 413–426. Handbook of Middle American Indians, Vol. 1. University of Texas Press, Austin.
1971 Speculation about How and Why Food Production and Village Life Developed in the Tehuacan Valley, Mexico. *Archaeology* 24(4): 307–315.
1992 *The Origins of Agriculture and Settled Life.* University of Oklahoma Press, Norman.

MacNeish, Richard S. (ed.)
1967 *Prehistory of the Tehuacan Valley, Vol. 1: Environment and Subsistence.* University of Texas Press, Austin.

Maffie, James
2014 *Aztec Philosophy: Understanding a World in Motion.* University Press of Colorado, Boulder.

Manel, Stéphanie, Michael Schwartz, Gordon Luikart, and Pierre Taberlet
2003   Landscape Genetics: Combining Landscape Ecology and Population Genetics. *Trends in Ecology and Evolution* 18(4): 189–197.

Manfredi, Marcello, Elisa Robotti, Greg Bearman, Fenella France, Elettra Barberis, Pnina Shor, and Emilio Marengo
2016   Direct Analysis in Real Time Mass Spectrometry for the Nondestructive Investigation of Conservation Treatments of Cultural Heritage. *Journal of Analytical Methods in Chemistry* 8: 1–11.

Mangelsdorf, Paul C.
1974   *Corn: Its Origin, Evolution, and Improvement.* Harvard University Press, Cambridge, Massachusetts.

Mangelsdorf, Paul C., Richard S. MacNeish, and Walton C. Galinat
1964   Domestication of Corn. *Science* 143(3606): 538–545.
1967   Prehistoric Maize, Teosinte, and Tripsacum from Tamaulipas, Mexico. *Botanical Museum Leaflets, Harvard University* 22(2): 33–63.

Mangelsdorf, Paul C., Richard S. MacNeish, and Gordon R. Willey
1964   Origins of Agriculture in Middle America. In *Natural Environment and Early Cultures,* edited by Robert C. West, pp. 427–445. Handbook of Middle American Indians, Vol. 1. University of Texas Press, Austin.

Mapes, Cristina, and Francisco Basurto
2016   Biodiversity and Edible Plants of Mexico. In *Ethnobotany of Mexico,* edited by Rafael Lira, Alejandro Casas, and José Blancas, pp. 83–129. Springer, New York.

Marcus, Joyce
2006   The Roles of Ritual and Technology in Mesoamerican Water Management. In *Agricultural Strategies,* edited by Joyce Marcus and Chip Stanish, pp. 221–254. Cotsen Institute of Archaeology, University of California, Los Angeles.

Mares, Teresa M.
2019   Cultivating Comida: What Maria Exposed to Us. *Journal of Agriculture, Food Systems, and Community Development* 9(1): 1–5.

Marler, Paris N., and Thomas E. Marler
2015   An Assessment of Red List Data for the Cycadales. *Tropical Conservation Science* 8(4): 1114–1125.

Marler, Thomas E., and April N. J. Cascasan
2018   Carbohydrate Depletion during Lethal Infestation of *Aulacaspis yasumatsui* on *Cycas revoluta. International Journal of Plant Sciences* 179(6): 497–504.

Marler, Thomas E., and John H. Lawrence
2012   Demography of *Cycas micronesica* on Guam Following Introduction of the Armoured Scale *Aulacaspis yasumatsui. Journal of Tropical Ecology* 28(3): 233–242.

Marler, Thomas E., Laura R. Snyder, and Christopher A. Shaw
2010   *Cycas micronesica* (Cycadales) Plants Devoid of Endophytic Cyanobacteria Increase in β-methylamino-L-alanine. *Toxicon* 56: 563–568.

Marler, Thomas E., and L. Irene Terry
2011   Arthropod Invasion Disrupts *Cycas micronesica* Seedling Recruitment. *Communicative and Integrative Biology* 4(6): 778–780.

Martin, Simon
2006   Cacao in Ancient Maya Religion: First Fruit from the Maize Tree and Other Tales from the Underworld. In *Chocolate in Mesoamerica: A Cultural History of Cacao,* edited by Cameron L. McNeil, pp. 154–183. University Press of Florida, Gainesville.

Martínez-Domínguez, Lilí, Fernando Nicolalde-Morejón, Francisco Vergara-Silva, and Dennis W. Stevenson
2018   Taxonomic Review of Ceratozamia (Zamiaceae) in the Sierra Madre Oriental, Mexico. *PhytoKeys* 100.

Martínez-Domínguez, Lilí, Fernando Nicolalde-Morejón, Francisco Vergara-Silva, Dennis W. Stevenson, and Enrique del Callejo
2017   Cryptic Diversity, Sympatry, and Other Integrative Taxonomy Scenarios in the Mexican *Ceratozamia miqueliana* Complex (Zamiaceae). *Organisms Diversity and Evolution* 17(4): 727–752.

Matsuoka, Yoshihiro, Yves Vigouroux, Major M. Goodman, Jesus Sánchez G., Edward Buckler, and John Doebley
2002   A Single Domestication for Maize, shown by Multilocus Microsatellite Genotyping. *PNAS* 99: 6080–6084.

Mayers, Marvin K.
1958   Pocomchí Texts with Grammatical Notes. Summer Institute of Linguistics No. 2. University of Oklahoma, Norman.

Mayle, Frances E., and José Iriarte
2014   Integrated Palaeoecology and Archaeology—A Powerful Approach for Understanding Pre-Columbian Amazonia. *Journal of Archaeological Science* 51: 54–64.

McAnany, Patricia A.
1992   Agricultural Tasks and Tools: Patterns of Stone Tool Discard Near Prehistoric Maya Residences Bordering Pulltrouser Swamp, Belize. In *Gardens of Prehistory: The Archaeology of Settlement Agriculture in Greater Mesoamerica,* edited by Thomas W. Killion, pp. 184–213. University of Alabama Press, Tuscaloosa.

McClung de Tapia, Emily
1992   The Origins of Agriculture in Mesoamerica and Central America. In *The Origins of Agriculture: An International Perspective,* edited by C. Wesley Cowan and Patty J. Watson, pp. 143–171. Smithsonian Institution Press, Washington, DC.

McGarry, Dan
2018   A Princely Title. *Vanuatu Daily Post,* April 9. Electronic document, https://dailypost.vu/news/a-princely-title/article_3b08dcb8-6286-51f6-bc85-ac5aee546242.html, accessed August 16, 2020.

Medina-Villarreal, Anwar, and Jorge González-Astorga
2016   Morphometric and Geographical Variation in the *Ceratozamia mexicana*

Brongn. (Zamiaceae) Complex: Evolutionary and Taxonomic Implications. *Biological Journal of the Linnean Society* 119: 213–233.

Medina-Villarreal, Anwar, Jorge González-Astorga, and Alejandro Espinosa de los Monteros

2019   Evolution of *Ceratozamia* Cycads: A Proximate-Ultimate Approach. *Molecular Phylogenetics and Evolution* 139: 106530. doi: 10.1016/j.ympev.2019.106530.

Meeks, John C.

1998   Symbiosis between Nitrogen-Fixing Cyanobacteria and Plants. *BioScience* 45(4): 266–276.

Meerow, Alan W., Javier Francisco-Ortega, Michael Calonje, M. Patrick Griffith, Tomás Ayala-Silva, Dennis W. Stevenson, and Kyoko Nakamura

2012   *Zamia* (Cycadales: Zamiaceae) on Puerto Rico: Asymmetric Genetic Differentiation and the Hypothesis of Multiple Introductions. *American Journal of Botany* 99(11): 1828–1839.

Meerow, Alan W., Dayana E. Salas-Leiva, Michael Calonje, Javier Francisco-Ortega, M. Patrick Griffith, Kyoko Nakamura, Francisco Jiménez-Rodríguez, John Lawrus, and Andreas Oberli

2018   Contrasting Demographic History and Population Structure of *Zamia* (Cycadales: Zamiaceae) on Six Islands of the Greater Antilles Suggests a Model for Population Diversification in the Caribbean Clade of the Genus. *International Journal of Plant Sciences* 179(9): 730–757.

Meerow, Alan W., Dayana E. Salas-Leiva, Javier Francisco-Ortega, M. Patrick Griffith, Michael Calonje, Dennis W. Stevenson, and Kyoko Nakamura

2018   Phylogeography and Conservation Genetics of the Caribbean Zamia Clade: An Integrated Systematic Approach with SSRs and Single Copy Nuclear Genes. *Memoirs of the New York Botanical Garden* 117: 278–296.

Meirmans, Patrick G.

2006   Using the AMOVA Framework to Estimate a Standardized Genetic Differentiation Measure. *Evolution* 60(11): 2399–2402.

Meléndez Guadarrama, Lucero, Felipe Trabanino, and Adriana Caballero Roque

2013   Tres perspectivas en torno al uso comestible de las inflorescencias de las palmas pacay(a) y chapay(a) en Chiapas, México: Enfoques paleoetnobotánico, nutricional y lingüístico. *Estudios de Cultura Maya* 41: 175–199.

Méndez Pérez, José

2012   La importancia comunitaria del sistema milpa Tojol-ab'al: el caso de la colonia 20 de Noviembre, las Margaritas, Chiapas. Unpublished undergraduate thesis, Universidad Intercultural de Chiapas, San Cristobal de las Casas, Mexico.

Mickleburgh, Hayley L., and Jaime R. Pagán-Jiménez

2012   New Insights into the Consumption of Maize and Other Food Plants in the Pre-Columbian Caribbean from Starch Grains Trapped in Human Dental Calculus. *Journal of Archaeological Science* 39(7): 2468–2478.

Miller, Walter S.

1956   *Cuentos Mixe*. Biblioteca de Folklore Indígena, Instituto Nacional Indigenista, Mexico City.

Minor, Emily S., and Dean L. Urban
2007	Graph Theory as a Proxy for Spatially Explicit Population Models in Conservation Planning. *Ecological Applications* 17(6): 1771–1782.

Miquel, Friedrich A. W.
1848	Over eenige nieuwe of zeldzame Cycadeen in den Hortus Botanicus te Amsterdam. *Tidschr. Wis-Natuurk. Wetensch. Eerste Kl. Ned. Inst. Wetensch.* 1: 33–43.

Mondragón Pichardo, Juana, and Heike Vibrans
2005	Ethnobotany of the Balsas Teosinte (*Zea mays* ssp. *parviglumis*). *Maydica* 50: 123–128.

Mora, Raymundo, Laura Yáñez-Espinosa, Joel Flores, and Nadya Nava-Zárate
2013	Strobilus and Seed Production of *Dioon edule* (Zamiaceae) in a Population with Low Seedling Density in San Luis Potosí, Mexico. *Tropical Conservation Science* 6(2): 268–282.

Morell-Hart, Shanti, Rosemary A. Joyce, and John S. Henderson
2014	Multi-proxy Analysis of Plant Use at Formative Period Los Naranjos, Honduras. *Latin American Antiquity* 25(1): 65–81.

Moreno de los Arcos, Roberto
1967	Los Cinco Soles Cosmogónicos. *Estudios de Cultura Náhuatl* 7: 183–210.

Moretti, Aldo, Paolo Caputo, Salvatore Cozzolino, Paolo De Luca, Luciano Gaudio, Gesualdo Siniscalco Gigliano, and Dennis W. Stevenson
1993	A Phylogenetic Analysis of *Dioon* (Zamiaceae). *American Journal of Botany* 80(2): 204–214.

Moretti, Aldo, and Sergio Sabato
1984	Karyotype Evolution by Centromeric Fission in *Zamia* (Cycadales). *Plant Systematics and Evolution* 146(3–4): 215–223.

Moretti, Aldo, Paulo Caputo, and Salvatore Cozzolino
1993	Karyotypes of New World Cycads. In *The Biology, Structure, and Systematics of the Cycadales: Proceedings of Cycad 90, The Second International Conference on Cycad Biology*, edited by Dennis W. Stevenson and Knut J. Norstog, pp. 263–269. Palm and Cycad Societies of Australia Ltd., Townsville, Australia.

Moretti, Aldo, Sergio Sabato, and Gesualdo Siniscalco Gigliano
1983	Taxonomic Significance of Methylazoxymethanol Glycosides in the Cycads. *Phytochemistry* 22(1): 115–117.

Morrone, Juan, Tania Escalante, and Gerardo Rodríguez-Tapia
2017	Mexican Biogeographic Provinces: Map and Shapefiles. *Zootaxa* 4277(2): 277–279.

Motshekga, Mathole Kherofo
2010	*The Mudjadji Dynasty. The Principles of Female Leadership in African Cosmology.* Kara Heritage Institute, Pretoria.

Mújica Vélez, Rubén
2007	*La rebelión de la semilla.* Editores Plaza y Valdés, Mexico City.

Nagalingum, Nathalie S., C. R. Marshall, Tiago B. Quental, H. S. Rai, D. P. Little, and S. Mathews
2011   Recent Synchronous Radiation of a Living Fossil. *Science* 334(6057): 796–799.

Nagoya, Sagenta
1984   *Nantō Zatsuwa: Bakumatsu Amami Minzokushi*. Toyo-Bunko, Tokyo.

Neff, Hector, Deborah M. Pearsall, John G. Jones, Barbara Arroyo, Shawn K. Collins, and David E. Freidel
2006   Early Maya Adaptive Patterns: Mid–Late Holocene Paleoenvironmental Evidence from Pacific Guatemala. *Latin American Antiquity* 17: 287–315.

Nei, Masatoshi, and Ronald K. Chesser
1983   Estimation of Fixation Indices and Gene Diversities. *Annals of Human Genetics* 47: 253–259

Neina, Dora
2019   The Role of Soil pH in Plant Nutrition and Soil Remediation. *Applied and Environmental Soil Science*. doi: 10.1155/2019/5794869.

Nelson, Mark, Chester R. Cooper, David E. Crowley, C. S. Paddy Reid, and Paul J. Szaniszlo
1988   An *Escherichia coli* Bioassay of Individual Siderophores in Soil. *Journal of Plant Nutrition* 11: 915–924.

Newell, Sandra J.
1985   Intrapopulational Variation in Leaflet Morphology of *Zamia pumila* L. in Relation to Microenvironment and Sex. *American Journal of Botany* 72(2): 217–221.

Nicholson, Henry B.
1971   Religion in Pre-Hispanic Central Mexico. In *Handbook of Middle American Indians, Volumes 10 and 11: Archaeology of Northern Mesoamerica*, edited by Gordon F. Ekholm and Ignacio Bernal, pp. 395–446. University of Texas Press, Austin.

Nicolalde-Morejón, Fernando, Jorge González-Astorga, Francisco Vergara-Silva, and Dennis W. Stevenson
2013   Biodiversidad de Zamiaceae en México. *Revista Mexicana de Biodiversidad* 84. doi: 10.7550/rmb.38114.

Nicolalde-Morejón, Fernando, Andrew P. Vovides, and Dennis W. Stevenson
2009   Taxonomic Revision of *Zamia* in Mega-Mexico. *Brittonia* 61(4): 301–335.

Nielsen, Erasmus, and Montgomery Slatkin
2013   *An Introduction to Population Genetics: Theory and Application*. Sinauer, Sunderland, Massachusetts.

Niklas, K. J., and Knut J. Norstog
1984   Aerodynamics and Pollen Grain Depositional Patterns on Cycad Megastrobili: Implications on the Reproduction of Three Cycad Genera (*Cycas, Dioon,* and *Zamia*). *Botanical Gazette* 145(1): 92–104.

Nishida, Harufumi, Kathleen B. Pigg, Kensuke Kudo, and John F. Rigby
2004   Zooidogamy in the Late Permian genus *Glossopteris*. *Journal of Plant Research* 117(4): 323–328.

Norstog, Knut J.
1987   Cycads and the Origin of Insect Pollination. *American Scientist* 75: 270–279.
Norstog, Knut J., and Priscilla K. S. Fawcett
1989   Insect Cycad Symbiosis and its Relation to the Pollination of *Zamia furfuracea* (Zamiaceae) by *Rhopalotria mollis* (Curculionidae). *American Journal of Botany* 76(9): 1380–1394.
1993   Hurricane Andrew and Fairchild Tropical Garden. *Encephalartos* 33: 26–28.
Norstog, Knut J., and Trevor J. Nicholls
1997   *The Biology of the Cycads*. Cornell University Press, Ithaca, New York.
Norstog, Knut J., and Dennis W. Stevenson
1980   Wind? Or Insects? The Pollination of Cycads. *Fairchild Tropical Garden Bulletin* 35: 28–30.
O'Brien, Charles W., and William Tang
2015   Revision of the New World Cycad Weevils of the Subtribe Allocorynina, with Description of Two New Genera and Three New Subgenera (Coleoptera: Belidae: Oxycoryninae). *Zootaxa* 3970(1): 1–87.
O'Donoghue, Alan C., Richard P. Evershed, and Terrence A. Brown
1996   Remarkable Preservation of Biomolecules in Ancient Radish Seeds. *Proceedings of the Royal Society of London B* 263: 541–547.
Ochoa, Lorenzo
2010   Topophilia: A Tool for the Demarcation of Cultural Microregions: The Case of the Huaxteca. In *Pre-Columbian Foodways: Interdisciplinary Approaches to Food, Culture, and Markets in Ancient Mesoamerica*, edited by John E. Staller and Michael D. Carrasco, pp. 535–552. Springer, New York.
Octavio-Aguilar, Pablo, Jorge Arturo González-Astorga, and Andrew P. Vovides
2008   Population Dynamics of the Mexican Cycad *Dioon edule* Lindl. (Zamiaceae): Life History Stages and Management Impact. *Botanical Journal of the Linnean Society* 157: 381–391.
2009   Genetic Diversity through Life History of *Dioon edule* Lindley (Zamiaceae, Cycadales). *Plant Biology* 11(4): 525–536.
Octavio-Aguilar, Pablo, Lourdes Georgina Iglesias-Andreu, Francisco Federico Núñez de Cáceres-González, and Dulce María Galván-Hernández
2017   Fine-Scale Genetic Structure of *Zamia furfuracea*: Variation with Life-Cycle Stages. *International Journal of Plant Sciences* 178(1): 57–66.
Oliver, José R.
2009   *Caciques and Cemí Idols. The Web Spun by Taíno Rulers Between Hispaniola and Puerto Rico*. University of Alabama Press, Tuscaloosa.
Olko, Justyna, and John Sullivan
2013   Empire, Colony, and Globalization. A Brief History of the Nahuatl Language. *Colloquia Humanistica* 2: 181–216.
Ordoñez Cabezas, Giomar
2004   *Pames. Pueblos indígenas del México contemporáneo*. Comisión Nacional para el Desarrollo de los Pueblos Indígenas, Programa de las Naciones Unidas para el Desarrollo, Mexico City.

Osborne, Roy
1989　　　Focus on: *Encephalartos transvenosus. Encephalartos* 20: 10–18.
Osborne, Roy, Andrew Grove, Peter Oh, Tom J. Mabry, Jack C. Ng, and Alan A. Seawright
1994　　　The Magical and Medicinal Usage of *Stangeria eriopus* in South Africa. *Journal of Ethnopharmacology* 43: 67–72.
Osborne, Roy, Ken D. Hill, Hiep T. Nguyen, and Loc Phan Ke
2007　　　*Cycads of Vietnam*. Brisbane.
Osborne, Roy, Andrew P. Vovides, and Dennis W. Stevenson
2006　　　What is *Ceratozamia fuscoviridis*? *Delpinoa* 48: 5–10.
Ostapkowicz, Joanna, Christopher Bronk, Fiona Brock, Tom Higham, Alex C. Wiedenhoeft, Erika Ribechini, Jeannette J. Lucejko, and Samuel Wilson
2012　　　Chronologies in Wood and Resin: AMS 14C Dating of Pre-Hispanic Caribbean Wood Sculpture. *Journal of Archaeological Science* 39(7): 2238–2251.
Pagán-Jiménez, Jaime R.
2005　　　Estudio Interpretativo de la Cultura Botánica de dos Comunidades Precolombinas Antillanas. La Hueca y Punta Candelero, Puerto Rico. Doctoral dissertation. Institute for Anthropological Research/Faculty of Philosophy and Literature, Universidad Nacional Autónoma de México, Mexico City.
2007　　　*De Antiguos Pueblos y Culturas Botánicas en el Puerto Rico Indígena: El Archipiélago Borincano y la Llegada de los Primeros Pobladores Agroceramistas*. Paris Monographs in American Archaeology 18, BAR International Series 1687, Archaeopress, Oxford.
2008　　　Envisioning Ancient Human Plant Use at the Río Tanamá Site 2 (AR-39) through Starch Analysis of Lithic and Clay Griddle Implements. In *A Multidisciplinary Approach to Site Testing and Data Recovery at Two Village Sites (AR-38 and AR-39) on the Lower Río Tanamá, Puerto Rico*, edited by Lisabeth A. Carlson, pp. 241–257. Southeastern Archaeological Research, Florida.
2009　　　Uso de Plantas en una Comunidad Saladoide Tardío del Este de Puerto Rico (Punta Candelero): Estudio de Residuos Vegetales (Almidones) en Artefactos Líticos, Cerámicos y de Concha. In *Mitigación Arqueológica de Punta Candelero, Palmas del Mar, Humacao, Puerto Rico*, edited by Marlene Ramos. Manuscript on file at the Council for the Protection of Terrestrial Archaeological Heritage of Puerto Rico, San Juan.
2011a　　Early Phytocultural Processes in the Precolonial Antilles: A Pan-Caribbean Survey for an Ongoing Starch Grain Research. In *Communities in Contact. Essays in Archaeology, Ethnohistory and Ethnography of the Amerindian Circum-Caribbean*, edited by Corinne L Hofman and Anne Van Duijvenbode, pp. 87–116. Sidestone Press, Leiden.
2011b　　Dinámicas Fitoculturales de un Pueblo Precolombino Saladoide Tardio (King's Helmet) en Yabucoa, Puerto Rico. *El Caribe Arqueológico* 12: 45–59.
2011c　　Assessing Ethnobotanical Dynamics at CE-11 and CE-33 through Analysis of Starch Grains, Plant Processing, and Cooking Artifacts. In *Phase III Data Recovery Investigations at Three Prehistoric Sites, Municipality of Ceiba, Naval*

	Activity Puerto Rico Volume 1: Final Report, edited by Lisabeth Carlson, pp. 325–367. Southeastern Archaeological Research, Florida.
2012	Atisbo a la Cultura Botánica de un Asentamiento Precolombino Agrocerámico en el Río San Juan, Sector Playa Grande, República Dominicana, por medio de Almidones Antiguos Recuperados en Utensilios Cerámicos (Ollas, Recipientes) y de Coral. In *El Sitio Arqueológico de Playa Grande, Río San Juan, María Trinidad Sánchez. Informe de las Excavaciones Arqueológicas Campaña 2011–2012*, edited by Adolfo López Belando. Manuscript on file at the Museo del Hombre Dominicano, Ministry of Culture, Santo Domingo.
2013	Human-Plant Dynamics in the Precolonial Antilles: A Synthetic Update. In *The Oxford Handbook of Caribbean Archaeology*, edited by William F. Keegan, Corinne L. Hofman, and Reniel Rodríguez Ramos, pp. 391–406. Oxford University Press, New York.
2015	*Almidones. Guía de Material Comparativo Moderno del Ecuador para los Estudios Paleoetnobotánicos en el Neotrópico.* Aspha Ediciones, Buenos Aires.

Pagán-Jiménez, Jaime R., Ana M. Guachamín-Tello, Martha E. Romero-Bastidas, and Angelo R. Constantine-Castro

2016	Late Ninth Millennium BP Use of *Zea mays* L. at Cubilán Area, Highland Ecuador, Revealed by Ancient Starches. *Quaternary International* 404: 137–155.

Pagán-Jiménez, Jaime R., Ana M. Guachamín-Tello, Martha E. Romero-Bastidas, and Pablo X. Vásquez-Ponce

2017	Cocción Experimental de Tortillas de Casabe (*Manihot esculenta* Crantz) y de Camote (*Ipomoea batatas* [L.] Lam.) en Planchas de Barro: Evaluando sus Efectos en la Morfometría de los Almidones desde una Perspectiva Paleoetnobotánica. *Americae: European Journal of Americanist Archaeology* 2: 26–49.

Pagán-Jiménez, Jaime R., and Julio C. Lazcano-Lara

2013	Toponymic Data Helps to Reveal the Occurrence of Previously Unknown Populations of Wild *Zamia pumila* L. on Volcanic Substrates in South Central Puerto Rico. *Ethnobiology Letters* 4: 52–58.

Pagán-Jiménez, Jaime R., and Hayley L. Mickleburgh

2015	Starchy Plant Food Consumption in the Precolonial Caribbean: New Evidence from Ancient Human Dental Calculus. Paper presented at the 26th Congress of the International Association for Caribbean Archaeology, Saint Martin.

Pagán-Jiménez, Jaime R., and José R. Oliver

2008	Starch Residues on Lithic Artifacts from Two Contrasting Contexts in North Central Puerto Rico: Los Muertos Cave and Vega Nelo Vargas Farmstead. In *Crossing the Borders: New Methods and Techniques in the Study of Archaeological Materials from the Caribbean*, edited by Corinne L. Hofman, Menno Hoogland, and Annelou L. van Gijn, pp. 137–158. University of Alabama Press, Tuscaloosa.

Pagán-Jiménez, Jaime R., Miguel A. Rodríguez-López, Luis A. Chanlatte-Baik, and Yvonne Narganes-Storde

2005	La Temprana Introducción y Uso de Algunas Plantas Domésticas, Silvestres

y Cultivos en las Antillas Precolombinas: Una Primera Revaloración desde la Perspectiva del "Arcaico" de Vieques y Puerto Rico. *Diálogo Antropológico* 3(10): 7–33.

Pagán-Jiménez, Jaime R., Reniel Rodríguez-Ramos, and Corinne L. Hofman
2019   On the Way to the Islands: The Role of Domestic Plants in the Initial Peopling of the Antilles. In *Early Settlers of the Insular Caribbean. Dearchaizing the Archaic,* edited by Corinne L. Hofman and Andrzej Antczak, pp. 89–106. Sidestone Press, Leiden.

Pagán-Jiménez, Jaime R., Reniel Rodríguez-Ramos, Basil A. Reid, Martijn van den Bel, and Corinne L. Hofman
2015   Early Dispersals of Maize and Other Food Plants into the Southern Caribbean and Northeastern South America. *Quaternary Science Reviews* 123: 231–246.

Pant, Divya Darshan
1973   *Cycas and the Cycadales.* Central Book Depot, Lucknow, India.

Paredes-Flores, Martin, Rafael Lira, and Patricia D. Davila Aranda
2007   Ethnobotanical Study of Zapotitlán Salinas, Puebla. *Acta Botánica Mexicana* 79: 13–61.

Pastor, Agustin, Gianni Gallello, M. Luisa Cervera, and Miguel de la Guardia
2016   Mineral Soil Composition Interfacing Archaeology and Chemistry. *TrAC Trends in Analytical Chemistry* 78: 48–59.

Patiño, Víctor Manuel
1989   Notas Preliminares sobre el Uso de las Zamiaceas por los Pueblos Primitivos y Aculturados del Intertrópico Americano. *Perez-Arbelaezia* 2(8): 429–442.

Pauling, Linus, and Emile Zuckerkandl
1963   Chemical Paleogenetics: Molecular Restoration Studies of Extinct Forms of Life. *Acta Chemica Scandinavica* 17(S1): 9–16. doi: 10.3891/acta.chem.scand.17s-0009.

Pearsall, Deborah M.
2002   Maize Is *Still* Ancient in Prehistoric Ecuador: The View from Real Alto, with Comments on Staller and Thompson. *Journal of Archaeological Science* 29: 51–55.

Pearsall, Deborah M., Karol Chandler-Ezell, K., and Alex Chandler-Ezell
2003   Identifying Maize in Neotropical Sediments and Soils Using Cob Phytoliths. *Journal of Archaeological Science* 30: 611–627.

Pearsall, Deborah M., Karol Chandler-Ezell, and James A. Zeidler
2004   Maize in Ancient Ecuador: Results of Residue Analysis of Stone Tools from the Real Alto Site. *Journal of Archaeological Science* 31: 423–442.

Pearson, Richard J.
1969   *Archaeology of the Ryukyu Islands: A Regional Chronology from 3000 B.C. to the Historic Period.* University of Hawaii Press, Honolulu.

Pérez-Farrera, Miguel Angel, and Andrew P. Vovides
2006   The Ceremonial Use of the Threatened 'Espadaña' Cycad (*Dioon merolae, Zamiaceae*) by a Community of the Central Depression of Chiapas. *Boletín de la Sociedad Botánica de México* 78: 107–113.

Pérez-Farrera, Miguel Angel, Andrew P. Vovides, and Sergio Avendaño
2014   Morphology and Leaflet Anatomy of the *Ceratozamia norstogii* (Zamiaceae, Cycadales) Species Complex in Mexico with Comments on Relationships and Speciation. *International Journal of Plant Sciences* 175: 110–121.

Pérez-Farrera, Miguel Angel, Andrew P. Vovides, Dolores González, Sergio López, Luis Hernández-Sandoval, and Mahinda Martínez
2017   Estimation of Genetic Variation in Closely Related Cycad Species in *Ceratozamia* (Zamiaceae: Cycadales) using RAPDs Markers. *Revista de Biología Tropical* 65(1): 305–319.

Pérez-Farrera, Miguel Angel, Andrew P. Vovides, Rubén Martínez-Camilo, Nayely Martínez-Meléndez, Héctor Gómez-Domínguez, and Sonia Galicia-Castellanos
2012   *Zamia grijalvensis* sp. nov. (Zamiaceae, Cycadales) from Chiapas, Mexico with Notes on Hybridization and Karyology. *Nordic Journal of Botany* 30(5): 565–570.

Pérez-Farrera, Miguel Angel, Andrew P. Vovides, Ruben Martínez-Camilo, Nayely Martínez-Meléndez, and Carlos Iglesias
2009   A Reassessment of the *Ceratozamia miqueliana* Species Complex (Zamiaceae) of Southeastern Mexico, with Comments on Species Relationships. *Systematics and Biodiversity* 7(4): 433–443.

Pérez-Farrera, Miguel Angel, Andrew P. Vovides, Pablo Octavio-Aguilar, Jorge González-Astorga, Jesús de la Cruz-Rodriguez, Rigoberto Hernández-Jonopá, and Susana Maza Villalobos-Méndez
2006   Demography of the Cycad *Ceratozamia mirandae* (Zamiaceae) under Disturbed and Undisturbed Conditions in a Biosphere Reserve of Mexico. *Plant Ecology* 187: 97–108.

Pérez-Farrera, Miguel Angel, Andrew P. Vovides, Christian Ruiz-Castillejos, Sonia Galicia, Angélica Cibrián-Jaramillo, and Sergio López
2016   Anatomy and Morphology Suggest a Hybrid Origin of *Zamia katzeriana* (Zamiaceae). *Phytotaxa* 270(3): 161.

Perry, Linda
2001   Prehispanic Subsistence in the Middle Orinoco Basin: Starch Analyses Yield New Evidence. PhD dissertation, Department of Anthropology, Southern Illinois University, Carbondale.
2005   Reassessing the Traditional Interpretation of "Manioc" Artifacts in the Orinoco Valley of Venezuela. *Latin American Antiquity* 16(4): 409–426.

Perry, Linda, Ruth Dickau, Sonia Zarrillo, Irene Holst, Deborah M. Pearsall, Dolores R. Piperno, Mary J. Berman, Richard G. Cooke, Kurt Rademaker, Anthony J. Ranere, J. Scott Raymond, Daniel H. Sandweiss, Franz Scaramelli, Kay Tarble, and James A. Zeidler
2007   Starch Fossils and the Domestication and Dispersal of Chili Peppers (*Capsicum* spp. L.) in the Americas. *Science* 315: 986–988.

Pfister, Barbara, and Samuel C. Zeeman
2016   Formation of Starch in Plant Cells. *Cellular and Molecular Life Sciences* 73(14): 2781–2807.

Phillips, Henry
1884   Notes upon the Codex Ramirez, with a Translation of the Same. *Proceedings of the American Philosophical Society* 21(116): 616–651.

Pinares, Ania, Jorge González-Astorga, Andrew P. Vovides, Julio Lazcano, and Wagner A. Vendrame
2009   Genetic Diversity of the Endangered Endemic *Microcycas calocoma* (Miq.) A. DC (Zamiaceae, Cycadales): Implications for Conservation. *Biochemical Systematics and Ecology* 37(4): 385–394.

Piperno, Dolores R.
2003   A Few Kernels Short of a Cob: On the Staller and Thompson Late Entry Scenario for the Introduction of Maize into Northern South America. *Journal of Archaeological Science* 30: 831–836.

2006a   Identifying Manioc (*Manihot esculenta* Crantz) and Other Crops in Pre-Columbian Tropical America through Starch Grain Analysis: A Case Study from Central Panama. In *Documenting Domestication: New Genetic and Archaeological Paradigms,* edited by Melinda A. Zeder, Daniel G. Bradley, Eve Emshwiller, and Bruce D. Smith, pp. 46–67. University of California Press, Berkeley.

2006b   *Phytoliths: A Comprehensive Guide for Archaeologists and Paleoecologists.* AltaMira Press, Lanham, Maryland.

2009   Identifying Crop Plants with Phytoliths (and Starch Grains) in Central and South America: A Review and an Update of the Evidence. *Quaternary International* 193: 146–159.

2011   The Origins of Plant Cultivation and Domestication in the New World Tropics: Patterns, Process, and New Developments. *Current Anthropology* 52: 453–470.

Piperno, Dolores R., and Thomas D. Dillehay
2008   Starch Grains on Human Teeth Reveal Early Broad Crop Diet in Northern Peru. *PNAS* 105(50): 19622-19627.

Piperno, Dolores R., and Kent V. Flannery
2001   The Earliest Archaeological Maize (*Zea mays* L.) from Highland Mexico: New Accelerator Mass Spectrometry Dates and Their Implications. *PNAS* 98(4): 2101–2103.

Piperno, Dolores R., and Irene Holst
1998   The Presence of Starch Grains on Prehistoric Stone Tools from the Humid Neotropics: Indications of Early Tuber Use and Agriculture in Panama. *Journal of Archaeological Science* 25: 765–776.

Piperno, Dolores R., Irene Holst, Linda Wessel-Beaver, and Thomas C. Andres
2002   Evidence for the Control of Phytolith Formation in *Cucurbita* Fruits by the Hard Rind (*Hr*) Genetic Locus: Archaeological and Ecological Implications. *PNAS* 99(16): 10923–10928.

Piperno, Dolores R., J. E. Moreno, Jose Iriarte, Irene Holst, M. Lachniet, John G. Jones, Anthony J. Ranere, and R. Castanzo
2007   Late Pleistocene and Holocene Environmental History of the Iguala Valley, Central Balsas Watershed of Mexico. *PNAS* 104(29): 11, 874–881.

Piperno, Dolores R., and Deborah M. Pearsall
1998   *The Origins of Agriculture in the Lowland Neotropics.* Academic, New York.
Piperno, Dolores R., Anthony J. Ranere, Irene Holst, Jose Iriarte, and Ruth Dickau
2009   Starch Grain and Phytolith Evidence for Early Ninth Millennium B.P. Maize from the Central Balsas River Valley, Mexico. *PNAS* 106(13): 5019–5024.
Piperno, Dolores R., Anthony J. Ranere, Irene Holst, and Patricia Hansell
2000   Starch Grains Reveal Early Root Crop Horticulture in the Panamanian Tropical Forest. *Nature* 407: 894–897.
Piperno, Dolores R., and Bruce D. Smith
2012   The Origins of Food Production in Mesoamerica. In *The Oxford Handbook of Mesoamerican Archaeology,* edited by Deborah L. Nichols and Christopher A. Pool, pp. 152–164. Oxford University Press, New York.
Piperno, Dolores R., and Karen E. Stothert
2003   Phytolith Evidence for Early Holocene *Cucurbita* Domestication in Southwest Ecuador. *Science* 299: 1054–1057.
Plukenet, Leonard
1691   In *Phytographia [. . .] pars altera,* pp. Tab. XCIX. Sumptibus Autoris, London.
Pohl, Mary E. D., Dolores R. Piperno, Kevin O. Pope, and John G. Jones
2007   Microfossil Evidence for Pre-Columbian Maize Dispersals in the Neotropics from San Andrés, Tabasco, Mexico. *PNAS* 104(16): 6870–6875.
Pohl, Mary E. D., Kevin O. Pope, John G. Jones, John S. Jacob, Dolores R. Piperno, Susan D. deFrance, David L. Lentz, John A. Gifford, Marie E. Danforth, and K. Kathryn Josserand
1996   Early Agriculture in the Maya Lowlands. *Latin American Antiquity* 7: 355–372.
Pool, Christopher A.
1997   The Spatial Structure of Formative Houselots at Bezuapan. In *Olmec to Aztec: Settlement Patterns in the Ancient Gulf Lowlands,* edited by Barbara L. Stark and Philip J. Arnold III, pp. 40–67. University of Arizona Press, Tucson.
Pope, Kevin O., Mary E. D. Pohl, John G. Jones, David L. Lentz, Christopher von Nagy, Francsico J. Vega, and Irvy R. Quitmyer
2001   Origin and Environmental Setting of Ancient Agriculture in the Lowlands of Mesamerica. *Science* 292: 1370–1373.
Prado, Alberto, Fret Cervantes-Díaz, Francisco G. Perez-Zavala, Jorge González-Astorga, Jacqueline C. Bede, and Angélica Cibrián-Jaramillo
2016   Transcriptome-Derived Microsatellite Markers for *Dioon* (Zamiaceae) Cycad Species. *Applications in Plant Sciences* 4(2). doi: 10.3732/apps.1500087.
Pritchard, Jonathan K., Matthew Stephens, Noah A. Rosenberg, and Peter Donnelly
2000   Association Mapping in Structured Populations. *American Journal of Human Genetics* 67(1): 170–181.
Pulido-Silva, María Teresa, Maricela Vargas-Zenteno, Aurelia Vite, and Andrew P. Vovides
2015   Range Extension of the Endangered Mexican Cycad *Ceratozamia fuscoviridis*

Moore (teosintle): Implications for Conservation. *Tropical Conservation Science* 8(3): 778–795.

Radha, P., and Rita Singh
2008   Ethnobotany and Conservation Status of Indian Cycas Species. *Encephalartos* 93: 15–21.

Rai, Amar Nath, E. Söderbäck, and B. Bergman
2000   Tansley Review No. 116. Cyanobacterium-Plant Symbioses. *New Phytologist* 147(3): 449–481.

Rai, Hardeep S., Heath E. O'Brien, Patrick A. Reeves, Richard G. Olmstead, and Sean W. Graham
2003   Inference of Higher-Order Relationships in the Cycads from a Large Chloroplast Data Set. *Molecular Phylogenetics and Evolution* 29(2): 350–359.

Ramírez, José, and Gabriel Alcocer
1902   *Sinonimia vulgar y científica de las plantas mexicanas.* Oficina Tipográfica de la Secretaría del Fomento, Mexico City.

Ranere, Anthony J., Dolores R. Piperno, Irene Holst, Ruth Dickau, and Jose Iriarte
2009   The Cultural and Chronological Context of Early Holocene Maize and Squash Domestication in the Central Balsas River Valley, Mexico. *PNAS* 106(13): 5014–5018.

Ravansari, Roozbeh, Susan C. Wilson, and Matthew Tighe
2020   Portable X-Ray Fluorescence for Environmental Assessment of Soils: Not Just a Point and Shoot Method. *Environment International* 134. doi: 10.1016/j.envint.2019.105250.

Rawal, Ashmita, Somsubhra Chakraborty, Bin Li, Katie Lewis, Maria Godoy, Laura Paulette, and David C. Weindorf
2019   Determination of Base Saturation Percentage in Agricultural Soils via Portable X-Ray Fluorescence Spectrometer. *Geoderma* 338: 375–382.

Reddy, Ch. Sudhakar, K. N. Reddy, Chiranjibi Pattanaik, and Vatsavaya S. Raju
2006   Ethnobotanical Observations on Some Endemic Plants of Eastern Ghats, India. *Ethnobotanical Leaflets* 10: 82–91.

Reichert, Edward T.
1913   *The Differentiation and Specificity of Starches in Relation to Genera, Species, etc.* Carnegie Institution of Washington, Washington, DC.

Reilly, F. Kent, III
1991   Olmec Iconographic Influences on the Symbols of Maya Rulership: An Examination of Possible Sources. In *Sixth Palenque Roundtable, 1986,* edited by Merle Green Robertson and Virginia M. Fields, pp. 151–166. University of Oklahoma Press, Norman.

Reinhart, Katrinka
2020   Analysis of Grooved Ceramic Vessels from the Early Chinese Bronze Age Site of Yanshi Shangcheng: Preliminary Evidence of Food Preparation Methods through Starch Analysis. *Journal of Archaeological Science: Reports* 29. doi: 10.1016/j.jasrep.2019.05.017.

Renfrew, Colin, and Katherine V. Boyle (eds.)
2000   *Archaeogenetics: DNA and the Population Prehistory of Europe.* McDonald Institute for Archaeological Research, Cambridge.

Rentería, Miroslava
2007   Breve Revisión de los Marcadores Moleculares. In *Ecología Molecular,* edited by Luis Eguiarte, Valeria Souza, and Xitlali Aguirre, pp. 499–516. SEMARNAT, UNAM, INECOL, CONABIO, Mexico City.

Reynolds, John, and Sean Cardenas
1973   Mound B-V-4: A Late Formative Ceremonial Structure and Middle Class Residence Complex. In *Pennsylvania State University Kaminaljuyu Project-1969, 1970 seasons* (No. 9), edited by Joseph Michels and William Sanders. Pennsylvania State University, Department of Anthropology, University Park.

Rice, Prudence M.
2018   Maya Crocodilians: Intersections of Myth and the Natural World at Early Nixtun-Ch'ich,' Petén, Guatemala. *Journal of Archaeological Method and Theory* 25: 705–738.

Rimski-Korsakov, Helena, Gerardo Rubio, and Raúl Silvio Lavado
2009   Effect of Water Stress in Maize Crop Production and Nitrogen Fertilizer Fate. *Journal of Plant Nutrition* 32(4): 565–578.

Rivera-Fernández, Andrés
2012   Diversidad y Estructura Genética Poblacional de *Ceratozamia mexicana* Brong. Unpublished doctoral thesis, Instituto de Biotecnología y Ecología Aplicada, Universidad Veracruzana. Xalapa, Veracruz, México.

Rodríguez-Ramos, Reniel
2010   *Rethinking Puerto Rican Precolonial History.* University of Alabama Press, Tuscaloosa.

Rodríguez-Ramos, Reniel, and Jaime R. Pagán-Jiménez
2006   Interacciones Multivectoriales en el Circum-Caribe Precolonial. Un Vistazo desde las Antillas. *Caribbean Studies* 34(2): 103–143.

Rodríguez-Ramos, Reniel, Jaime R. Pagán-Jiménez, and Corinne L. Hofman
2013   The Humanization of the Insular Caribbean. In *The Oxford Handbook of Caribbean Archaeology,* edited by William F. Keegan, Corinne L. Hofman, and Reniel Rodríguez-Ramos, pp. 126–140. Oxford University Press, New York.

Rodríguez-Suárez, Roberto, and Jaime R. Pagán-Jiménez
2008   The Burén in Precolonial Cuban Archaeology: New Information Regarding the Use of Plants and Ceramic Griddles during the Late Ceramic Age of Eastern Cuba Gathered through Starch Analysis. In *Crossing the Borders: New Methods and Techniques in the Study of Archaeological Materials from the Caribbean,* edited by Corinne L. Hofman, Menno L. P. Hoogland, and Annelou L. van Gijn, pp. 159–169. University of Alabama Press, Tuscaloosa.

Rosas Guerra, Oswaldo Saúl
2016   La creacion del maíz en las culturas del Golfo de Mexico: Los mitos de Chicomexochitl, Dhipaak y Homshuk. *Revista de Estudios Interculturales* 4: 86–101.

Rosenberg, Noah A., and Magnus Nordborg
2002   Genealogical Trees, Coalescent Theory, and the Analysis of Genetic Polymorphisms. *Nature Reviews Genetics* 3: 380–390.
Rosenswig, Robert M.
2015   A Mosaic of Adaptation: The Archaeological Record for Mesoamerica's Archaic Period. *Journal of Archaeological Research* 23: 115–162.
Rosenswig, Robert M., Deborah M. Pearsall, Marilyn A. Masson, Brendan J. Culleton, and Douglas J. Kennett
2013   Archaic Period Settlement and Subsistence in the Maya Lowlands: New Starch Grain and Lithic Data from Freshwater Creek, Belize. *Journal of Archaeological Science* 41: 308–321.
Rosenswig, Robert M., Amber M. VanDerwarker, Brendan J. Culleton, and Douglas J. Kennett
2015   Is it Agriculture Yet?: Intensified Maize-Use at 1000 cal BC in the Soconusco and Mesoamerica. *Journal of Anthropological Archaeology* 40: 89–108.
Rubio-Tobón, Cuauhtémoc, and Pablo Octavio-Aguilar
2019   Evaluación de la diversidad y estructura genética de *Ceratozamia kuesteriana* Regel., (Zamiaceae), en Tamaulipas, México: propuesta de conservación. Master's thesis, Universidad Autónoma del Estado de Hidalgo, Pachuca, Mexico.
Ruiz-García, Noe, Brenda Yesenia Méndez-Pérez, Mario Valerio Velasco-García, Guillermo Sánchez-de la Vega, and Juana Laura Rivera-Nava
2015   Distribución, ciclo biológico y tabla de vida de *Eumaeus toxea* (Lepidoptera: Lycaenidae) en la provincia fisiográfica Costa de Oaxaca, México. *Revista Mexicana de Biodiversidad* 86(4): 998–1003.
Ruiz-Mallén, Isabel, Christoph Schunko, Esteve Corbera, Matthias Rös, and Victoria Reyes-García
2015   Meanings, Drivers, and Motivations for Community-Based Conservation in Latin America. *Ecology and Society* 20: 33.
Rzedowski, Jerzy
1991   Diversidad y Orígenes de la Flora Fanerogámica de México. *Acta Botánica Mexicana* 14: 3–21.
Sabato, Sergio, and Paolo De Luca
1985   Evolutionary Trends in *Dion* (Zamiaceae). *American Journal of Botany* 72(9): 1353–1363.
Sahagún, Bernardino de
1577   *Historia general de las cosas de Nueva España*. Medicea Florenziana Library, Florence. Electronic document, http://teca.bmlonline.it/TecaRicerca/index.jsp, accessed April 5, 2022.
Salas-Leiva, Dayana E., Alan W. Meerow, Michael Calonje, M. Patrick Griffith, Javier Francisco-Ortega, Kyoko Nakamura, Dennis W. Stevenson, Carl E. Lewis, and Sandra Namoff
2013   Phylogeny of the Cycads Based on Multiple Single-Copy Nuclear Genes: Congruence of Concatenated Parsimony, Likelihood, and Species Tree Inference Methods. *Annals of Botany* 112(7): 1263–1278.

Salas-Morales, Silvia H., Jeffrey Chemnick, and Timothy J. Gregory
2019 A New Cycad Species in the Genus *Dioon* (Zamiaceae) from the Mixteca Region of Oaxaca, Mexico. *Cactus and Succulent Journal* 88(1): 35–42.

Sánchez Jiménez, José
2005 El Relato del maíz. Implicaturas y desencadenantes de sentido. *Alteridades* 15: 99–112.

Sandstrom, Alan R., and Enrique Hugo García Valencia
2005 *Native Peoples of the Gulf Coast of Mexico.* University of Arizona Press, Tucson.

Sandstrom, Alan R., and Pamela E. Sandstrom
1986 *Traditional Papermaking and Paper Cult Figures of Mexico.* University of Oklahoma Press, Norman.

Santi, Carole, Didier Bogusz, and Claudine Franche
2013 Biological Nitrogen Fixation in Non-Legume Plants. *Annals of Botany* 111: 743–767.

Santos Vecino, Gustavo, Carlos Alberto Monsalve Marín, and Luz Victoria Correa Salas
2014 Alteration of Tropical Forest Vegetation from the Pleistocene Holocene Transition and Plant Cultivation from the End of Early Holocene through Middle Holocene in Northwest Colombia. *Quaternary International* 363: 28–42.

Sass, Chodon, Damon P. Little, Dennis W. Stevenson, and Chelsea D. Specht
2007 Barcoding in the Cycadales: Testing the Potential of Proposed Barcoding Markers for Species Identification of Cycads. *PLOS ONE* 2(11): e1154. doi: 10.1371/journal.pone.0001154.

Schellhas, Paul
1897 *Die Göttergestalten der Mayahandschriften: Ein mythologisches Kulturbild aus dem alten Amerika.* Verlag von Richard Bertling, Dresden.
1904 Representation of Deities of the Maya Manuscripts. Papers of the Peabody Museum of American Archaeology and Ethnology 4(1). Harvard University, Cambridge.

Schlumbaum, Angela, Marrie Tensen, and Viviane Jaenicke-Després
2008 Ancient Plant DNA in Archaeobotany. *Vegetation History and Archaeobotany* 17: 233–244.

Schutzman, Bart, Russell S. Adams, Jody L. Haynes, and Loran M. Whitelock
2008 A New Endemic *Zamia* from Honduras (Cycadales: Zamiaceae). *Cycad Newsletter* 31(2/3): 22–26.

Schutzman, Bart, Andrew P. Vovides, and Bijan Dehgan
1988 Two New Species of *Zamia* (Zamiaceae, Cycadales) from Southern Mexico. *Botanical Gazette* 149(3): 347–360.

Schwendemann, Andrew B., Thomas N. Taylor, and Edith L. Taylor
2009 Pollen of the Triassic Cycad *Delemaya spinulosa* and Implications on Cycad Evolution. *Review of Palaeobotany and Palynology* 156(1–2): 98–103.

Secretaría de Medio Ambiente y Recursos Naturales (SEMARNAT)
2000 *Protección, Conservación y Recuperación de la Familia Zamiaceae (Cycadales) de México.* Instituto Nacional de Ecología, Mexico City.

Seler, Eduard
1902–1923   *Gesammelte Abhandlungen zur Amerikanischen Sprach- und Alterthumskunde.* 5 vols. (1908: *Die Ruinen von Chichen Itza in Yucatan,* Vol. 5.) A. Asher and Co., Berlin.
1963   *Comentarios al Códice Borgia.* Fondo de Cultura Económica, Mexico City.
1976   Observations and Studies in the Ruins of Palenque, 1915. Translated by Gisela Morgner, edited by Thomas Bartman and George Kubler. Robert Louis Stevenson School, Pebble Beach, California.
Sheets, Payson D.
1982   Prehistoric Agricultural Systems in El Salvador. In *Maya Subsistence: Studies in Memory of Dennis E. Puleston,* edited by Kent V. Flannery, pp. 99–118. Academic, New York.
Sheets, Payson D., Christine Dixon, Monica Guerra, and Adam Blanford
2011   Manioc Cultivation at Cerén, El Salvador: Occasional Kitchen Garden Plant or Staple Crop? *Ancient Mesoamerica* 22(1): 1–11. doi: 10.1017/S0956536111000034
Sherzer, Joel
1990   *Verbal Art in San Blas: Kuna Culture through Its Discourse.* Cambridge University Press, Cambridge.
Siemens, Alfred H.
1983   Wetland Agriculture in Pre-Hispanic Mesoamerica. *Geographical Review* 73(2): 166–181.
Sifuentes de Ortiz, María S.
1983   *Importancia Económica del Chamal* Dioon edule *Lindl. (Cycadaceae) en el Estado de Nuevo Leon, México.* Unpublished undergraduate thesis, Facultad de Ciencias Biológicas, Universidad Autónoma de Nuevo León, Monterrey.
Sharma, Ish K., David L. Jones, and Paul I. Forster
2004   Genetic Differentiation and Phenetic Relatedness among Seven Species of the *Macrozamia plurinervia* Complex (Zamiaceae). *Biochemical Systematics and Ecology* 32(3): 313–327.
Sharma, Ish K., David L. Jones, Paul I. Forster, and Andrew G. Young
1998   The Extent and Structure of Genetic Variation in the *Macrozamia pauliguilielmi* Complex (Zamiaceae). *Biochemical Systematics and Ecology* 26(1): 45–54.
1999   Low Isozymic Differentiation among Five Species of the *Macrozamia heteromera* Group (Zamiaceae). *Biochemical Systematics and Ecology* 27(1): 67–77.
Simms, Stephanie R.
2014   Prehispanic Maya Foodways: Archaeological and Microbotanical Evidence from Escalera al Cielo, Yucatán, Mexico. PhD dissertation, Department of Archaeology, Boston University.
Slater, Sam S., and Charles H. Wellman
2015   A Quantitative Comparison of Dispersed Spore/Pollen and Plant Megafossil Assemblages from a Middle Jurassic Plant Bed from Yorkshire, UK. *Paleobiology* 41(4): 640–660.

Sluyter, Andrew
1994 Intensive Wetland Agriculture in Mesoamerica: Space, Time, and Form. *Annals of the Association of American Geographers* 84(4): 557–584.

Sluyter, Andrew, and Gabriela Dominguez
2006 Early Maize (*Zea mays* L.) Cultivation in Mexico: Dating Sedimentary Pollen Records and Its Implications. *PNAS* 103(4): 1147–1151.

Smalley, John, and Michael Blake
2003 Sweet Beginnings: Stalk Sugar and the Domestication of Maize. *Current Anthropology* 44(5): 675–703.

Smartt, Joseph
1988 Morphological, Physiological, and Biochemical Changes in *Phaseolus* Beans under Domestication. In *Genetic Resources of Phaseolus Beans,* edited by Paul Gepts, pp. 143–162. Kluwer, Boston.

Smith, Bruce D.
1997 Reconsidering the Ocampo Caves and the Era of Incipient Cultivation in Mesoamerica. *American Antiquity* 8(4): 342–383.
1998 *The Emergence of Agriculture.* Scientific American Library, New York.
2005 Reassessing Coxcatlan Cave and the Early History of Domesticated Plants in Mesoamerica. *PNAS* 102(27): 9438–9445.

Smith, C. Earle
1967 Plant Remains. In *Prehistory of the Tehuacan Valley, Vol. 1: Environment and Subsistence,* edited by Richard S. MacNeish, pp. 220–255. University of Texas Press, Austin.

Smith, Hale G.
1951 The Ethnological and Archeological Significance of Zamia. *American Anthropologist* 53(2): 238–244.

Smith, Moya
1982 Late Pleistocene Zamia Exploitation in Southern Western Australia. *Archaeology in Oceania* 17(3): 117–121.

Smouse, Peter E., and Rod Peakall
1999 Spatial Autocorrelation Analysis of Individual Multiallele and Multilocus Genetic Structure. *Heredity* 82(5): 561–573.

Song, Shuhui, Dongmei Tian, Zhang Zhang, Songnian Hu, and Jun Yu
2018 Rice Genomics: Over the Past Two Decades and into the Future. *Genomics, Proteomics and Bioinformatics* 16(6): 397–404.

Spiekermann, Rafael, André Jasper, Anelise Marta Siegloch, Margot Guerra-Sommer, and Dieter Uhl
2021 Not a Lycopsid but a Cycad-like Plant: *Iratinia australis* gen. nov. et sp. nov. from the Irati Formation, Kungurian of the Paraná Basin, Brazil. *Review of Palaeobotany and Palynology* 289. doi: 10.1016/j.revpalbo.2021.104415.

Spinden, Herbert Joseph
1913 *A Study of Maya Art: Its Subject Matter and Historical Development.* Memoirs of the Peabody Museum of American Archaeology and Ethnology 6. Harvard University, Cambridge. Reprinted in 1975 by Dover Publications, New York.

Stafleu, Frans A.
1966   F.A.W. Miquel, Netherlands Botanist. *Wentia* 16: 1–95.
1970   The Miquel-Schechtendal Correspondence: A Picture of European Botany, 1836–1866. *Regnum Vegetabile* 71: 295–341.

Staller, John
2018   Origin, Evolution, Biogeography, and Cultivation of Maize: The Role of Nixtamal to Nutrition and Economic Importance in Ancient and Contemporary Food and Culture. In *The Origin and Evolution of Food Production and Its Impact on Consumption Patterns,* edited by Nuria Sanz, pp. 179–195. Center for Research and Advanced Studies, Mexico City.

Standley, Paul Carpenter, and Julian Alfred Steyermark
1958   Flora of Guatemala. Cycadaceae. *Fieldiana Botany* 24: 11–20.

Standley, Paul Carpenter, and Louis Otho Williams
1950   *Dioon mejiae,* a New Cycad from Honduras. *Ceiba* 1: 36–38.

Stevenson, Dennis W.
1980   Observations on Root and Stem Contraction in Cycads (Cycadales) with Special Reference to *Zamia pumila. Botanical Journal of the Linnean Society* 81(4): 275–281.
1987   Again the West Indian Zamias. *Fairchild Tropical Garden Bulletin* 42(3): 23–27.
1990   Morphology and Systematics of Cycadales. *Memoirs of the New York Botanical Garden* 57: 8–55.
1991   The Zamiaceae in the Southeastern United States. *Journal of the Arnold Arboretum.* Supplementary Series 1: 367–384.
1992   A Formal Classification of the Extant Cycads. *Brittonia* 44(2): 220–223.

Stevenson, Dennis W., Aldo Moretti, and Luciano Gaudio
1998   A New Species of *Zamia* (Zamiaceae) from Belize and the Yucatan Peninsula of Mexico. *Delpinoa* 37–38: 3–8.

Stevenson, Dennis W., and Sergio Sabato
1986a   Typification of Names in *Ceratozamia* brongn, *Dion* lindl, and *Microcycas* A. DC. (Zamiaceae). *Taxon* 35(3): 578–584.
1986b   Typification of Names in *Zamia* L. and *Aulacophyllum* Regel (Zamiaceae). Note. *Taxon* 35(1): 134–144.

Stevenson, Dennis W., Andrew P. Vovides, and Jeffrey Chemnick
2003   Regional Overview: New World. In *Cycads: Status, Survey, and Conservation Action Plan,* edited by John S. Donaldson, pp. 31–38. IUCN/SSC Cycad Specialist Group. IUCN, Gland, Switzerland.

Stewart, Wilson N.
1983   *Paleobotany and the Evolution of Plants.* Cambridge University Press, New York.

Stoner, Wesley D.
2017   Risk, Agricultural Intensification, Political Administration, and Collapse in the Classic Period Gulf Lowlands: A View from Above. *Journal of Archaeological Science* 80: 83–95.

Storfer, Andrew, Melanie A. Murphy, Stephen F. Spear, Rolk Holderegger, and Lisette P. Waits
2010   Landscape Genetics: Where Are We Now? *Molecular Ecology* 19(17): 3496–3514.

Stothert, Karen E., and Amelia Sánchez Mosquera
2011   Culturas del Pleistoceno Final y el Holoceno Temprano en el Ecuador. *Boletín de Arqueología de la Pontificia Universidad Catolica del Perú* 15: 81–119.

Stross, Brian
1994   Maize and Fish: The Iconography of Power in Late Formative Mesoamerica. *RES: Anthropology and Aesthetics* 25: 10–35.
2006   Maize in Word and Image in Southeastern Mesoamerica. In *Histories of Maize: Multidisciplinary Approaches to the Prehistory, Linguistics, Biogeography, Domestication, and Evolution of Maize,* edited by John E. Staller, Robert H. Tykot, and Bruce F. Benz, pp. 577–598. Academic, New York.

Stuart, David S.
2005   *The Inscriptions from Temple XIX at Palenque: A Commentary.* Precolumbian Art Research Institute, San Francisco.
2017   Gods of Heaven and Earth: Evidence of Ancient Maya Categories of Deities. In *Del saber ha hecho su razón de ser . . . : Homenaje a Alfredo López Austin,* 2 vols, edited by Eduardo Matos Moctezuma and Ángela Ochoa, pp. 247–267. Secretaría de Cultura, INAH, UNAM, Mexico.

Sturtevant, William C.
1969   History and Ethnography of some West Indian Starches. In *The Domestication and Exploitation of Plants and Animals,* edited by Peter J. Ucko and G. W. Dimbleby, pp. 177–199. Gerald Duckworth, London.

Suárez-Moo, Pablo de Jesús, Andrew P. Vovides, M. Patrick Griffith, Francisco Barona-Gómez, and Angélica Cibrián-Jaramillo
2019   Unlocking a High Bacterial Diversity in the Coralloid Root Microbiome from the Cycad Genus *Dioon. PLOS ONE* 14(2): e0211271. doi: 10.1371/journal.pone.0211271.

Summerhayes, Glenn
2018   Coconuts on the Move: Archaeology of Western Pacific. *Journal of Pacific History* 53(4): 375–396.

Sutherland, Cyril Hardy Nelson
1986   *Plantas comunes de Honduras.* Vol. 2. Editorial Universitaria, Tegucigalpa, Honduras.

Swanton, John R.
1913   Coonti. *American Anthropologist* 15: 141–142.

Szpak, Paul
2014   Complexities of Nitrogen Isotope Biogeochemistry in Plant-Soil Systems: Implications for the Study of Ancient Agricultural and Animal Management Practices. *Frontiers in Plant Science,* 5(288): 1–19.

Tadeu Costa Junior, Geovani, Lidiane Cristina Nunes, Marcos Henrique Feresin Gomes, Eduardo de Almeida, and Hudson Wallace Pereira de Carvalho
2019   Direct Determination of Mineral Nutrients in Soybean Leaves under Vivo

Conditions by Portable X-Ray Fluorescence Spectroscopy. *X-Ray Spectrometry* 49(2): 274–283.

Takamiya, Hiroto
2013 南島中部圏先史時代遺跡出土の植物遺体 [Plant Remains Excavated from Prehistoric Sites in the Central South Island]. In ナガラ原東貝塚の研究―5世紀から7世紀前半の沖縄伊江島 ["Nagarahara Higashi-Kaizuka" Research: Okinawa-Iejima from the 5th Century to the First Half of the 7th Century], edited by Naoko Kinoshita, pp. 317–325. Kinoshita Laboratory, Faculty of Letters, Kumamoto University, Kumamoto

Tang, William
1987 Insect Pollination in the Cycad *Zamia pumila* (Zamiaceae). *American Journal of Botany* 74(1): 90–99.

Tang, William, Guang Xu, Charles W. O'Brien, Michael Calonje, Nico M. Franz, M. Andrew Johnston, Alberto Sidney Taylor Blake, Andrew P. Vovides, Miguel Angel Pérez-Farrera, Silvia H. Salas-Morales, Julio C. Lazcano-Lara, Paul Skelley, Cristina Lopez-Gallego, Anders Lindström, and Stephen Rich
2018 Molecular and Morphological Phylogenetic Analyses of New World Cycad Beetles: What They Reveal about Cycad Evolution in the New World. *Diversity* 10(2): 38.

Tate, Carolyn E.
2012 *Reconsidering Olmec Visual Culture: The Unborn, Women, and Creation.* University of Texas Press, Austin.

Taube, Karl A.
1985 Classic Maya Maize God: A Reappraisal. In *Fifth Palenque Round Table,* 1983, Vol. 7, edited by Merle Green Robertson and Virginia M. Fields, pp. 171–181. Pre-Columbian Art Research Institute, San Francisco.
1989 Maize Tamale in Classic Maya Diet; Epigraphy and Art. *American Antiquity* 54(1): 31–51.
1992 *The Major Gods of Ancient Yucatan.* Dumbarton Oaks, Washington, DC.
1996 The Olmec Maize God: The Face of Corn in Formative Mesoamerica. *RES: Anthropology and Aesthetics* 29/30: 39–81
2000 Lightning Celts and Corn Fetishes: The Formative Olmec and the Development of Maize Symbolism in Mesoamerica and the American Southwest. In *Olmec Art and Archaeology in Mesoamerica,* edited by John E. Clark and Mary E. Pye, 297–331. Studies in the History of Art, No. 58. National Gallery of Art and Yale University Press, Washington, DC.
2004 *Olmec Art at Dumbarton Oaks.* Pre-Columbian Art at Dumbarton Oaks, No. 2. Dumbarton Oaks, Washington, DC.

Tawada, Shinjun
1975 沖縄先史、原始時代の主食材料について [On Plant Use during Pre-/Proto-Historic Times in Okinawa]. *Nantō Kōko* 4: 25–28.

Taylor Blake, Alberto Sidney, Jody Haynes, and Gregory Holzman
2008 Taxonomical, Nomenclatural and Biogeographical Revelations in the *Zamia skinneri* Complex of Central America (Cycadales: Zamiaceae). *Bo-*

tanical Journal of the Linnean Society 158: 399–429. doi: 10.1111/j.1095-8339.2008.00886.x.

Taylor Blake, Alberto Sidney, and Gregory Holzman
2012   A New *Zamia* Species from the Panama Canal Area. *Botanical Review* 78(4): 335–344.

Tedlock, Dennis
1985   *Popol Vuh*. Simon & Schuster, New York.

Terrell, John Edward, John P. Hart, Sibel Barut, Nicoletta Cellinese, L. Antonio Curet, Tim Denham, Chapurukha M. Kusimba, Kyle Latinis, Rahul Oka, Joel Palka, Mary E. D. Pohl, Kevin O. Pope, Patrick Ryan Williams, Helen Haines, and John E. Staller
2003   Domesticated Landscapes: The Subsistence Ecology of Plant and Animal Domestication. *Journal of Archaeological Method and Theory* 10(4): 323–368.

Terry, Irene
2001   Thrips and Weevils as Dual, Specialist Pollinators of the Australian Cycad *Macrozamia communis* (Zamiaceae). *International Journal of Plant Sciences* 162(6): 1293–1305.

Terry, Irene, William Tang, and Thomas E. Marler
2012   Pollination Systems of Island Cycads: Predictions Based on Island Biogeography. *Memoirs of the New York Botanical Garden* 106: 102–132.

Terry, Irene, William Tang, Alberto Sidney Taylor Blake, John S. Donaldson, Rita Singh, Andrew P. Vovides, and Angélica Cibrián-Jaramillo
2012   An Overview of Cycad Pollination Studies. *Memoirs of the New York Botanical Garden* 106: 352–394.

Terry, Irene, Gimme H. Walter, Chris Moore, Robert Roemer, and Craig Hull
2007   Odor-Mediated Push-Pull Pollination in Cycads. *Science* 318(5847): 70.

Thieret, John
1958   Economic Botany of the Cycads. *Economic Botany* 12(1): 3–41.

Thiselton-Dyer, William Turner
1884   Cycadaceae. In *Biología Centrali-Americana, Botany,* Vol. 3, edited by William Botting Hemsley, pp. 190–195. Published for the editors by R. H. Porter and Dulau, London.

Tomasco, Ivanna, and Enrique Lessa
2015   Variación Genética. Equilibrio Hardy-Weinberg y Mutación. In *Evolución Orgánica,* edited by Arturo Becerra, América Castañeda, and Daniel Piñero, pp. 5–45. UNAM, Mexico City.

Torrence, Robin, and Huw Barton (eds.)
2006   *Ancient Starch Research*. Left Coast Press, Walnut Creek, California.

Townsend, Richard
1992   *The Aztecs*. Thames and Hudson, New York.

Tōyama, Masanao, and Takako Ankei
2015   *Sotetsu o minaosu—Amami Okinawa no sotetsu bunka-shi tankōbon*. Bōdāinku, Tokyo.

Trembath-Reichert, Elizabeth, Jonathan Paul Wilson, Shawn E. McGlynn, and Woodward W. Fischer
2015   Four Hundred Million Years of Silica Biomineralization in Land Plants. *PNAS* 112(17): 5449–5454.

Tristán, Elvia
2012   Aprovechamiento alimentario de Dioon edule Lindl. (Chamal) en comunidades de la región Xi'iuy del estado de San Luis Potosí. Undergraduate thesis, Universidad Autónoma de San Luis Potosí, Mexico.

Tristán Martínez, Elvia, Javier Fortanelli Martínez, and Mark A. Bonta
2020   Toxic Harvest: Chamal Cycad (*Dioon edule*) Food Culture in Xi'Iuy Indigenous Communities of San Luis Potosi, Mexico. *Journal of Ethnobiology* 40: 519–534.

Tsuji, Seiichiro, Shinobu Omatsu, and Keiko Tsuji
2007   伊礼原遺跡の植物遺体群 [Plant Remains of Ireihara Ruins]. In 伊礼原遺跡 [*Ireihara Ruins*], pp. 433–444. Chatan Town Board of Education, Chatan Town, Okinawa.

Turner, Monica G., and Robert H. Gardner
2001   *Landscape Ecology in Theory and Practice: Pattern and Process*. Springer, New York.

Ulloa Hung, Jorge
2013   *Arqueología en la Línea Noroeste de La Española. Paisaje, cerámicas e interacciones*. Doctoral dissertation. Caribbean Research Group, Faculty of Archaeology, Leiden University, Leiden.

Valdez, Ulises
2009   La flor de espadaña en Terán: ofrenda de los hojeros a la Santa Cruz. Universidad Politécnica de Chiapas, Suchiapa.

van den Bel, Martijn, Sebastiaan Knippenberg, and Jaime R. Pagán-Jiménez
2018   From Cooking Pits to Cooking Pots: Changing Modes of Food Processing during the Late Archaic Age in French Guiana. In *The Archaeology of Caribbean and Circum-Caribbean Farmers (6000 BC–AD 1500)*, edited by Basil A. Reid, pp. 391–418. Routledge, London.

van der Ent, Antony, Guillaume Echevarria, A. Joseph Pollard, and Peter D. Erskine
2019   X-Ray Fluorescence Ionomics of Herbarium Collections. *Scientific Reports* 9(1): 4746. doi: 10.1038/s41598-019-40050-6.

van der Ent, Antony, Wojciech J. Przybyłowicz, Martin D. de Jonge, Hugh H. Harris, Christopher G. Ryan, Grzegorz Tylko, David J. Paterson, Alban D. Barnabas, Peter M. Kopittke, and Jolanta Mesjasz-Przybyłowicz
2018   X-Ray Elemental Mapping Techniques for Elucidating the Ecophysiology of Hyperaccumulator Plants. *New Phytologist* 218: 432–452.

VanDerwarker, Amber M.
2005   Field Cultivation and Tree Management in Tropical Agriculture: A View from Gulf Coastal Mexico. *World Archaeology* 37(2): 274–288.
2006   *Farming, Hunting, and Fishing in the Olmec World*. University of Texas Press, Austin.

VanDerwarker, Amber M., and Robert P. Kruger
2012   Regional Variation in the Importance and Uses of Maize in the Early and Middle Formative Olmec Heartland: New Archaeobotanical Data from the San Carlos Homestead, Southern Veracruz. *Latin American Antiquity* 23(4): 509–532.

van't Hooft, Anuschka
2008   Chikomexochitl y el origen del maíz en la tradición oral Nahua de la Huasteca. *Revista Destiempos* 15: 53–60.

Vázquez Torres, Mario
1990   Algunos datos etnobotánicos de las cicadas en Mexico. *Memoirs of the New York Botanical Garden* 57: 144–147.

Vázquez Torres, Mario, L. Torres Hernández, and Luis H. Bojórquez Galván
2001   Distribución, abundancia, estructura poblacional y potencial reproductor de *Zamia furfuracea* L. Universidad Veracruzana. Instituto de Investigaciones Biológicas. Informe Final SNIB-CONABIO Proyecto No. R176. Mexico City.

Vekemans, Xavier, and Olivier J. Hardy
2004   New Insights from Fine-Scale Spatial Genetic Structure Analyses in Plant Populations. *Molecular Ecology* 13(4): 921–935.

Veloz Maggiolo, Marcio
1973   Aspectos Etnológicos Aborígenes y Actuales del uso de la Guáyiga y sus Derivados en Santo Domingo. *Revista del Instituto de Cultura Puertorriqueña* 59: 33–39.
1992   Notas sobre la Zamia en la Prehistoria del Caribe. *Revista de Arqueología Americana* 6: 125–138.

Velsko, Irina M., Katherine A. Overmyer, Camilla Speller, Matthew Collins, Louise Loe, Laurent A.F. Frantz, Juan Rodriguez Martinez, Eros Chavez, Lauren Klaus, Krithivasan Sankaranarayanan, Cecil M. Lewis, Joshua J. Coon, Greger Larson, and Christina Warinner
2017   The Dental Calculus Metabolome in Modern and Historic Samples. *Metabolomics* 13(11): 134.

Vite-Reyes, Aurelia
2010   Etnobotánica de Cícadas en Hidalgo y Algunos Aspectos Demográficos de *Ceratozamia fuscoviridis* D. Moore. Unpublished undergraduate thesis, Area Académica de Biología, Universidad Autónoma del Estado de Hidalgo, Mexico.
2012   Etnobotánica de cícadas en comunidades nahuas y mestizas de Tlanchinol, Hidalgo. Master's thesis, Universidad Autónoma del Estado de Hidalgo, Mexico.

Vite-Reyes, Aurelia, María Teresa Pulido-Silva, and Juan Carlos Flores
2010   Aspectos Etnobotánicos de las Cícadas en Algunas Zonas de Hidalgo, México. In *Sistemas Biocognitivos Tradicionales: Paradigmas en la Conservación Biológica y el Fortalecimiento Cultural*, edited by Ángel Moreno, María Teresa Pulido, Ramón Mariaca, Raúl Valadéz Azúa, Paulina Mejía Correa, and

Tania V. Gutiérrez Santillán, pp. 481–486. Asociación Etnobiológica Mexicana, Mexico City.

Viveiros de Castro, Eduardo
2009 The Gift and the Given: Three Nano-Essays on Kinship and Magic. In *Kinship and Beyond: The Genealogical Model Reconsidered*, edited by Sandra C. Bamford and James Leach, pp. 237–67. Berghahn Books, New York.

Voorhies, Barbara
2004 *Coastal Collectors in the Holocene: The Chantuto People of Southwest Mexico*. University Press of Florida, Gainesville.

Vovides, Andrew P.
1983 Systematic Studies on the Mexican Zamiaceae. I. Chromosome Numbers and Karyotypes. *American Journal of Botany* 70(7): 1002–1006.
1985 Systematic Studies on Mexican Zamiaceae II. Additional Notes on *Ceratozamia kuesteriana* from Tamaulipas, Mexico. *Brittonia* 37: 226–231.
1989 Problems of Endangered Species Conservation in Mexico: Cycads an Example. *Encephalartos* 20: 35.
1990 Spatial Distribution, Survival, and Fecundity of *Dioon edule* (Zamiaceae) in a Tropical Deciduous Forest in Veracruz, Mexico, with Notes on its Habitat. *American Journal of Botany* 77(12): 1532–1543.
1991a Cone Idioblasts of Eleven Cycad Genera: Morphology, Distribution, and Significance. *Botanical Gazette* 152(1): 91–99.
1991b Insect Symbionts of Some Mexican Cycads in their Natural Habitat. *Biotropica* 23(1): 102–104.
2000 México: Segundo Lugar Mundial en Diversidad de Cícadas. *Biodiversitas* 6(31): 6110.

Vovides, Andrew P., Sergio Avendaño, Miguel Angel Pérez-Farrera, and Dennis W. Stevenson
2012 What is *Ceratozamia brevifrons* (Zamiaceae)? *Brittonia* 64(1): 35–42.

Vovides, Andrew P., James A. R. Clugston, José Said Gutiérrez-Ortega, Miguel A. Pérez-Farrera, M. Ydelia Sánchez-Tinoco, and Sonia Galicia
2018 Epidermal Morphology and Leaflet Anatomy of *Dioon* (Zamiaceae) with Comments on Climate and Environment. *Flora* 239: 20–44.

Vovides, Andrew P., John R. Etherington, P. Quentin Dresser, Andrew Groenhof, Carlos Iglesias, and Jonathan Flores Ramírez
2002 CAM-cycling in the Cycad *Dioon edule* Lindl in Its Natural Tropical Deciduous Forest Habitat in Central Veracruz, Mexico. *Botanical Journal of the Linnean Society* 138: 155–162.

Vovides, Andrew P., and Sonia Galicia
2016 G-fibers and Florin Ring-like Structures in *Dioon* (Zamiaceae). *Botanical Sciences* 94: 263–268.

Vovides, Andrew P., Dolores González, Miguel Angel Pérez-Farrera, Sergio Avendaño, and Cristina Bárcenas
2004 A Review of Research on the Cycad Genus *Ceratozamia* Brongn. (Zamiaceae) in Mexico. *Taxon* 53(2): 291–297.

Vovides, Andrew P., and Carlos G. Iglesias
1994   An Integrated Conservation Strategy for the Cycad *Dioon edule* Lindl. *Biodiversity and Conservation* 3: 137–141.

Vovides, Andrew P., Carlos G. Iglesias, Victor Luna, and Teodolinda Balcázar
2013   Los Jardines Botánicos y la Crisis de la Biodiversidad. *Botanical Sciences* 91: 239–250.

Vovides, Andrew P., and Nancy P. Moreno
1983   Proposal to Conserve the Spelling *Dioon* against *Dion* (Zamiaceae). *Taxon* 32(3): 484–485.

Vovides, Andrew P., Knut J. Norstog, Priscilla K. S. Fawcett, Mark W. Duncan, Robert J. Nash, and Dian V. Molsen
1993   Histological Changes During Maturation in Male and Female Cones of the Cycad *Zamia furfuracea* and Their Significance in Relation to Pollination Biology. *Botanical Journal of the Linnean Society* 111(2): 241–252.

Vovides, Andrew P., and Mariana Olivares
1996   Karyotype Polymorphism in the Cycad *Zamia loddigesii* (Zamiaceae) of the Yucatan Peninsula, Mexico. *Botanical Journal of the Linnean Society* 120(1): 77–83.

Vovides, Andrew P., Miguel Angel Pérez-Farrera, José Said Gutiérrez-Ortega, Sergio Avendaño, Anwar Medina-Villarreal, Jorge González-Astorga, and Sonia Galicia
2020   A Revision of the *Ceratozamia miqueliana* (Zamiaceae) Species Complex Based on Analyses of Leaflet Anatomical Characters. *Flora* 270. doi: 10.1016/j.flora.2020.151649.

Vovides, Andrew P., Miguel Angel Pérez-Farrera, and Carlos Iglesias
2010   Cycad Propagation by Rural Nurseries in Mexico as an Alternative Conservation Strategy: 20 Years On. *Kew Bulletin* 65: 603–611.

Vovides, Andrew P., Miguel Angel Pérez-Farrera, Bart Schutzman, Carlos Iglesias, Luis Hernández-Sandoval, and Mahinda Martínez
2004   A New Species of *Ceratozamia* (Zamiaceae) from Tabasco and Chiapas, Mexico. *Botanical Journal of the Linnean Society* 146(1): 123–128.

Vovides, Andrew P., Miguel Angel Pérez-Farrera, Mario Vázquez Torres, and Uwe Schippmann
2002   Peasant Nurseries: A Concept for an Integrated Conservation Strategy for Cycads in Mexico. In *Plant Conservation in the Tropics: Perspectives and Practice,* edited by Mike Maunder, Colin Clubbe, Clare Hankamer, and Madeline Groves, pp. 421–444. Royal Botanic Gardens, Kew.

Vovides, Andrew P., and C. M. Peters
1987   *Dioon edule,* la planta más antigua de México. *Ciencia y Desarrollo* 13: 19–24.

Vovides, Andrew P., John D. Rees, and Mario Vásquez Torres
1983   Zamiaceae. In *Flora de Veracruz,* Vol. 26, edited by Arturo Gómez-Pompa, pp. 1–31. INIREB, Xalapa, Mexico.

Vovides, Andrew P., Dennis W. Stevenson, Miguel Angel Pérez-Farrera, Sergio López, and Sergio Avendaño
2016   What is *Ceratozamia mexicana* (Zamiaceae)? *Botanical Sciences* 94(2): 419–429.

Wahl, David, Roger Byrne, Thomas Schreiner, and Richard Hansen
2006   Holocene Vegetation Change in the Northern Petén and Its Implications for Maya Prehistory. *Quaternary Research* 65(3): 380–389.

Wang, Jiajing, Li Liu, Terry Ball, Linjie Yu, Yuanqing Li, and Fulai Xing
2016   Revealing a 5,000-year-old Beer Recipe in China. *PNAS* 113(23): 6444–6448.

Watanabe, Iwao, Corazon R. Espinas, Nilda S. Berja, and B. V. Alimagno
1977   *The Utilization of the Azolla-Anabaena Complex as a Nitrogen Fertilizer for Rice*. IRRI Research Papers No. 11. International Rice Research Institute, Los Baños, Philippines.

Wendt, Carl J., Henri N. Bernard, and Jeffery Delsescaux
2014   A Middle Formative Artifact Excavated at Arroyo Pesquero, Veracruz. *Ancient Mesoamerica* 25(2): 309–316.

Wendt, Thomas
1987   Las Selvas de Uxpanapa, Veracruz-Oaxaca, México: Evidencia de Refugios Florísticos Cenozoicos. *Anales del Instituto de Biología, Serie Botánica* 58: 29–54.

Weyrich, Laura S, Keith Dobney, and Alan Cooper
2015   Ancient DNA Analysis of Dental Calculus. *Journal of Human Evolution* 79: 119–24.

Weyrich, Laura S., Sebastian Duchene, Julien Soubrier, Luis Arriola, Bastien Llamas, James Breen, Alan G. Morris, Kurt W. Alt, David Caramelli, Veit Dresely, Milly Farrell, Andrew G. Farrer, Michael Francken, Neville Gully, Wolfgang Haak, Karen Hardy, Katerina Harvati, Petra Held, Edward C. Holmes, John Kaidonis, Carles Lalueza-Fox, Marco de la Rasilla, Antonio Rosas, Patrick Semal, Arkadiusz Soltysiak, Grant Townsend, Donatella Usai, Joachim Wahl, Daniel H Huson, Keith Dobney, and Alan Cooper
2017   Neanderthal Behaviour, Diet, and Disease Inferred from Ancient DNA in Dental Calculus. *Nature* 544(7650): 357–361.

Whitaker, Thomas W., Hugh C. Cutler, and Richard S. MacNeish
1957   Cucurbit Materials from Three Caves near Ocampo, Tamaulipas. *American Antiquity* 22(4): 352–358.

Whitelock, Loran M.
2002   *The Cycads*. Timber Press, Portland, Oregon.

Whiting, Marjorie Grant
1963   Toxicity of Cycads. *Economic Botany* 17(4): 270–302.
1989   Neurotoxicity of Cycads: An Annotated Bibliography for the Years 1829–1989. *Lyonia—Occasional Papers of the Harold L. Lyon Arboretum* 2(5): 201–270.

Wichmann, Søren
1995   *The Relationship Among the Mixe–Zoquean Languages of Mexico*. University of Utah Press, Salt Lake City.

Wilkes, Hilbert Garrison
1967   *Teosinte: The Closest Relative of Maize*. Bussey Institute, Harvard University, Cambridge, Massachusetts.

Winterhalder, Bruce, and Douglas J. Kennett
2006   Behavioral Ecology and the Transition from Hunting and Gathering to Agriculture. In *Behavioral Ecology and the Transition to Agriculture,* edited by

Douglas J. Kennett and Bruce Winterhalder, pp. 1–21. University of California Press, Berkeley.

Witthoft, John
1946   The Cherokee Green Corn Medicine and the Green Corn Festival. *Journal of the Washington Academy of Sciences* 36(7): 213–219.

Wright, Sewell G.
1943   Isolation by Distance. *Genetics* 28: 114–138.
1951   The Genetical Structure of Populations. *Annals of Eugenics* 15: 323–354.

Wu, Chung-Shien, and Shu-Miaw Chaw
2015   Evolutionary Stasis in Cycad Plastomes and the First Case of Plastome GC-Biased Gene Conversion. *Genome Biology and Evolution* 7(7): 2000–2009.

Wu, Yaqiong, Qi Zhou, Shujing Huang, Guibin Wang, and Li-an Xu
2019   SNP Development and Diversity Analysis for *Ginkgo biloba* Based on Transcriptome Sequencing. *Trees Structure and Function* 33(2): 587–597.

Xiao, Longqian, Xun Gong, Gang Hao, and Si-xiang Zheng
2004   ISSR Variation in the Endemic and Endangered Plant *Cycas guizhouensis* (Cycadaceae). *Annals of Botany* 94: 133–138.

Xiao, Longqian, Xun Gong, Gang Hao, Xuejun Ge, Bo Tian, and Sixiang Zheng
2005   Comparison of the Genetic Diversity in Two Species of Cycads. *Australian Journal of Botany* 53(3): 219–223.

Xiao, Siyue, Yunheng Ji, Jian Liu, and Xun Gong
2019   Genetic Characterization of the Entire Range of *Cycas panzhihuaensis* (Cycadaceae). *Plant Diversity* 42(1): 7–18.

Xu, Meng, Xin Liu, Jian-Wen Wang, Shi-Yuan Teng, Ji-Qing Shi, Yuan-Yuan Li, and Min-Ren Huang
2017   Transcriptome Sequencing and Development of Novel Genic SSR Markers for *Dendrobium officinale*. *Molecular Breeding* 37: 18.

Yang, Si-Lin, and Alan W. Meerow
1996   The *Cycas pectinata* (Cycadaceae) Complex: Genetic Structure and Gene Flow. *International Journal of Plant Sciences* 157(4): 468–483.

Yew, Joanne Y.
2019   Natural Product Discovery by Direct Analysis in Real Time Mass Spectrometry. *Mass Spectrometry* 8: 1–8.

Zapata Avendaño, Luis Enrique
2018   El chamal, el venado y el jaguar. Aproximaciones etnográficas en la Reserva de la Biosfera Sierra Gorda de Querétaro. Master's thesis, Facultad de Ciencias Sociales y Humanidades, Universidad Autónoma de San Luis Potosí, Mexico.

Zarate Escamilla, Jaime
2007   *Huamelula, pueblo danzante*. CDI, Mexico City.

Zarrillo, Sonia
2012   Human Adaptation, Food Production, and Cultural Interaction during the Formative Period in Highland Ecuador. PhD dissertation. Department of Archaeology, University of Calgary, Calgary.

Zarrillo, Sonia, Nilesh Gaikwad, Claire Lanaud, Terry Powis, Christopher Viot, Isabelle Lesur, Olivier Fouet, Xavier Argout, Erwan Guichoux, Franck Salin, Rey Loor Solorzano, Olivier Bouchez, Hélène Vignes, Patrick Severts, Julio Hurtado, Alexandra Yepez, Louis Grivetti, Michael Blake, and Francisco Valdez
2018 The Use and Domestication of *Theobroma cacao* during the Mid-Holocene in the Upper Amazon. *Nature Ecology and Evolution* 2(12): 1879–1888.

Zarrillo, Sonia, Deborah M. Pearsall, J. Scott Raymond, Mary Ann Tisdale, and Dugane J. Quon
2008 Directly Dated Starch Residues Document Early Formative Maize (*Zea mays* L.) in Tropical Ecuador. *PNAS* 105(13): 5006–5011.

Zeder, Melinda A., Eve Emshwille, Bruce D. Smith, and Daniel G. Bradley
2006 Documenting Domestication: The Intersection of Genetics and Archaeology. *TRENDS in Genetics* 22(3): 139–155.

Zhou, Hui Ping, and Jin Chen
2010 Spatial Genetic Structure in an Understorey Dioecious Fig Species: The Roles of Seed Rain, Seed and Pollen-Mediated Gene Flow, and Local Selection. *Journal of Ecology* 98(5): 1168–1177.

Zier, Christian J.
1992 Intensive Raised-Field Agriculture in a Posteruption Environment, El Salvador. In *Gardens of Prehistory: The Archaeology of Settlement Agriculture in Greater Mesoamerica,* edited by Thomas W. Killion, pp. 217–233. University of Alabama Press, Tuscaloosa.

Ziesemer, Kirsten A., Allison E. Mann, Krithivasan Sankaranarayanan, Hannes Schroeder, Andrew T. Ozga, Bernd W Brandt, Egija Zaura, Andrea Waters-Rist, Menno Hoogland, Domingo C. Salazar-García, Mark Aldenderfer, Camilla Speller, Jessica Hendy, Darlene A. Weston, Sandy J. MacDonald, Gavin H. Thomas, Matthew J. Collins, Cecil M. Lewis, Corinne L. Hofman, and Christina Warinner
2015 Intrinsic Challenges in Ancient Microbiome Reconstruction using 16S rRNA Gene Amplification. *Scientific Reports* 5: 16498.

Zizumbo-Villarreal, Daniel, and Patricia Colunga-García Marín
2010 Origin of Agriculture and Plant Domestication in West Mesoamerica. *Genetic Resources and Crop Evolution* 57(6): 813–825.

Zizumbo-Villarreal, Daniel, Alondra Flores-Silva, and Patricia Colunga-García Marín
2012 The Archaic Diet in Mesoamerica: Incentive for Milpa Development and Species Domestication. *Economic Botany* 66(4): 328–343.

Zonneveld, B. J. M., and Anders Jan Lindstrom
2016 Genome Sizes for 71 Species of *Zamia* (Cycadales: Zamiaceae) Correspond with Three Different Biogeographic Regions. *Nordic Journal of Botany* Early View: 1-EV.

Contributors

Francisco Barona-Gómez is full professor at Langebio, Cinvestav, Mexico, where he leads the Evolution of Metabolic Diversity Laboratory. He is Level 3 of Conacyt's Sistema Nacional de Investigadores, Mexico, and holds a Newton Advanced Fellowship from the Royal Society of the UK. His research aims to decipher the evolutionary mechanisms at different scales—from atoms to microbial communities—underlying the evolution of metabolism during bacterial adaptation. To approach this problem he integrates diverse interdisciplinary conceptual frameworks and technologies, from analytical chemistry, evolutionary analysis, and genomics, and focuses on cycads as an ancestral microbial reservoir with metabolically diverse bacterial communities that have intimately adapted to their plant host.

Emanuel Bojórquez Quintal is a native of Hunucmá, Yucatán, Mexico. He is a chemist, biologist, and bromatologist with a BA from the Faculty of Chemistry, Universidad Autónoma de Yucatán (2008) and a master's (2010) and doctorate (2015) in plant biological sciences with a specialization in biochemistry and molecular biology from the Centro de Investigación Científica de Yucatán. Since 2015, he has been research professor–Cátedra CONACYT assigned to the Laboratorio de Análisis y Diagnóstico del Patrimonio (LADiPA) of El Colegio de Michoacán in La Piedad, Mexico. His research interests include the design of restoration products from organic materials present in cultural heritage; the characterization and identification of materials and organic molecules of plants by direct analysis techniques; and mineral nutrition, metal toxicity, and phytoremediation in higher plants. He is a member of the CONACYT Sistema Nacional de Investigadores.

Mark A. Bonta is a cultural geographer with degrees from Penn State, The University of Texas at Austin, and Louisiana State University. He has published widely in ethno-ornithology, the philosophy of geography, cultural ecology, and the ethnobotany of cycads, focusing particularly on Honduras and Mexico. His scholarship on the human geography of cycads began in 1999 through a chance encounter with *Dioon mejiae,* known locally as "teocinte," in northeastern Honduras. This inspired in-depth collaborations with botanists and local people across Mesoamerica to both document and preserve cycad culture, which led eventually to this volume, as well as to the initiation of several successful cycad conservation efforts in Honduras, including an annual cycad festival in Gualaco. Bonta has taught at Penn State Altoona and Delta State University, as well as in China.

Edder D. Bustos-Díaz is a PhD student at Langebio, Cinvestav, Mexico, where he studies the role of the bacterial communities associated with symbiotic cyanobacterial strains from the Nostocales order associated with plants like cycads. His research explores the composition and role of these communities through a combination of (meta)genomic data analysis and microbiological experiments.

Dánae Cabrera-Toledo is an evolutionary biologist by training and an expert in ecology and biological systematics in plants. She is currently affiliated with the Department of Botany and Zoology of the University Center for Biological and Agricultural Sciences of the University of Guadalajara. She obtained her doctorate in science (Ecology and Natural Resource Management) from the same institution, with a thesis titled "Population Biology of Two Endemic Cycads That Differ in Their Levels of Rarity." She has worked as a research assistant in the Laboratory of Marine Ecosystem Ecology and as laboratory technician of genetics in animal production, and she has taught classes in viability analysis of wild organisms in the master's graduate program BIMARENA.

Michael Calonje is the cycad biologist at Montgomery Botanical Center in Coral Gables, Florida. His main research interests include taxonomy, nomenclature, phylogenetics, and ex-situ conservation of cycads. He has conducted extensive fieldwork throughout Latin America and the Caribbean, focused particularly on the New World genus *Zamia.* He currently serves as deputy chair for the IUCN SSC's Cycad Specialist Group, coordinator of The World List of Cycads, and board member of the International Cycad Society and the Colombian Cycad Society. He holds a PhD in biology from Florida International University and an MSc from Humboldt State University, California.

Michael D. Carrasco is associate dean for academic affairs and research in the College of Fine Arts at Florida State University and an associate professor of the visual cultures of the Americas in the Department of Art History. His scholarship draws on diverse, interdisciplinary perspectives to elucidate the origins of writing in the Americas and examine Indigenous aesthetics, theology, and epistemologies. Over the last decade his research has expanded to encompass ethnobiology, cultural heritage, and the dynamic interaction between folk traditions and the global art system, specifically in Japan. The fruits of his scholarship have appeared in numerous journal publications and chapters, as well as in the edited volumes *Interregional Interaction in Ancient Mesoamerica*, *Parallel Worlds: Genre, Discourse, and Poetics in Contemporary, Colonial, and Classic Maya Literature*, and *Pre-Columbian Foodways: Interdisciplinary Approaches to Food, Culture, and Markets in Ancient Mesoamerica*.

Angélica Cibrián-Jaramillo has a degree in biology from the Universidad Nacional Autónoma de México, where she graduated with honors working on the genetics of vanilla populations. She completed a master's degree and a doctorate in ecology, evolution, and environmental biology at Columbia University with a Fulbright-Robles fellowship, and a certificate in environmental policy also at Columbia University. During this stage she worked to help develop the United Nations Millennium Goals and carried out a research stay at Bioversity International in Rome. She had postdoctoral stays as a Lewis and Dorothy Cullman Postdoctoral Fellow at the New York Botanical Garden and the American Museum of Natural History, in the Harvard University Department of Organismic and Evolutionary Biology and the Harvard Medical School, and at New York University working on plant phylogenomics and the genetic basis of plant ecotypes. She is currently head of the Ecological and Evolutionary Genomics lab at the National Genomics Laboratory for Biodiversity of CINVESTAV, Level II of the SNI and has more than 30 publications in prestigious journals and publishing houses. Her current work focuses on understanding the genomic bases of domestication in Mesoamerican plants.

Joshua D. Englehardt is research professor of archaeology at El Colegio de Michoacán, Mexico, and a Level I National Investigator of the Mexican National Council on Science and Technology. He specializes in Mesoamerican archaeology and epigraphy, with a research focus on the development of Mesoamerican writing systems in the Formative period. He is the co-director of *The Origins of Writing in Early Mesoamerica*, an NEH-funded interdisciplinary project that is developing print and digital materials for the investigation and presentation of early Mesoamerican visual cultures. He is also a member of a multidisciplinary team currently exploring the cultural uses of cycads and their significance as biocultural heritage. Recent publications include *Archaeological Paleography*, the edited volumes *Nuevos Enfoques en la Arqueología de la Región de Tequila*, *Interregional Interaction in Ancient Mesoamerica*, and *Ancient West Mexicos* (University Press of Florida, 2019), as well as more than 20 articles and book chapters that stem from his fieldwork throughout Mexico and 25 years of living, working, or studying in the Global South.

Jorge González-Astorga has been a researcher since 1999 at the Instituto de Ecología A. C. in Xalapa, Veracruz, Mexico, in the population genetics laboratory of the department of Evolutionary Biology. He has published about 50 papers, 30 of which are on cycads of the American continent on various topics: genetics and population ecology, phylogenetic systematics, phylogeography, taxonomy and conservation, in some species of the Zamiaceae family of the genera *Dioon, Ceratozamia, Zamia,* and *Microcycas*. His scientific research is based on hard work in the field, in the laboratory, and in collaboration with his students. He is a member of the IUCN group of specialists for the Cycadales. He has a degree in biology from the Universidad Nacional Autónoma de México and received a Doctor of Science from the Universidad Complutense de Madrid, Spain.

Naishla M. Gutiérrez-Arroyo is a master's student in the integrative biology program at the Center for Research and Advanced Studies of the National Polytechnic Institute (CINVESTAV, UGA-LANGEBIO), Mexico. She worked with the variability and genetic structure of an endemic cycad from San Luis Potosí, Mexico, and she is also interested in the estimation of coralloid root age and the relationship between their anatomy and the distribution of endophytic organisms in *Dioon edule* individuals.

José Said Gutiérrez-Ortega is a Mexican biologist interested in plant evolution, especially the processes concerning biogeography, phylogenetics, population genetics, ecology, and speciation. He started his formation as a researcher in Universidad de Sonora, Mexico, in 2005, and has continued his career in Japan since 2012. As a cycad specialist, he has contributed to understanding the evolutionary history of Mexican cycads and the macroevolutionary mechanisms that have allowed their persistence and diversification. To date, he has coauthored the description of three cycad species and expects the implications of his research to help the conservation of threatened species. He is a member of the Cycad Specialist Group of the IUCN Species Survival Commission. Currently, he works as assistant professor in the Institute for Excellence in Educational Innovation, Chiba University, Japan, where he also coordinates a science education program aimed at forming young, highly talented researchers.

Jaime R. Pagán-Jiménez is a Puerto Rican paleoethnobotanist and archaeologist who received his doctoral degree in anthropology at Universidad Nacional Autónoma de México (2005). For the past 22 years he has been studying the paleoethnobotany of the Caribbean islands, and more recently of French Guiana, Ecuador, and Venezuela. He has published extensively on anti- and postcolonial archaeology in the context of archaeological praxis in Latin America and the Caribbean, and on the paleoethnobotany of the Caribbean islands and northern

South America. Among his main published works are *De antiguos pueblos y culturas botánicas en el Puerto Rico indígena*, "Early dispersals of maize and other food plants into the southern Caribbean and northeastern South America," and "Late ninth millennium B.P. use of *Zea mays* L. at Cubilán area, highland Ecuador, revealed by ancient starches." He is currently senior researcher and archaeologist at Cultural Heritage & Plantscape Research (Leiden, Netherlands), and affiliated senior researcher in paleoethnobotany at the Royal Netherlands Institute of Southeast Asian and Caribbean Studies (Netherlands).

Francisco Pérez-Zavala is a doctoral student at the Center for Research and Advanced Studies of the National Polytechnic Institute's Laboratorio Nacional de Genómica para la Biodiversidad (Langebio, Cinvestav). His research focuses on plant biotechnology and the genetic structure of plant populations, particularly cycads. He received his MSc from the Departamento de Ingeniería Genética at Langebio in 2017, and his work has been published in the journals *Applications in Plant Sciences* and *Ecology and Evolution*.

Luis Rojas Abarca is research assistant at the Laboratorio de Análisis y Diagnóstico del Patrimonio (LADiPA) of El Colegio de Michoacán in La Piedad, Mexico. He has a bachelor's degree in bio-pharmaceutical chemistry from the Universidad Michoacana de San Nicolás Hidalgo (2006). He is currently assigned to the operation of LADiPA's mass spectrometer by direct analysis in real time (DART), carrying out the analysis of cultural and natural heritage samples. Recently he has been involved in a variety of research projects on cultural and natural heritage, focused specifically on the molecular analysis and chemical characterization of organic compounds in *Indigofera tinctoria*, *Phaseolus vulgaris*, *Dioon edule*, *Capsicum* spp., *Opuntia ficus indica* mucilage, and *Salvia hispanica* oil.

Esteban Sánchez Rodríguez is an electrical engineer by training, with a degree from the Universidad Michoacana de San Nicolás de Hidalgo. Since 2010, he has worked as a research associate and technician at the Laboratorio de Análisis y Diagnóstico del Patrimonio of El Colegio de Michoacán, where he serves as the scanning electron microscopist. His principal research centers on elemental and chemical analyses, via dispersed energy spectrometry and scanning electron microscopy, of samples of both natural heritage (water, soil, vegetal materials)—with particular emphasis on agricultural applications—and cultural patrimony, such as stratigraphic sections from paintings and sculpture.

Dennis William Stevenson received his BS and MS degrees in botany from Ohio State University and his doctorate in botany from the University of California. He was a postdoctoral fellow at Harvard University from 1975 through 1980, and during that time spent a year at the Royal Botanic Gardens, Kew, as a NATO Fellow. Prior to joining The New York Botanical Garden in 1987, he was on the faculty of Barnard College of Columbia University. His research has focused on the biology of the cycads from collecting to pollination biology to neurobiology to fossils, supported by an active field program, particularly in the Neotropics. His other area of expertise is the evolution and classification of the monocots, which include plants such as lilies, grasses, and gingers. Dr. Stevenson has published more than 350 peer-reviewed papers, with 170 of them on cycads, and edited 11 books ranging from horticulture to paleobotany to genomics. He holds adjunct faculty positions at City University of New York, Columbia University, Cornell University, New York University, and Yale University. He also serves as editor of the journals *Botanical Review* and *Cladistics,* in addition to being one of 50 honorary foreign members of the Linnaean Society of London.

Amber M. VanDerwarker is professor of anthropology at the University of California, Santa Barbara, where she directs the Integrative Subsistence Laboratory. VanDerwarker received her PhD from the University of North Carolina, Chapel Hill, in 2003. Her research examines plant domestication and agriculture, and the ways in which ancient foodways articulate with conditions of warfare, the development of complex political institutions, gendered labor and power, and culture contact situations. VanDerwarker focuses her research questions in the Americas, with particular emphasis in Mesoamerica and the Eastern Woodlands of the United States of America.

Luis R. Velázquez Maldonado is research assistant at the Laboratorio de Análisis y Diagnóstico del Patrimonio (LADiPA) of El Colegio de Michoacán in La Piedad, Mexico, specializing in the study of cultural patrimony via noninvasive analytic techniques, particularly X-ray fluorescence. He has undergraduate (2004) and master's (2011) degrees in radiochemistry from the Facultad de Ciencias y Tecnologías Nucleares (FCTN) of the Instituto Superior de Tecnologías y Ciencias Aplicadas (INSTEC), Havana, Cuba, as well as a master's degree (2017) in archaeology from the Centro de Estudios Arqueológicos of El Colegio de Michoacán. His work has been published in academic journals such as *Ancient Mesoamerica, International Journal of South American Archaeology, Restaurierung und Archäologie, Nucleus,* and *CENIC Ciencias Químicas.*

Andrew P. Vovides works on the systematics of Mexican cycads, mainly morphology, taxonomy, and anatomy, as well as conservation aimed at sustainable management. He has coauthored 17 new cycad species for Mexico and has experience in conservation in botanic gardens and establishing and curating living collections. He is a founding member of the Francisco Javier Clavijero Botanic

Garden in Xalapa and curator of the Living National Collection of Cycads, and founding member of the Asociación Mexicana de Jardines Botánicos, A.C. (Mexican Association of Botanic Gardens). His collaborations include with researchers from UNAM in Mexico City, and foreign institutions such as the Fairchild Tropical Botanic Garden (with which the Clavijero Botanic Garden has enjoyed a sister garden relationship since 1989), and the Montgomery Botanical Center, Miami, Florida. He is also a member of the Cycad Specialist Group for the IUCN Species Survival Commission, and he proposed a cycad consortium during the 4th Global Botanic Gardens Congress held in Dublin, Ireland, in 2010.

Index

Page numbers followed by the letters *f* and *t* indicate figures and illustrations

Aach-Eagle grandmother, 176, 177
A-amylase, 107. *See also* Enzymatic digestion and starch survival
Abbad y Lasierra, Fray Iñigo, 98–99
Acecentli, 174f
Acorns, 174f
Acrobat, 190t
Acrocomia mexicana, 179, 180. *See also* Palms
aDNA, 160f
Africans in the Caribbean, 123
Agave. See Maguey
"Age of cycads," 33
Agroceramic age sites, 115
Agroecological mythological symbolic regimes: in Chiapanec cycad traditions, 186; cycads in, 170f
Agroforestry: systems of, in ancient Mesoamerica, 125; as threat to cycads, 24
Aguán Valley (Honduras), 218
Agua Prieta (Chiapas), 87
Ahaatik (a) eem, 180, 226, 227t
Aján, 227t
Ajätic, 227t
Alcorn, Janis, 168
Alimentation, 167–168
Alleles, 64, 65; dissimilar male/female frequencies of, in *Dioon,* 84
Alligator (day sign), 181, 198f
Allocorynina weevils, 32
Allozymes: in cycad genetics studies, 76; in *Dioon caputoi* studies, 84; as markers of genetic variation and diversity, 65
Alma de cintli, 235, 236t
Alma del maíz, 235, 236t
Almagre phase (Sierra de Tamaulipas), 145t
ALS, 51
Amami Oshima, 12, 25

Ammonia (NH_3), 130
Ampara mines (Jalisco), 220
Amuzgo (people), cycad use by, 240
Amylopectin, 101
Amyloplasts, 101
Amylose, 101
Amyotrophic lateral sclerosis. *See* ALS
Anabaena, 130
Ancient food (cycad name), 239t
Anemophily in cycads, 48
Anse à la Gourde (Guadeloupe), 110t, 115
Anthropocene, 61
Ants in Xi'iuy cycad-origin story, 233
Apogeotrophic roots. *See* Coralloid roots
Aquatic resources: in Mesoamerican foodways, 170; storage of, vs. maize, 195; in symbolism, 207f
Aradopsis thaliana, 253
Archaeobotanical evidence of cycad use: further research needed in, 139; overview of, in Mexico, 90
Archaeogenetics, 158
Archaeo-historical research in cycads, 91
Archaeological methods: integrative/interdisciplinary, for cycads, 151–152, 160f, 164–165; lack of, for detecting cycads, 142
Archaeology, experimental, 152–154, 160f; and Teenek people, 153–154
Archaeology, precolonial, 120
Archaeometric methods in archaeology, 154
Archaic, late (Caribbean period), 114
Archaic (Mesoamerican period), 148; burning and cultivation in, 133–136; cycads used in, in Mexico, 90, 149, 171; food eaten during, 170, 173–174; Late, 135, 206; symbolism focused on, 170
Archegonial chamber, 32
Archegonium, 32
Archives in cycad research, 142–143
Arecibo 39 (Puerto Rico), 110t

Arepa de guáyiga, 123
Argyle 2 (Saint Vincent) archaeological site with zamia starch, 110t, 113, 115
Aridification affecting *Dioon,* 39, 83
Aroids in San Bartolo murals, 208
Arrowroot, 179
Arroyo Pesquero celt, 196, 197f; as portraying maize deity, 198
Arroyo Pesquero jadeite sculpture, 203–205, 205f; as a cycad, 203, 205f; as maize, 203
Art historical studies, 167
Artist/s: Corn Master as, 187; TMG/GCMH and, 188; Xi'iuy, as creators of Tancoyol mission church façade, 230
Aruba: archaeological sites with zamia starches in, 108, 110t, 111t, 115; no natural distribution of *Zamia* in, 108; underground stems of zamia eaten in, 115
Ash for cycad seed detoxification, 90, 91f, 127, 158, 209n2
Atoles as cycad drinks, 128
Atractosteus spatula. See Gar
Aulacaspis yasumatsui scale insect. *See* Cycad Asian Scale (CAS)
Australia, cycad species richness of, 35
Australian Aboriginal relationships to cycads, 9, 10f, 245
Avocado: domestication of, 125; (*Persea americana*) as dominant in Early Formative period, 210n9
Axe, lightning, 190t
Aztec (empire). *See* Mexica

Bahamas, The: plant evidence in late ceramic age sites of, 115; *Zamia* in, 108; *Zamia* as foodstuff in, 97, 123
Baizabal Cenozoic floristic refuge, 42
Ballgame/ball player, 190t
Balobedu people, 15
Balsas River valley/Balsas drainage basin, 132, 133, 212, 216
Banana leaves in *Zamia* food preparation, 121f
Banwari Trace (Trinidad and Tobago), 112
Bats, fruit, 248
Beaches. *See* Coastal habitat
Beans (*Phaseolus* spp.), 141, 154; common, 133; common, in Tamaulipas caves, 145t; and cycads, 125–140; intercropping with maize, 136–138; overview of domestication of, 133; runner, 133; starch grains of, in Puerto Rican archaeological sites, 114, 115; tepary, 133; year-long, 133

Beaucarnea recurvata, 56, 57
Becerra, Professor Marcos E., 48
Beetles as cycad pollinators, 32
Belize, 47, 242
Bennettitales. *See* Cycadeoidales
Berlin Tripod Vase, 200, 201f
β-Nmethylamino-L-alanine. *See* BMAA
Beverages (cycad), 127, 128
Bezuapan (Veracruz), 137
Bioarchaeology related to cycad microorganisms, 153
Biological nitrogen fixation. *See* BNF
Biomarkers, 160; in cycads, 161, 164–165
Birds in Xi'iuy cycad-origin story, 232, 233
Blue-green algae. *See* Cyanobacteria
BMAA, 27, 51, 253; as a biomarker, 161, 162f; detected via DART-MS; and MAM, 121f; produced by cyanobacteria/other microorganisms, 161; produced by cycads, 161
BNF: comparison of, in cycads and legumes, 130–131; cycads and, 154, 155, 157f, 160f; effect of controlled burning on, in cycads, 131, 134; as explanation for Nahua cycad belief, 235; general overview of, 126, 129–130; in Late Archaic maize fields, 134; overview of, in cycads, 130–131
Bo'jor, 227t, 228
Bones, human: as artifacts to look for ancient zamia evidence, 121f; collagen of, in Mokaya, 210n8; isotopic nitrogen in, 155; preservation of, 139
Book of Chilam Balam of Chumayel, 182
Botanical Garden of the University of Florence (Italy), 46
Botanic gardens: importance of, for living cycad collections, 50; role of, in ex situ cycad conservation, 55, 61
Bowenia, 9; bipinnate leaves of, 30
Brains (human): *chamal* cycads derived from, 232, 233; cycads and, 185; discussion of, as cycad origin, 233. *See also* Head
Braun, Alexander, 245
Brazil, *Zamia* in, 43
Bread, cycad, 98–99, 121f, 150; in the Dominican Republic, 123; of Xi'iuy, 231
Breadnut tree, 170, 207; as associated with Aach-Eagle, 177–178; *Brosimum*-based religion, 178; defeated by cycad, 207; as dominant in Early Formative period, 210n9; as embodying Thipaak's grandmother, 226; fruiting of, influencing Teenek maize harvest, 178; and maize, 177, 178; in Teenek

world ages, 177t; as *iximte'* in Tzotzil/Yucatec, 191; as relic foodway, 196
Bromeliad. See *Guapilla*
Brosimum alicastrum. See Breadnut tree
Brosimum-based religion. See Breadnut tree
Buddhism, cycads in, 14–15, 16f
Bull, William, horticultural catalogues of, 47, 48
Bull (name for cycad), 228
Bundles: of cycad leaves in pilgrimages, 128; in Formative/Olmec imagery, 196; in Vegetal Bundle motif, 205f, 206
Burning. See Fire/burning
Butterflies, 174f
Butterflies, cycad. See *Eumaeus*

Cabeza. See Head
Cabo, El (Dominican Republic), 110t
Cacao, 142; and Maya Maize God, 190–192, 191f; pods/fruits of, 190–192, 191f; (*Theobroma cacao*) as dominant in Early Formative, 210n9
Cacti as Chicomoztoc, 208
Cahua, 239t
Calendrical systems, Mesoamerican, 180
Calocitta formosa. See Magpie-jay, white-throated, as *Dioon* seed disperser
CAM photosynthetic pathway, 129
Canashito (Aruba), 110t, 114
Canavalia, starch grains of, in Puerto Rican archaeological sites, 114
Cannibal/cannibalism: in Homshuk story, 184; role of, in Teenek mythology, 176, 177, 200; in Xi'iuy cycad origin story, 232; in Xi'iuy ogress narrative, 185, 196, 232–233
Carcinogenic properties of cycads, 127, 241; known to Chontal de Oaxaca, 241
Cardboard palm. See *Zamia furfuracea*
Caribbean *Zamia* clade, 35, 40–41, 76–77
Casita de Piedra (Panama), 117
Cataphylls of cycads, 30
Catemaco: Lake, 200, 202f; region, 236t
Catfish, 181
Cattle, raising of, affecting *Ceratozamia fuscoviridis* populations, 81
Cave: in maize god iconography, 192; in Xi'iuy cycad-origin story, 232
Cave sites with cycad remains, 90, 144f. See also Rock shelters with cycad remains
Cayman Islands: admixture with Jamaican cycads in, 77; *Zamia* in, 108
Ceiba 11 (Puerto Rico), 111t
Ceiba 33 (Puerto Rico), 111t
Cencocopi as *teocentli,* 174

Cenozoic, 33; floristic refuges of, 42
Centeuctli (maize deity), 235
Centli/Cintli, 52, 169, 236t; as compared to *ixim,* 192; cycads as, representing primordial foodstuff, 195; as "dried ear of maize," 174; as ear of maize, 212; as food appropriate for humans, 174–175, 212; as maize/beyond maize, 195; as maize ear/cycad cone homology, 220; as Middle Formative maize/cycad category, 208; as "spindles," 174f; in Tenochtitlan 5 ages sequence, 174, 208
Central America: early maize in, 132–133; raised fields in, 138
Ceratozamia, 20, 27, 35, 36f, 63, 126; absence of starch reference database for, 165n4; clades of, 37; cultural relationships of, with maize, 225f; description of seeds of, in the Tuxtlas, 203; distribution of, 215f; evolutionary origins of, 78–79; as food, 51; habitats of, 37; leaves and cataphylls of, 30, 78; longevity of, 79; Mesoamerican hot spots of, 41–42; Nahua names for, 236t–237t; overview of, 37–38; population genetics of, 75; seeds of, as resembling eggs, 203; seeds of, as insecticide, 53; stems of, 30, 37, 78; as SMA, 219, 234; as *tepecentli* in Totonac/Nahua, 213–214; *Thipaak* term applied to, 178; Totonac names for, 239t; Totonac term *kun* used for, 179; undescribed species of, 37
Ceratozamia alvarezii, 35, 37
Ceratozamia becerrae, 37; in honor of Becerra, 48; synonymization of, with *C. zoquorum,* 48
Ceratozamia brevifrons, 45f; diversification of, 78–79; reestablishment of, as viable species, 44; as synonymous with *C. mexicana,* 44
Ceratozamia chamberlainii: Nahua names for, 236t–237t; as Teenek *konlib/konlif,* 176, 178–179
Ceratozamia chimalapensis, 37
Ceratozamia decumbens, 37; diversification of, 78–79
Ceratozamia euryphyllidia, 37
Ceratozamia fuscoviridis, 5f, 6f, 37, 80f; as *Chicōmexōchitl/Chicome-Sintli,* 175; decorative use of, 52; designation of epitype of, from Molango, 46; diversification of, 78–79; in European botanical gardens, 46; genetic diversity and structure of, 72t, 79; genetic variation in, 81; as Huayacocotla teocentli in 1560s Hernández text, 213–214; leaves of, used in Michaelmas, 237; Nahua names for, 236t–237t; as published by William Bull, 46;

Ceratozamia fuscoviridis—continued
range expansion of, 46; as SMA, 234, 235, 236t, 237t; taxonomic confusion surrounding, 46; "tentative" description of (as "*Ceratozamia fusca-viridis*"), by Moore, 45–46; as *teosinte*, 172; as Teosintle/San Miguel, 237
Ceratozamia hildae, 37; diversification of, 78–79
Ceratozamia hondurensis, 37
Ceratozamia huastecorum, 37
Ceratozamia kuesteriana, 37, 80f; diversification of, 78–79; genetic diversity and structure of, 72t, 75, 79; in IBD models, 81
Ceratozamia latifolia, 37; as described by Miquel, 44; female strobilus of, compared to Arroyo Pesquero jadeite sculpture, 203; female strobilus of, compared to Olmec imagery, 205f; as Teenek *konlib/konlif*, 176, 178–179, 226, 228–229; Teenek terms for, 228
Ceratozamia longifolia, 44
Ceratozamia matudae, 37; community cycad nursery with, 57–58; sympatry of, with Z. *soconuscensis*, 48; as *tapacapon/tapacarbón*, 187
Ceratozamia mexicana, 37; confusion about identity of, 49; diversification of, 78–79; eaten as food in Chiapas/Oaxaca, 128; genetic diversity and structure of, 72t; megasporophyll/microsporophyll/strobili of, 204f; other species under synonymy with, 44; population of, in El Mirador, Veracruz, 49; as *tapacapon/tapacarbón*, 187
Ceratozamia mexicana var. *tenuis*, 49
Ceratozamia microstrobila, diversification of, 78–79
Ceratozamia miqueliana, 37; as described by Wendland, 45; identification of *C. becerrae* as, by Becerra, 48; rediscovery of, 45
Ceratozamia mirandae, 37; community nursery project with, 57–58
Ceratozamia mixeorum, 37
Ceratozamia "Molango." See *Ceratozamia fuscoviridis*
Ceratozamia morettii, 37; diversification of, 78–79
Ceratozamia norstogii, 37; complex, population genetics of, 79
Ceratozamia robusta, 35; as synonymous with *Ceratozamia mexicana*, 44
Ceratozamia robusta, 37
Ceratozamia sabatoi, 37; diversification of, 78–79

Ceratozamia santillanii, 37
Ceratozamia subroseophylla, 37
Ceratozamia tenuis, 37; as *C. mexicana* var. *tenuis*, 49; community cycad nursery of, 57
Ceratozamia totonacorum, Nahua names for, 236t–237t
Ceratozamia vovidesii, 37; ovulate strobili of, 204f
Ceratozamia whitelockiana, 37
Ceratozamia zaragozae, 37, 80f; diversification of, 78–79; genetic diversity and structure of, 72t, 75, 79; illegal trade in, 79
Ceratozamia zoquorum, 37; synonymization of *C. becerra* with, 48
Cereal grasses, 214
Ceremonial bar, 190t
Cerén (El Salvador), 137
Cerro Tepezcuintle (Tuxtepec, Oaxaca), 240
C-4 pathway plants, 210n8
Chak waj: among Tojol-ab'al, 241–242; among Yucatec Maya, 242
Chalcatzingo Vase, 199f
Chalchiuhtlicue, 174f
Chamaedorea palms, in Chiapas community cycad nurseries, 57–58
Chamaedorea tepejilote, 179, 180
Chamal, 52, 236t; cones of, on Tancoyol church façade, 230; *dameu* as derived from, 230; *Dioon angustifolium* as, 236t; *D. edule* as, 236t, 237t; *D. purpusii* as, 236t; *D. spinulosum* as, 236t; *D. tomasellii*, 236t as; Nahua relation to, 234; name of, derived from *Tzamaal*, 229, 236t; in NE Mexico, 221; as preferred to maize, 232; seeds of, as derived from human brains, 232; term discussed, 238; as term used in mestizo culture, 238; in Xi'iuy culture, 167, 185, 233t
Chamal chico, 52, 236
Chamalillo, 52, 236; *palmilla*, 236
Chamberlain, Charles J., 47–48; *C. mexicana* location recognized by, 49
Chaneques, 236
Chapulhuacán (Hidalgo), 52
Chavarrillo (Veracruz), 48
Chemín-Bassler, Heidi, 231
"Chewing" tools, 115
Chiapanec (culture), 52, 128; cycad rituals in, 186; as formerly Chiapanec-speaking, 186; herald presented in language of, 186
Chiapas, 40, 45, 48, 82f; community cycad nurseries in, 57–58; sales of cycads from community nurseries in, 59; SMA absent from, 241

Chiapas subgraph of *Dioon,* 86f, 87; as solely containing *Dioon merolae,* 87
Chichimeca, 230
Chickens, 172
Chicomesintli. See *Chicome-Sintli*
Chicome-Sintli: as cycad, 175, 235, 236t; in Huastecan Nahua thought, 235; as maize-cycad deity, 234, 235; as maize deity, 175, 235
Chicōmexōchitl, 207f, 234, 235, 236t; association of, with writing, 190t; as *Chicomexochil,* 236; as Corn Master narrative, 167; as cultural hero, 177; as cycad, 175; as fertility hero, 188; in Huastecan Nahua thought, 235; as maize/cycad deity like Teosintle, 221, 229; as Mesoamerican maize deity, 175
Chicomoztoc, 208
Chigua, 247; Indigenous poem about, 247
Chikomexochitl. *See* Chicōmexōchitl
Child deities, cycad/maize, 187; in Corn Master narratives, 200; in Ch'orti' Maya, 209n3
Child-maize god, 176
Children in maize/cycad stories: antagonism of, with grandmother, 187; cycad seedlings as, 235, 236t; and Homshuk, 194; maize ancestors as, among Nahua, 235; as sprouting from split bottle gourd, 208; in Xi'iuy culture, 232–233
Chili pepper, starches of, in archaeological context, 111
China, *Cycas* in, 252
Chinantec (people), no SMA among, 240
Chloroplast DNA, 81, 158; of leaves, 101; used for detecting spread of *Cycas revoluta,* 159
Chocó region (Colombia), 247
Chola (cycad food), 123
Ch'ol Mayan, 172
Chontal de Oaxaca: cycad terms in, 239t; more research needed among, 243; SMA among, 241
Chorro de Maíta (Cuba), 111t
Ch'orti' (Maya), 190; spirit of maize in, 209n3
Christianity, evangelical, 226
Chromoplasts of leaves, 101
Chromosomal fissions in cycads, 49
Chukmal, 227t, 228
Chum, 227t
Ciénega del Sur (Veracruz), cycad nursery in, 57
Ciliate antherozoids, 32, 34
Cincocop, 174f; as *cincocop/cintrococopi,* 174; as *cincocopi,* in reference to *Zea,* 214
Cinnamon in *Zamia* food, 121f
Cintektli, 235

Cintli. See *Centli/Cintli*
Cintli i nana, 235, 236t
Cintrococopi. See *Cincocop*
Cipactli, 180–184, 187, 188, 196, 207f; as "alligator," 180; as ancestor, 200; as aquatic monster, 181–182; as calendrical day, 180; Classic/Post-Classic mythology of, 181; codical/glyphic/sculptural depictions of, 183f; as crocodilian deity, 180, 198, 226; as cycad, 207f; as equivalent to Imix, 180; as maize/cycad deity, 198; at Palenque, 182–183; shark/piscine creature as, 181, 198; as source for new life, 200; starry-deer-crocodile similar to, 182; and *zipac* alligator gar, 181
Cipacuatli. *See* Cipactli
Çipaqli. *See* Cipactli
Circinate leaves, emergence of, 8
CITES, 53; enforcement of, and cycad barcoding, 252
Classic, Early (Mesoamerica), 134; Maya Maize God in iconography of, 191f
Classic, Late (Mesoamerica), 117, 134
Classic, Terminal/Late (Yucatán/Maya), 136, 152
Classic period (Mesoamerica), 138–139; Cipactli mythology in, 181–182; Corn Master traditions of, 167; images from, related to Formative period images, 192; images from, depicting humans sprouting from trees, 201f; Maya Maize God in, 187–192
Clavijero Botanic Garden, 50, 50f, 55–57, 61; sales of cycads in, 59. *See also* National Cycad Collection at Xalapa
Climate change, 140; affecting *Dioon,* 39; affecting *Dioon* evolution, 84; and *Ceratozamia,* 78; as reflected in cycad morphology, 34; in relation to cycad hot spots, 42; research on, affected by cycad research, 60; as threat to cycads, 24; as threat to food security vis-à-vis *Zamia* consumption in the Caribbean, 120
Coastal habitat: beaches as, for *Dioon edule,* 83; sand dunes as, for Mesoamerican *Zamia,* 127; sand dunes as, for *Zamia furfuracea,* 57; scrub as, for Caribbean *Zamia* habitat, 41
Coatzacoalcos (Veracruz), cycads as ornamentals in, 59
Cocijo, 186
Coco, Río. *See* Segovia, Río
Coconuts, 62
Cocopatic, 174f
Cocoyam: secondary role of, in pan-Caribbean foodways, 122; *Xanthosoma* spp. of, as dominant in Early Formative (Mesoamerica), 210n9

Cocultivation of cycads and other cultivars, 160f
Codex, Borgia, 183
Codex, Dresden, 210n7
Codex, Madrid, 210n7
Codex Fejérváry-Mayer, 183f, 208
Colds, common, and cycad medicine, 128
Colombia, 53; cycads in, 245; maize-related terms for cycads in, 242; northwest, and early maize, 132
Colón (Honduras), 90
Comida antígua, 239t, 241
Comida de gentil, 239t, 241
Comida de los abuelos, 239t
Compañero de maíz, 226, 227t
Condif as variant of *konlib,* 179, 227t
Cones, cycad, 30–32: female, as head, 233; female, in Teenek cycad-maize terminology, 227t; male, as phallus, 186; male, in Teenek cycad-maize terminology, 227t; pollen, 31f; seed, 31f
Conifers, 8, 32
Conlif, 227t; as variant of *konlib,* 179
Containers, ceramic: as artifacts to look for ancient zamia evidence on, 121f; for reception/storage of zamia dough, in archaeological context, 115, 116
Convention on Biological Diversity, 54–55, 57
Convention on International Trade in Endangered Species of Wild Fauna and Flora. *See* CITES
Cooking of *Zamia,* archaeological evidence of, 112; ceramic pots as, in ancient contexts, 116; fire-altered rocks as, 112
Coontie, 97, 123
Copper in siderophores, 155
Coprolites, plant starches found in human, 102
Coral: grater teeth of, 115; tools made from, 102, 109f
Coralloid roots, 7f, 30, 130–131, 156, 157f, 160f; better understanding needed of, 155; coralloid masses on, 130; cyanobacteria on, 7f, 30, 130, 155; effects of fire/burning on, 131; functions of, 30
Cormorants, 190t
Corn. *See* Maize
Corn Master: definition of, as Popolucan expression, 209n1; among Mije, 184; narratives of, 167, 169, 184–185, 187, 200; ovogenesis in, stories of, 200; among Popoluca, 184; Thipaak as, 176–178; among Xi'iuy, 184–185
Corn palm (cycad), 242

Cortés, Hernán de, 217; expedition of, to Honduras, 217–218
Coscomatepec (Veracruz), 220
Cosmogenesis, 186
Costa Rica, 77; northern, as possibly part of *teosinte-*as-grass region, 218, 221
Coulter, Thomas, 45
Coxcatlán (Mexico): in context of Greater Caribbean ancient zamia sites, 117; rockshelter containing ancient *Dioon* and other plants, 147, 148, 148f, 149
CpDNA. *See* Chloroplast DNA
Cress, thale. *See Aradopsis thaliana*
Cretaceous period, 9, 61
Crocodile. *See* Crocodilian
Crocodilian, 170, 184, 188, 207; connected to scepter/staff, 204; cycads as, 206; as Mesoamerican mythological entity, 181, 182; in representations of Olmec Maize God, 196–198; teeth of, becoming roots, 200; in Young Lord imagery, 196
Crossozamia, 8; compared to Cycad *revoluta,* 33, 34f
Crystal in Teenek cycad/maize mythology, 178–179
C_3 plant, 129
Cu. *See* Copper in siderophores
Cuacintle, 236t
Cuacintli, 236t
Cuauhmochitl, 171
Cuautitla, cintli, 236t
Cuautitla, sintli (de), 235, 236t
Cuba: archaeological sites with zamia starches in, 111t, 112; as likely genetic origin of Caribbean *Zamia,* 77; Mexican cycad seeds sent to, 45; *Zamia* as foodstuff in, 97; *Zamia* in, 108
Cucurbita. See Squash
Cucurbits. *See* Squash
Cueva de los Muertos (Puerto Rico), 110t
Cueva Ventana (Puerto Rico), 110t, 114
Cuicateco (people), cycad use of, 240
Culinary evolution, 167, 173–175
Culinary identities and ancient Caribbean zamias, 117, 122
Culinary-ideological assets, zamia as, 117
Culinary practices in zamia: as blended by African slaves in Dominican Republic, 123; as related to exchange networks, 117
Cultural ecology, 151
Cultural-symbolic significance of cycads. *See* Symbolic uses of cycads
Cunelcintli, 235, 236t
Cyanide glycosides, 99

Cyanobacteria, 130, 131, 154, 155; as BMAA producers, 161; DNA of, 159
Cycadaceae, 27, 99
Cycadales, 5, 27, 33, 61, 92, 126–127, 170; genetic patterns in, 95; as nitrogen-fixing, 125; pollen of, in archaeological context, 118; slow reproduction rate of, 70
Cycad anatomy: evolution, 8, 33; overview, 5, 28f, 29–30; as reflective of climate, 33–34
Cycad Asian Scale (CAS), 54
Cycad biology overview, 6, 27–43
Cycad conservation, 23, 24, 53–60; biocultural conservation efforts, 53; and community nurseries, 55–60; demographic studies' roles in, 60; ex situ, as best occurring in country of origin, 57; gender issues in community nurseries, 58; in Honduras, 53; as inclusive of cultural traditions, 124; and issues with commercialization, 58–59; landscape genetics studies as useful for, 68; in Mexico, 53, 54–60; population genetics as useful for, 65, 70, 91–92; risk and threat categories in, 24, 27, 53–54; roles of local and Indigenous communities in, 24–25, 53; sustainable utilization as strategy for, 55–60
Cycad distribution, 7f, 33; in Belize, 35, 37, 40; in the Caribbean region, 35, 40–41; in the Central American Isthmus, 35; in El Salvador, 35; in Florida, 35; in the Greater Antilles, 35; in Guatemala, 35, 37, 40; in Honduras, 35, 37, 38, 40; in the Isthmian-South American region, 35; in the Mesoamerican region, 35; in the New World, 35, 36f; and range restriction of New World species, 35
Cycadeoidales, 9, 33
Cycad (folk) terminology: in Mexico, 53; in Mexico and Central America, 63
Cycad-grass *teosinte,* 222
Cycad harvesting camps, 90
"Cycad hell." See *Sotetsu jigoku*
Cycad leaves: for decoration of church facades in Mexico, 52; for decoration of village in Mexico, 52; for decoration of Rain Queen's royal *kraal,* 15; [foliage] as archive of isotopes of atmospheric nitrogen, 155; on floral arches with dried maize cones, 237; as green manure, 11; in Mexican pilgrimages, 52, 64, 87, 128; morphology of, 30; in rituals, 15, 17–18, 240; as roofing, 52; in temples and shrines, 15, 17–18; in term "sacred leaf," 220, 236
Cycad longevity, 8; in *Dioon,* 38; of *Dioon edule,* 48; in Japan, 14
Cycad morphology, 27, 28f, 29–30; in

Ceratozamia, 78; convergent evolution of, in Caribbean *Zamia,* 41
Cycadophytes, 33
Cycad pollination, overview of, 5, 32
Cycad reproduction, 30–31; erroneously thought to be wind-pollinated, 48; evolution of, 32; fertilization overview, 32; reproductive anatomy, 31f; work done on, 48
Cycad roots: contractile, 30; overview of, 7; system, 30
Cycads: abundance and, 18; in archaeological research, problems of, 141–142; associations with poverty in the Ryukyu Islands, 25; branching of, 30; as crocodilians, 206; cultural identity/heritage and, 26, 52; defined, 5–8, 27; distribution of, 35–36; fish and, 181; hallucinatory effects of, 128; human longevity and, 10, 15, 18; in Indigenous lexicons, 18; in Mesoamerica and the Americas, 20–22; in the New World, pollination of, 32; overview of evolution and distribution worldwide, 32–34; overview of human-cycad relationships, 1–2, 7, 9–20, 23–24, 26; resistance to fire, 30; as "the Rosetta Stone," 33, 34; sense of place and, 26, 52; sexual potency/fertility and, 10, 15, 18; and snake deities, 15, 17f; snakes and, 16, 18
—fossil: in the Mesozoic Era, 8–9, 33; morphology of, 8; in the Paleozoic, 8
—rain and: and Chiapanec cycad pilgrimage, 186–187; and rainmaking, 15; related to pulling down cloud moisture, 129; Teenek associations of, 179. See also Thunder
—resistance to drought, 30, 33–34; in *Dioon,* 38
—as toys: as bullroarers (*zumbadores*), 53; as children's toys/whistles, 128; as potential, in Tamaulipas cave, 128
—traditional knowledge/use of: disappearance of, 53; as threats to cycad conservation, 53
Cycads in agriculture: cultivation of, in ancient Honduras, 218; defined as "wild" plants, 3–4; in hedged fields, 12; in managed landscapes, 11–12, 12f; for pest control, 10, 129; related to development of social complexity, 24; related to transitions to agriculture, 11, 24, 142
Cycads in art/architecture: design motifs on Catholic church facades, 17–18, 230, 256; in folklore, 51–52; iconography of, in Formative Gulf Coast, 200–206; in Japanese art, 15, 16f, 17f; in Japanese literature, 14–15, 17f; in music, 12; as national monuments in Japan, 15; in rock art, 9, 10; in South Africa, 15, 18f; in textiles, 12

334 · Index

Cycads as food: in Australia, 9; in the Caribbean, 25, 97–101; in Colombia, 53, 247; consumption of leaves, 9; consumption of sarcotestas, 9–10, 64, 213; consumption of trunks (stems), 9, 51, 98–99; as dominant in Early Formative, 210n9; as famine food, 52, 62–63, 150, 231; in 1560s Hernández text, 213; graphic representations of, 13; in Honduras, 219; importance of, in ancient Honduras, 218; in Jamaica, 248; loss of knowledge about, 243; as luxury food, 232; in the Mexican archaeological record, 53–54; in miso, 12, 13f; overview, 51–52; for purchase online, 12, 13f; related to cycad genetic structure, 63; in the Ryukyu Islands, 12, 13f; as starchy staple, 62–63, 150; at Tamaulipas cave sites, 143–147; among Xi'iuy, 231

Cycads in religion: sacred sites and, 128
—in Hinduism: in Bali, 19f; in India, 19f
—in Roman Catholicism: in Chiapanec cycad pilgrimage, 186; in Mexico, overview of, 52; in the Philippines, 17; in Thipaak stories, 226
—in temples and other religious landscapes, 15, 17–18; in Bali, 19f; in India, 15; in Japan, 14–15; in the Philippines, 17; in Vietnam, 19f

Cycad species, increase in number known of, 35, 251–252; in Mesoamerica/Mexico, 43, 126

Cycad starch: analysis of, 100; as food, 51; in stems, used as food, 51, 97–101

Cycad taxonomy: in the Caribbean, 40–41; controversy surrounding, in Caribbean *Zamia*, 41; history of, in Mexico, 43–49; in Mesoamerica, 37–40; in Mexico, overview of, 20; world overview of, 6, 27–28

Cycads, uses of: as insecticides, 53; magical, 10–11, 51; medicinal, 9, 10, 52, 127; medicinal, *teocentli* as, in 1560s Hernández text, 213; among Honduran cycads, 219; in Japanese sake, 12; as narcotics, 10–11, 128; as ornamentals, 52, 53, 59, 59f, 127; as poison, 52; for preventing bad dreams, 16, 18; as seabreaks, 11; as symbolic mediators in Pan-Mesoamerican mythology, 206, 207f, 208; in tourism in Japan, 16f; as utilitarian items, 9, 10; in weddings, 15; as windbreaks, 11

Cycads, regional overviews of human relationships to: in southern Africa, human uses of, 10–11, 15, 18f; in Vanuatu, 12–14
—in Mesoamerican foodways: ancestral relation to ferns, 32; as early wild food resource, 2–3, 23; relationships to palms and ferns, 6, 27, 127; in symbolic systems, 3

—in the Ryukyu Islands, 11–12; archaeology, 11; for food, 12; human uses, 11–12; threats to, 25

Cycas, 7f, 27; increase in species numbers of, 252; leaf morphology, 30; as "living fossils," 32–33; population genetics of, 75; pseudocone of, 32

Cycas balansae, genetic diversity and structure of, 73t

Cycas beddomei in medicine, 10

Cycas circinalis: as decoration, 15; in medicine, 10

Cycas debaoensis: genetic diversity and structure of, 74, 74t; in IBD models, 81

Cycas elongata, 19f

Cycas guizhouensis: genetic diversity and structure of, 73t; pollen and seed dispersal in, 70

Cycasin, 27, 51, 99, 127; compared to MAM, 121f; as proxy for ancient zamia detection, 121f

Cycas micronesica: effect of CAS on, 54; genetic diversity and structure of, 73t, 77; moth pollination of, 32

Cycas multipinnata, genetic diversity and structure of, 74t

Cycas multipinnata, pollen and seed dispersal in, 70

Cycas orixensis as decoration, 15, 19f

Cycas panzhihuensis, genetic diversity and structure of, 74t

Cycas parvula, genetic diversity and structure of, 73t

Cycas pectinata, 246f; as decoration, 15; genetic diversity and structure of, 73t; pollen dispersal of, 79; for sexual potency, 10, 11f

Cycas revoluta, 8; genetic variation of, 74; human role in spread of, 159; in Japan, 11–12, 14–15, 16f, 17f; megasporophyll of, on Basilica di San Marco facade, 256, 256f; as out-competing *Dioon edule* in German horticultural market, 58–59; in the Philippines, 17; similarity of, to fossil cycads, 33, 34f

Cycas rumphii: in Bali, 19; in ceremony, 14f; genome sequencing of, 253

Cycas seemannii, genetic diversity and structure of, 73t

Cycas siamensis, genetic diversity and structure of, 73t

Cycas simplicipinna, genetic diversity and structure of, 74t

Cycas taitungensis: genetic diversity and structure of, 73t, 74; pollen and seed dispersal in, 70

Index · 335

Dameu, 230; as miraculous plant, 231; stories about origin of, 231–233; term, as derived from *chamal/tzamaal*, 230, 233t
Dance encompassed by TMG/GCMH, 188, 190t
DART-MS: in cycad studies, 161, 162f; defined, 165n5; to detect BMAA in cycads, 162
Day of the Dead: cycad leaves in, 64; cycads in, 52
Day of the Holy Cross. See Día de la Santa Cruz
De Luca, Paola, 49
Defecation. See Excrement as food for humans
Deforestation and early maize cultivation, 134
Dehesa, Theodore, 48
Deities, Mesoamerican, definitions of, 168
De las Casas, Fray Bartolomé, account of *guáyiga* cycad food by, 98
δ15N. See Nitrogen, isotopic, measurement as cycad-detection method
Demographic events: affecting evolution of cycads, 42
Demographic movements: of humans, affecting cycads, 93
Demographic studies, importance of, for cycad conservation, 60
Dendrobium officinale, SNPs used for assessing genetic diversity of, 66
Dental calculus (dog), as artifact to look for ancient zamia evidence in, 121f
Dental calculus (human): research needed to detect cycads in, 139; starches preserved in, 102; *Zamia* starches in ancient, 114, 115, 116, 136
Dental calculus (hutia), as artifact to look for ancient zamia evidence in, 121f
Deserts: as *Ceratozamia* habitat, 78; as *Dioon* habitat, 42, 127; as *Zamia* habitat, 127
Detoxification (toxin removal) of cycads, 9, 51, 64, 127; in Australia, 152; convergence of methods of, in Mexico and Honduras, 90, 218; ethnoarchaeological research on, 152; in Honduras, as like methods in the Caribbean basin 219; Honduran methods of, as similar to Mexican methods, 219; in Mexico, 229
Devil in Teenek cycad terminology, 228t
Dhipak, amigo de, 227t
Dhipak, ayudante de, 227t
Dhipak. See Thipaak
Dhipak (as Teenek term for cycad), 228t, 229
Dhókob, 227t
Diablo, planta del, 228t, 229–230
Diablo Cave (Sierra de Tamaulipas), 144
Diablo (Christian "Devil"), 228t

Diablo phase (Sierra de Tamaulipas), 145t
Día de la Santa Cruz, cycads used during, 52, 186
Diadem on Tenaspi Egg, 201
Diarrhea and cycad medicine, 128, 213
Diaspora, Nahua-speaking, 219
Dicotyledons, 193
Dien Tien Hoang Temple complex, 19f
Dinitrogen gas, 129
Dioecious. See Dioecism
Dioecism (dioecious reproduction) of cycads, 30–31, 169; in *Ceratozamia*, 79; cycad-grass *teosinte* and, 222; and maize/cycad deity, 222; related to gender duality in Mesoamerican religion, 185–187; as related to gene flow and variation, 70; and sexual union, creating maize in Teenek thought, 229; in Teenek mythology, 185–186
Dioecy. See Dioecism
Dion as alternative to *Dioon*, 44–45. See also *Dioon*
Dion edule. See *Dioon edule*
Dioon, 20, 27, 35, 36f, 63, 126; absence of reference database of starches for, 165n4; as absent from Central American *teosinte*-as-grass region, 218; arborescent stems of, 30, 38; beetle/thrips pollination of, 32, 83; biogeographic events affecting, 81; as *chamal*, 229; clades of, 38; cones of, 38; conservation of spelling of, 44–45; crown age of, 39; cultural relationships of, with maize, 225f; distribution of, 215f; edible sarcotesta of, 83; epidermal anatomy of, 33, 38; as food, 51, 53, 91f; as food, in early Tehuacán Valley caves/rockshelters, 149; hot spot of, 42; human-induced dispersal of, 88–89; leaves of, 38; leaves and cataphylls of, 30; longevity, of, 38, 59; low vagility of pollinators in, 87–88; molecular phylogeny of, published, 49; morphology overview of, 83; overview of distribution of, 82; population structure of, 63; seed dispersal of, 83; as SMA, in NE Mexico, 219; as *teosinte* in western Mexico, 215f; transplantation of, for ritual/food reasons, 218; trunks of, related to gar scales, 181
Dioon angustifolium, 39, 83; as *chamal* among mestizos, 238; in dry cave sites in Tamaulipas, 171; exclusion of, from population graph, 86; genetic diversity and structure of, 72t; Nahua names for, 236t–237t
Dioon argenteum, 39, 42, 84; exclusion of, from population graph, 86

336 · Index

Dioön as alternative to *Dion,* 44
Dioon califanoi, 39, 42, 84; discovery of, 49; genetic diversity and structure of, 71t
Dioon caputoi, 35, 39, 42, 82f, 84; discovery of, 49; genetic diversity and structure of, 71t, 72t, 84; populations of, excluded from Oaxaca and Northeast subgraph, 87, 93
Dioon edule, 5f, 6f, 28f, 39, 44f, 82f, 83, 129; attempted introduction of nursery-grown plants of, to German horticultural trade, 58–59; as *chamal,* 167; as *chamal* among mestizos, 238; conservation of, through nurseries and sustainable utilization, 55–57, 58–60; domestic market in, for nursery-grown plants, 59; export of seedlings of, to the US, 59; female seed-bearing strobili of, resembling corn fetish iconography, 204–206; first taxonomic description of, 43; as food, 127, 229; found in ancient Mexican caves, 135; genetic diversity and structure of, 71t, 72t; genetic flow of, connecting to *D. mejiae* and *D. holmgrenii,* 88; growth rates of, in nurseries, 59–60; lack of genetic connection between Querétaro and Veracruz populations of, 88; longevity of wild plants of, 59, 75; Nahua names for, 236t–237t; Nahua relationship to, 234; as ornamental, 59f; poaching of crowns of, for sale, 54; reintroduction of nursery-grown plants of, to wild, 58, 59–60; relationships between seed availability and seed size of, 81; reproductive rate of, 70; SMA concept among Teenek largely lacking for, 230; in Tamaulipas cave sites, 144–149, 171; Teenek ter ms for, 228; in Tehuacán Valley caves/rockshelters, 149, 171; threats to, 54; as *tiotamal,* 219; type locality rediscovery of, 48; in Xi'iuy cycad mythology, 185
Dioon holmgrenii, 39, 84; discovery of, 49; as eaten by Zapotecs, 240; genetic diversity and structure of, 71t, 72t; genetic flow of, connecting to *D. edule* and *D. mejiae,* 88
Dioon mejiae, 38, 83, 218, 221, 231, 242; distance of, from other *Dioon* species, 83; genetic diversity and structure of, 71t; genetic flow of, connecting to *D. edule* and *D. holmgrenii,* 88; only one local name for, 217, 219; origin of name, 216; significance of genetic connection of, with Querétaro, 88; SMA and, 242; *tiusinte* region of, in Honduras, 215f
Dioon merolae, 39, 82f, 84; Chiapas populations of, excluded from Oaxaca and Northeast subgraph, 87, 93; Chontal de Oaxaca and, 239t, 241; community nursery project with, 57–58; discovery of, 49; Día de la Santa Cruz pilgrimage involving, 52, 128, 186–187; distinctiveness of Agua Prieta population of, 87; genetic diversity and structure of, 71t, 72t, 84, 87, 88
Dioon oaxacensis, 39, 42, 84
Dioon planifolium, 39, 84, 85; exclusion of, from population graph, 86; recent discovery of, 86
Dioon purpusii, 39, 42, 84; genetic diversity and structure of, 71t; Mazatec (Oaxaca) names for, 239t; Nahua names for, 236t–237t
Dioon rzedowskii, 38, 42, 83; discovery of, 49; genetic diversity and structure of, 71t; Mazatec (Oaxaca) names for, 239t
Dioon sonorense, 38, 39, 85; bottleneck events and low vagility of pollinators in, 87–88; genetic diversity and structure of, 71t, 72t; used for alcoholic drink, 128
Dioon spinulosum, 38, 42, 83; genetic diversity and structure of, 71t; Mazatec (Oaxaca) names for, 239t, 240; Nahua names for, 236t–237t; seed germination of, 203f; seeds of, used as toys/bullroarers, 53, 128
Dioon stevensonii, 38, 39, 85; genetic connection of, to *D. vovidesii,* 88; genetic diversity and structure of, 71t
Dioon tomasellii, 35, 38, 85; discovery of, 49; genetic diversity and structure of, 71t, 72t; Nahua names for, 236t–237t; paucity of ethnobotanical knowledge of, 220; as possibility for *teosinte,* 220
Dioon vovidesii, 38, 39, 85; bottleneck events and low vagility of pollinators in, 87–88; genetic connection of, to *D. stevensonii,* 88; genetic diversity and structure of, 72t
Dioscorea. See Yams, domestic, secondary role of, in pan-Caribbean foodways; Yams, wild
Direct analysis real-time mass spectrometry. *See* DART-MS
Diversification, factors influencing, 42
Divine ancestry, 237
Divine food, cycads as, 232
DNA barcoding and cycads, 252
DNA (human), related to cycad population genetics in Mesoamerica, 94, 158
DNA (mitochondrial), used for detecting spread of *Cycas revoluta,* 159
DNA (zamia), analysis of ancient, as potential research avenue, 120
Dogs, 121f, 174f
Domesticated plants: artificial selection for

desirable traits related to, 62–63; early, in Tamaulipas caves, 143–147; human-induced movement related to, 62

Dominican Republic: archaeological sites with zamia starches in, 110t, 111t; high genetic flow of cycads in, 77; plant evidence in late ceramic age sites in, 115; *Zamia* in, 108; *Zamia* as foodstuff in colonial and more recent times, 97, 123

Dough (zamia): processing of in archaeological context, 115, 121f; discarded, in modern Dominican Republic zamia preparation, 123

Drum, 190t; carapace, 190t

Duality, gender, 185–187

Eagle. *See* Aach-Eagle grandmother

Earspools, on Tenaspi Egg, 201

Earth monster as name for maize/cycads, 173

Easter Week: cycads used in, 52; cycad leaves used in, 64

Ecuador and early maize, 132

Edges (population genetics): defined, 68; in *Dioon* population graphs, 86f

Edilio Cruz (Dominican Republic), 111t

Edule clade, 39, 83; population graph of, 86

Effective population size (N_e), 69

Eggs: and *Ceratozamia*, 204; in Corn Master narratives, 184, 200, 202; cycad seeds as, 206; forms of, 202. *See also* Ovogenesis; Tenaspi Egg

Ehatal, 176

Ehecatl-Quetzalcoatl, 184

Ejatal em, 227t

Ejattaläb, 227t

Ekmito'tol, 227t, 228

Elolcintli, 236t

Elote (cycad), 236t

Elote (maize), 232

Elotillo, 236t

Elotsintli, 235, 236t

Eloyo, 236t

El Salvador, 137; *teosinte* in toponyms in, 214

ēm, 227t

Embryos/fetuses, Olmec imagery of aborted, 194

"Emergence" (iconographic theme), 181, 188, 189f, 192f

Encephalartos: human uses of, 10–11; population genetics of, 75

Encephalartos barteri ssp. *barteri*, genetic diversity and population differentiation in, 74

Encephalartos transvenosus, 15

Encephalartos woodii, 24

Endemism: in cycads, 35; in Mexican cycads, 63

Endotesta (cycad) in Teenek cycad terminology, 228t

England, cycads cultivated in, 43

Enzymatic digestion and starch survival, 102

Epidermis, 33

Epistemologies: Indigenous, 167; Teenek, as compared to Western, 168

Epistomatal chamber, 33

Erotylidae, 32

Escalero al Cielo (Yucatán), cycad starch on tools at, 136

Eslabones phase (Sierra de Tamaulipas), 145t

Espadaña. *See Dioon merolae*

Ethnoarchaeology, 152–154, 160f; in Australian cycads, 152

Ethnographic investigations in cycads, 91, 95, 150; correlated with archaeology, 95; correlated with specific archaeological objects containing zamia, 119; historical, in Greater Antilles, 98–99; of SMA, 223–243; urgency of, in Puerto Rico, 124; value of, in Caribbean zamia research, 120

Etiological creation narratives in Mesoamerica, 167, 173

Etruscans, 245

Euchlaena. *See Zea luxurians*

Eumaeus: caterpillars of, in milpa, 226; DART-MS used to detect BMAA in, 161, 162f

Eva 2 (French Guiana) archaeological site with zamia, 110t, 111–112, 113f, 114, 115, 117

Exchange networks, ancient pan-Caribbean: maritime, 122; in relation to movement of domesticates, 116–117; in relation to zamia, 117, 122

Excrement as food for humans, 176; in Flat Earth Teenek world age, 177

Expected and observed heterozygosity (H_e, H_o), 67; in cycads, 71t–74t

Extinction cross. *See* Maltese cross

Extinction of cycads, 24

Extraction of cycads from the wild, 53–54; of Tomasellii clade species, 85

Eyes (human) becoming *guapilla*, 233

Fabaceae. *See* Legumes as nitrogen-fixing

Family pedigrees, 66

Famine foods. *See* Hunger/famine

Farmers, subsistence: as damaging cycad habitat, 55; as participants in community cycad nurseries, 55

Female maize spirit, 209n3

Fermentation of maize, 195

338 · Index

Ferns, 8
Fertility: as associated with TMG/GCMH, 188, 190; in Mesoamerican stories, 187
Fertility deities, Mesoamerican, 166; definitions of, 168; "maize" gods as, 167
Fertility Deity Complex, 169
Fields, raised. *See* Fields, relict drained (systems)
Fields, relict drained (systems), 138, 139
Fields, terraced, 137, 138, 139
Fincas (cycad), 218
Fine Scale Genetic Structure. *See* FSGS
Fire/burning: as affecting *Dioon merolae* in Chiapas, 87; agricultural, effect on cycad growth and BNF, 131; in agroecological mythological symbolic regimes with cycads, 170; in the Archaic period in relation to agriculture, 134; and "charcoal" in cycad names, 187; and early foragers, 134; regeneration of cycads after, in mythology, 187; in Xi'iuy cycad-origin stories/mythology, 185, 232–233
Fischeri clade, 40, 77
Fish, 207f; and crocodilians, 182; as human ancestor, 210n6; humans becoming, in Teenek mythology, 210n6; piscine imagery, in Formative iconography, 196; piscine-monster tails, 199f *Popol Vuh* Hero Twins as fish people, 181; in Teenek world ages, 176, 177t, 210n6; in Tenochtitlan 4 age sequence, 174f. *See also* Gar
Fissures in zamia starch grains, 104t
Flacco phase (Ocampo), 145t; *Dioon edule* remains found in caves of, 144
Flat Earth, 177t
Flesh (human), cycads as, 238, 239t
Flood: in *Chilam Balam,* 182; great, in Tenochtitlan 5 ages sequence, 174f; in Teenek world ages, 176, 177t, 210n5
Flora de Veracruz, 45
Flor de Dhipak, 227t
Flor de konlif, 227t
Florida, La, phase (Ocampo), 145t; *Dioon edule* remains found in caves of, 144
Florida, zamia storage starch used for food in, 123. *See also* Coontie
Floristic refugia in Southeast Mexico, 42
Flour (cycad), 90; of *Zamia,* processing of in the Caribbean, 121f
Flying foxes, 248
Foliation, 202
"Foodness," 167–168. *See also* Alimentation

Food security: and *Zamia,* in the Caribbean, 120–122
Foodways, Mesoamerican, pre-maize, 173–174
Foragers, 134. *See also* Hunter-gatherers
Forest, cloud: as *Ceratozamia* habitat, 37, 127; as cycad habitat, 127, 171
Forest, evergreen: as *Ceratozamia* habitat, 37, 42; as *Dioon* habitat, 42; humid tropical, as Mexican cycad habitat, 127; as *Zamia* habitat, 42
Forest, oak: as *Ceratozamia* habitat, 127; stunted, as *Dioon* habitat, 127; as *Zamia* habitat, 40
Forest, pine: as Caribbean *Zamia* habitat, 41; as *Ceratozamia* habitat, 127; as *Dioon edule* habitat, 83; landscapes of, in Honduras, as resource-poor, 218
Forest, pine-oak: as *Ceratozamia* habitat, 37; as *Dioon* habitat, 38; as *Microcycas calocoma* habitat, 41; as *Zamia* habitat, 40
Forest, semideciduous, as habitat for *Microcycas calocoma,* 41
Forest, sub-deciduous tropical, as *Zamia* habitat, 40
Forest, tropical deciduous: as *Dioon* habitat, 38; as *Zamia* habitat, 40
Forest, tropical dry, as *Zamia* habitat, 40
Forest, tropical seasonally dry: as *Dioon edule* habitat, 44f; as Mexican cycad habitat, 127
Forest, tropical thorn, conserved in Mexico as cycad habitat, 55
Formative Gulf Coast: art/ideation of, theories on, 187; cycads in iconography of, 200–206; imagery from, reliant on *Popol Vuh,* 194
Formative period, Early (Mesoamerica), 134, 166, 171, 206; foodways/diet of, 173–174, 194–196; imagery in, 194; resources dominant in, 210n9
Formative period, Late/Terminal (Mesoamerica), 134, 136; sculpture of, 202f
Formative period, Middle (Mesoamerica), 134, 136, 166; change of foodways/mythologies in, 196; diet of, 195–196; imagery in, 194, 208; maize imagery in, 195
Formative period (Mesoamerica), 136–138; art and thought of, reconsidered, 208–209; complex of fertility/maize gods in, 196; crocodilian in imagery of, 181; Maya Maize God in, 187
Founder effects: of *Ceratozamia norstogii* complex, 79; indicating migration of people/cycads, 158; related to cycad diversity, 70, 74

Francisco Clavijero Botanic Garden. *See* Clavijero Botanic Garden
Frass, caterpillar, as helpful in Teenek milpa, 226
French Guiana: archaeological site with zamia starches in, 110t, 113f, 114; archaeological *Zamia* in, 108, 122; and early maize, 132; why no natural distribution of *Zamia* in, 108
Fruit trees associated with TMG/GCMH, 190t, 190–191; ancestor figures as, compared to crocodilian trees, 200; in Thipaak narrative, 200; on vessel, 191
FSGS: definition of, 67; as useful for cycads, 67
Fungi, 155, 156
Furfuracea subclade, 40

Gamagrass. See *Tripsacum*
Games associated with TMG/GCMH, 190t
Gametophytes (cycad): Chamberlain's description of, 48; as food, 51
Ganesh (Hindu god), 19f
Gar, 170, 173, 207; alligator gar (*zipac*) as related to Thipaak; and Arroyo Pesquero celt imagery, 198; as Cipactli, 198 and *Dioon* cycads, 181
Garbanzos, 231
Gar fish. *See* Gar
GCMH, 188, 207f; equated with TMG, 188, 189–190; not equated with Hun-Hunahpu, 188; youthfulness as characteristic of, 190t
Gene flow, 64; barriers to, in *Dioon*, 81, 85; in *Ceratozamia*, 79; in *Cycas*, 74; in *Dioon edule*, 84; fine-scale, in *D. caputoi*, 84; patterns of, 68; patterns of, in cycads, 70, 86–88; population graphs and subgraphs demonstrating, 86, 87
General Law of Ecological Balance and Protection of the Environment (Mexico), 55
Genetic differentiation (*Fst*), 67, 74, 75; in *Ceratozamia*, 79; in *Dioon*, 88, 89f; in *Dioon caputoi* and *D. merolae*, 84
Genetic drift: of *Ceratozamia norstogii* complex, 79; in *Dioon*, 81
Genetic neighborhoods, 67
Genome, cycad, 159, 164
Gentil, 239t; as maize/cycad, 241
Geochemistry, 151
Geographic isolation: as affecting evolution of cycads, 42; as affecting population genetics of *Encephalartos barteri* ssp. *barteri*, 74; in *Dioon*, 81, 84
Gestation, human, in Olmec deity representations, 193

GI, 177; in Palenque panel, 182, 184
Giants: in Tenochtitlan 5 suns sequence, 174, 174f; in Teenek world creation, 176, 177, 177t
Gigante, El (Honduras), *Zamia* in, 135, 163
GII, 193
Ginkgo. See *Ginkgo biloba*
Ginkgo biloba, 8, 32; SNPs used for assessing genetic diversity of, 66
Glottochronological. *See* Glottochronology
Glottochronology, 93
Glycoside, 99. *See also* Cycasin
Gnetophytes, 8
God (Christian): and cycads, 178–179, 227t; in Teenek mythology, 178–179; in Xi'iuy cycad origin story, 232
God E, 210n7
God II. *See* Olmec Maize God
Godpowers: cycads as incarnating, 169; in Teenek epistemology, 167, 168
Gourd, bottle. *See* Squash
Grandfather in maize/cycad stories, 241
Grandmother in maize/cycad stories, 187; and antagonism with young boy motif, 209n4–210n4
Grandparents' food (cycad term), 239t, 241
Grass, 175, 221–22; *cincocopi* as, 214; *teosinte* term as applied to, 212–213, 214
Grasshoppers, 231
Grasslands as cycad habitat, 127
Graters, 115, 116, 121f; as artifacts to look for ancient zamia evidence on, 121f; coral branch, 121f; coral/lithic/shell, 115; lithic-coral slab, 121f; teeth of, 121f
"Great Dying." *See* Permian-Triassic extinction event
Greater Antilles and cycads, 97, 116, 121f
Greater Caribbean, map of cycads as food in, 100f
Griddles: as artifacts to look for ancient zamia evidence on, 121f; clay, for ancient zamia cooking, 116; lithic/clay, as work tables
Grinder, edge, for grinding/pounding zamia dough, 115, 116
Groves, cycad, 218
Guachumpoyo, 237t
Guadeloupe (island) archaeological site with zamia starches in, 110t, 115
Guam: effects of *Aulacaspis* scale in, 54; effects of cycad consumption on people in, 51, 248
Guam Dementia, 248
Guanajuato: *chamal* in, 236t, 238; Nahua names for cycads in, 236t–237t

Guánica (Puerto Rico), 99
Guapilla, 185; as derived from womb or eyes in Xi'iuy mythology, 232, 233
Guatemala, 137, 221; Nahua transportation of term *teosinte* to SW part of, 221; *teosinte* in toponyms of, 214, 216; *teosinte* as term for cycad in N part of, 219; Veracruz *Tripsacum teosinte* introduced from, 220; *Zea luxurians* teosinte derived from, 220
Guáyiga, 97, 98
Guerra phase (Ocampo), 145t; *Dioon edule* remains found in caves of, 144
Guerrero (Mexico), 132, 212; SMA absent from, 241
Guilá Naquitz (Oaxaca) cave, 149
Gulf Coast: imagery of, 181; maize/cycad deities in, 187; Corn Master mythologies in, 188
Gulf Coast Maize Hero. *See* GCMH
Gulf (of Mexico coast/al) lowlands, 39; as SMA origin region, 223, 243
Gunung Kawi temple, 19f
Gusano de konlif, 226
Gymnosperms, 32; largest cone of, 38; phytoliths of, 118–119
Gymnosperms: Structure and Evolution, 48

Habitat destruction: as affecting Tomasellii clade gene flow, 85; as threat to cycads, 24, 27, 53, 61
Hardwood hammocks as Caribbean *Zamia* habitat, 41
Havana, 45
Head: *cabeza* (cycad term), 233; "crazy in the head," 233; discussion of, in cycad terminology, 233; of evil woman in Xi'iuy mythology, 232; as origin of cycads, 233; in Teenek cycad-maize terminology, 228t; Tzompoyo as "evil head/hair," in Nahua terminology, 237t, 237; as Xi'iuy cycad term, 233t
Headband, 190t
Headdress, tripartite, in Tenaspi Egg, 201
Heart of maize. *See Ichi-ich*
Heat and starch survival, 102, 107
Hechtia glomerata. *See Guapilla*
Hermano de Dhipak, 227t
Hernández, Francisco, 172, 213, 214
Heroes, 187, 243
Herons, 190t
Hero Twins, 181; defeat of Zipacna, 181; Hun-Hunahpu as father of, 188
Heterozygosity, 66, 67; in *Ceratozamia*, 79; in *Dioon caputoi, D. holmgrenii*, and *D. merolae*, 84; expected, in cycads, 71t–74t; observed, in cycads, 71t–74t
Heterozygotes, 67; excess of, in *Dioon edule*, 84
Heterozygous genotypes, 65
Hidalgo (Mexican state), 45; *chamal* term among mestizos in, 238; Nahua names for cycads in, 236t–237t; nixtamalization of cycads in, prior to maize, 209n2; *teosinte* as referring to cycads in, 175
Higüey (Hispaniola), 98, 123
Hills, sacred: and cycads, 214; as maize origin places, 214. *See also* Nambiyugua
Hilum, 101
Hiroshige (Edo period artist), 15
Hispaniola, *Zamia* as food in, 97
Historia de los Mexicanos por sus Pinturas, 167, 173–174, 182
Historical ecology: research directions of, related to *Zamia*, 118–122; research problems in, related to *Zamia*, 116–117
Hoeiacocotla. *See* Huayacocotla
Hojaldre, 123
Holocene, 142
Holy Week, 242
Homozygosity, 66
Homozygotes, 66
Homozygous genotypes, 65
Homshuk, 207f, 240; as Corn Master narrative, 167, 184; as fertility hero, 188; life cycle of, as equivalent to maize, 194; and ovogenesis, 200; as potentially related to cycads, 240
Honduras: ancient *Zamia* in, 135, 163; connections of, with Mexico, 88–90, 93; cycads as food in, 51; cycad harvesting camps in, 90; *Dioon* in, 63; endemic cycads in, 219; nixtamalization in, prior to maize, 209n2; similarities of TMG/GCMH to Ch'orti' Kumix story in, 190; *teosinte* grasses in toponyms of W part of, 214, 221; *tiusinte* region in, 215f. *See also Dioon mejiae*
Hornitos (Panama), archaeological site with zamia, 117
Horticulture: ancient, as wild food management strategy, 142; ancient, role of cycads in, 154; modern, as threat to cycads, 24, 27
Horto Botanico (Naples, Italy), 49, 61
Hot spots of cycad diversity, 41–42
Huamelula (Oaxaca), 243
Huasteca region: archaeobotanical evidence of cycad use in, 90; beliefs in cycad cones as head in, 233; culinary mythologies in, 173; cycad seeds sold in markets of, 90;

hidalguense, 224f; multiethnic mix in, 221; origin of cycad as maize ancestor idea in, 88; paper cutout figures in, 209n3; potosina, 231; relatedness of cycads in, with Honduras, 93; as *teosinte* cycad origin area, 214, 221
Huastec Maya. *See* Teenek
Huautla (Hidalgo), 235
Huave (people), cycad use by, 240
Huayacocotla (Veracruz), 213
Huazalingo, San Juan (Hidalgo), 175; maize-cycad deity in, 234, 237
Hueca, La (Puerto Rico), 110t
Human-induced dispersal: in Caribbean zamia, 119; of cycads, 64, 88, 90; of *Cycas revoluta*, 159; of *Dioon*, 93; of domesticated plants, 62–63; of japonica and indica rice varieties, 63; of maize, as related to cycads, 93–94; of plants, 62; of plants, as affecting biogeography, 62; of plants, as obscuring evolutionary history, 62; unintentional or intentional, of cycads, 158. *See also* Human-mediated cycad seed dispersal
Human-induced gene flow, 63
Human-mediated cycad seed dispersal, 90
Humans involved in cycad distribution: in Japan, 11; in Mexico and Honduras, 88, 90
Hunger/famine: and cycads, 229, 231, 240; in Chontal de Oaxaca beliefs about cycads/maize, 241; and cycads, in Jamaica, 248; foods, among the Zapotec, 240; in Xi'iuy cycad origin story, 232
Hun-Hunahpu, 188, 189
Hunter-gatherers: as cycad consumers, 53; diets of, in Sierra de Tamaulipas caves, 147t
Hurricane María, 122
Hurricanes, 99; related to food scarcity and *Zamia* in the Puerto Rico, 120, 122, 123; in Tenochtitlan 5 ages sequence, 174
Hutia, 121f
Hybridity, 190, 208; exemplified in emergence of maize god from bottle gourd, 192; in Formative imagery of crocodilians, 198
Hybrid zones in population genetics, 67
Hypocotyledonous stem. *See* Hypocotyls
Hypocotyls, 253
Hypodermis, 33

IBD models, 81
Ichi-ich, 176
Iconography, 188; Classic Maya, 189f; Maya, 168; Formative Gulf Coast, cycads in, 200–206; and origin of Formative interpretations in Classic Maya interpretations, 193; and question of early Olmec depictions, 196–200, 211
Iconological analyses beyond iconography, 208
ICP. *See* Inductively coupled plasma (ICP) technologies
Ijben ixik, 209n3
Ijben winik, 209n3
Ikiapacinti, 239t
Ilye'e dameaw kon garbants, 233t
Imix, 180, 181; day sign, 198f
India, cycad use in, 10; Andhra Pradesh, 10; Assam, 10, 15, 17; Kerala, 10; Meghalaya, 10, 11f; Tamil Nadu, 10
Indigenous Caribbeans, 97
Inductively coupled plasma (ICP) technologies, 154
Infiernillo Canyon (Tamaulipas), 143–144
Infiernillo phase (Ocampo), 145t
Inik, 185. *See also* *Konlif*
Insects, siderophore-producing bacteria in guts of cycad-associated, 156
Integrative research: model of, for cycad studies, 160f; value of, in cycad studies, 95–96, 151; value of multiproxy research, 101
Intensification (agricultural), 134; in Formative Period, 136–138
International Convention on the Trade in Endangered Species, 24
International Plant Science Center, 164
Inter Simple Sequence Repeats. *See* ISSRs as markers of genetic variation and diversity
Introduced (alien) species, effects on cycads of, 54
Ipomoea batatas. *See* Potato, sweet
Irati Formation, 8
Ireland, 45
Iron: in cycad tissues, 156; and Nahua cycad beliefs, 235; in siderophores, 155
Irrigation: canal, 137; field, 139
Islas Marías Federal Penal Colony, 254
Isolation by Distance models. *See* IBD models
ISSRs as markers of genetic variation and diversity, 65–66; advantages and disadvantages of, 66; in *Ceratozamia* studies, 81; in *Dioon caputoi*, 84. *See also* Allozymes
Isthmus clade, 77
Isthmus de Tehuantepec, 39, 243; as cycad hot spot, 41–42
Isthmus of Panama, 77
Itzam Cab Ain, 182. *See also* Cipactli
Itzamna, 190t
IUCN Red List of Threatened Species, 53

342 · Index

Ixim, uncertain etymology of, 192
Iximte', 191, 192
Ixtepec (Oaxaca), 239
Izapa, Stelae 25 and 5 at, 196, 197f, 200

Jadeite, 116, 203
Jaguar, 174f
Jaguar-skin cushion, 189f
Jalisco (Mexico): *chamal* in, 236t, 238; loss of cycad knowledge in, 243; Nahua names for cycads in, 236t–237t; *teosintes* in, 220–221
Jalpan (Querétaro), 230
Jamaica: cycad population genetics of, 77; *Zamia* in, 108; *Zamia erosa* as famine food in, 248
Jango, 240
Jealousy as cycad trait, 241
Jobal, 227t
Jojo, 227t
Jojobal, 227t
Jomxuc. *See* Homshuk
Joya, La (Veracruz), 137
Juana Díaz (Puerto Rico), 121f
Juan Dolió (Dominican Republic), 110t
Jun Ixim, 189f
Jun Ye Nal Chak. *See* GI
Jurassic period, 8, 126

Kab (earth signs), 200
Kakaw. *See* Cacao
Kaminaljuyu (Guatemala), 137
Kanau namew, 233t
Kanaw namew, 233t
Katun, 182
Kene, 239t
Kerr vessel 5761, 192f
Kew Herbarium, 49
K'iche' Maya, 167, 180, 188; and maize origins in the *Popol Vuh*, 181
K'id, 227t
Kidney function and cycad medicine, 128
King's Helmet (Puerto Rico), 110t
Kingship, 190t; (rulership) in Tenaspi Egg iconography, 202
K'inich Janaab' Pakal sarcophagus reliefs, 200, 201f
Kin lihuaykan, 238, 239t
Knowledge loss of cycads, 243
Kombi as variant of *konlib*, 179, 227t
Kombil as variant of *konlib*, 179, 227t
Kon/kum. *See* Kon/m
Kon/m: and cycads, 179; as element of shared terminology, 187; etymology of, 179, 184; as food palm, 180; as "king," ruler, 179; as *kom*, in Soteapan Gulf Sokean, 179; as *kom/kum*, coyol palm, 179; and *koon-mook*, 179; as **kuma* in Proto-Mije-Sokean, 179; and Mije *kon*, Kondoy, 179; as Mije *kon/konk* plus *oí* or *kon* plus *toi*, 210n5; as Soke *kom*, related to Teenek *kon*, 179; and Totonac *kun*, 179
Konb as variant of *konlib*, 179, 227t
Kondif as variant of *konlib*, 179, 227t
Kondoy, 179, 207f; as Corn Master, 179–180; as Corn Master narrative, 167, 184; as fertility hero, 188; and ovogenesis, 200; suggested etymology of, 210n5; and Zapotecs, 210n5
Kone, 239t
Konfi as variant of *konlib*, 179
Konlib, elote de, 227t
Konlib, elotito de, 227t, 228
Konlib: etymology of, 179–180; as maize ancestor, 178–179; shared identity of, with Thipaak, 179; as Teenek "spirit of the milpa," 129
Konlif, 129, 226–227, 228–229; as attracting rain, 226; as brother/friend/helper of Thipaak, 226; elderly knowledge of, 226; as flesh of maize/milpa, 226; *inik* and *uxum*, 226, 227, 228, 229; as king of maize/milpa, 226; as life-force, 226; as maize ancestor, 178–179; as maize cob, 178; as maize companion, 226; as maize ear's uncle, 226; as maize-guardians, 226; as predator trap, 226; as "sand-plant," 226, 227t; as Teenek "spirit of the milpa," 176. *See also Ceratozamia latifolia*
Konlif, olote de, 228t
Konlif, ushu, 228t
Konlif, ushum, 228t
Koon-mook, 179
K-Pg Extinction, 8–9, 61
K-T Extinction. *See* K-Pg Extinction
Kuasintli, 236t
*Kuma. *See Kon/m*
Kumix story, 190, 209
Kun: as Sierra Norte Totonac term for cycad, 179, 238, 239t; as related to Mije *kon*, 179

Lacandón Maya, 241
La-fané-tejuá, 239t, 241
Lagenaria siceraria. *See* Squash
Laguna de Limoes (Cuba), 111t
Laguna de los Cerros, 210n9
Laguna phase (Sierra de Tamaulipas), 145t; diets of hunter-gatherers in, 147t; *Dioon edule* remains found in caves of, 144, 147t, 147
Lamb, Sister Alice, cycad research by, 47
Lamellae in zamia starch grains, 104t, 111–112

Landesque capital, 138
Landscape ecology: definition of, 68; as related to population genetics, 68
Landscape genetics: definition of, 68–69; examples of, 68
Landscaping: as threat to cycads, 54, 55; trade as related to spread of *chamal* term, 238
Lan-zi-lé, 239t
Lan-zi-li, 239t
Larvae. *See* Maggots
Late Archaic (Caribbean), 112
Lavoutte (Saint Lucia), archaeological site with zamia starch, 111t, 115
Laws protecting cycads, 53, 55, 57
Leaf bases (cycad), 30
Leaf (cycad), spines on, in Teenek cycad-maize terminology, 227t
Leaflets (cycads): variation of, 41; variation of, in *Ceratozamia,* 78
Leaf traces (cycad), 29–30, 29f; functions of, 30
Legend of the Suns. *See* Leyenda de los Soles
Legumes as nitrogen-fixing, 125, 130, 154
Lenca, 217
Lepisosteus tristoechus. See Gar
Lerio, Giovanni, 43
Lerma phase (Sierra de Tamaulipas), 145t
Lesions, skin, and cycad medicine, 128
Lesser Antilles, 97; archaeological *Zamia* in, 108, 114, 115; why no natural distribution of *Zamia* in, 108, 116
Leyenda de los Soles, 167, 173, 214
Lightning, TMG/GCMH associated with, 190t
Likeness, role in meaningful association, 202
Limestone: hills of, as *Zamia* habitat, 99f; as substrate for *Ceratozamia* habitat, 127; as *Z. erosa* and other zamia habitat in Puerto Rico, 114
Lime used in cycad nixtamalization, 209n2
Limpopo Province, 15
Lindley, John, 43
Linguistic shifts correlated to demographic movements of humans/cycads/maize, 94
Links. *See* Nodes (population genetics)
Linnaeus and binomial cycad nomenclature for *Zamia furfuracea,* 43
Lints'i', 176, 210n6
Lipids. *See* Biomarkers
Liver damage from cycads, 127
Lizard: in Popolucan narrative, 185; in Xi'iuy mythology, 185, 232–233
Lizard-hero, 185
Loaves. *See* Bread, cycad
Lord, Maize, 180; *konlif* as, 226, 227t

Lord 9 Wind, 177
Los Angeles phase (Sierra de Tamaulipas), 145t; diets of hunter-gatherers in, 147t; *Dioon edule* remains in caves of, 146, 147, 147t
Lotus, 180

Maam, 179
Maam, planta del, 228t
MacNeish, Richard, 90, 143–149, 164, 171
Macrozamia, 252
Macrozamia: moth pollination of, 32; population genetics of, 75
Macrozamia communis, genetic diversity and structure of, 73t
Macrozamia heteromera, genetic diversity and structure of, 73t
Macrozamia heteromera, pollen and seed dispersal in, 70
Macrozamia parcifolia, genetic diversity and structure of, 73t
Macrozamia pauli-guilielmi, genetic diversity and structure of, 73t
Macrozamia plurinerva, genetic diversity and structure of, 73t
Macrozamia riedlei, genetic diversity and structure of, 73t
Macrozamin, 27, 51, 127
Madjedbebe/Malakunanja II archaeological site. *See* Madjedbebe rock shelter
Madjedbebe rock shelter, 9, 10f
Maggots: in *Zamia* food, 98, 121
Magpie-jay, white-throated, as *Dioon* seed disperser, 83
Maguey, 149; (*Agave* sp.) in La Perra horizon, Tamaulipas, 171; as dominant in Early Formative, 210n9; as part of Archaic diet, 172; (agaves) on Tancoyol church façade, 230
Maicito (cycad), 242
Maisabel (Puerto Rico), 110t
Maize, 141; becoming humans/human flesh, 173; domestication of, evidence for, in Tamaulipas caves, 143–147; as dominant interpretation of Olmec Maize God, 193; in Early Formative Gulf Coast, 194–195; Early/Middle Formative use of, tied to sugary stalks; found in La Perra horizon, Tamaulipas, but not eaten, 171; intercropping of, with beans, 136–138; maize deity, TMG/GCMH as, 190t; male spirit of, 209n3; Maya Maize God identification with, 188; as monoecious in Nahua culture, 235; monoecious, as birthed by dioecious cycad, 222; as non-staple in Mokaya sites, 210n8;

Maize—*continued*
  in Olmec imagery, 195; over-emphasis on, in Mesoamerican research, 125, 139–140; overview of domestication of, 132–133; pollen of, in Teenek mythology, 177; in the *Popol Vuh*, 181; ritual significance of, 169; in Teenek cycad mythology, 186; in Teenek world ages, 177t; Tojol-ab'al and, 241–242; as tribute food, 195; as unimportant in diet but important in symbolism, 195
—early: in Mesoamerica, and burning, 134; in Panama, 132; in Tamaulipas caves, 144, 145t, 147t; in Tehuacán Valley caves/rockshelters, 148
—starch grains of, 111; in ancient dental calculus, 114; compared to cycads, 107; in Puerto Rican archaeological sites, 114

Maize, cycads and, 2, 21, 51–52, 127, 129, 224f, 225f; cycad-maize-ancestor complex, 188; cycad/maize culture region (map), 224f, 225f; cycads as forest maize, 235; cycads as masters of maize, 235; cycads as wild maize, 235, 236t, 237t; history of investigation into similarities between, 212; iconology of, 166–211; interrelatedness of diversification and spread of, 93–94; maize as monoecious vs. dioecious cycads, 186; maize-cycad complex, 172; maize-cycad plant, 235; maize as cycad, 243; maize found inside *teocintle* cycad, 175; mental template for, 150, 165n3, 169; resemblance between, 129, 150, 151f, 170, 212; stories of, as having common origin, 187. *See also* Maize; SMA
—in agriculture: cycads and BNF in early Mesoamerican fields of, 134–135; cycads left intentionally in fields of maize, 242; cycads as maize/milpa guardians/shepherds, 209 (Huasteca), 240 (Oaxaca); cycads as only plants surviving in maize fields after burns, 131; intercropping of maize and cycads, 135, 138, 154, 231; intercropping of maize, cycads, and beans, 139; maize and cycads in agricultural terraces, 139; planting of maize in cycad groves, 242; seed setting of cycads predicting good maize crop, 178
—in association with: crocodilians, 206; fish, 208–209; fish and Cipactli, 184; fish and crocodilians, in Pan-Mesoamerican mythology, 206, 207f; floral arches representing Teosintle/San Miguel, 237; Formative Gulf Coast iconography, 200–206; palms, 179, 180; piscine-maize-cycad relationship, 181; seed-eggs, 206
—association found in: Belize, 242; Colombia, 242; Tamaulipas caves, 147–149
—complementarity of: among Xi'iuy, 231; in Honduras, 242
—culture of, in: Chontal de Oaxaca, 241; Maya, 241–242; Mazatec (Oaxaca), 240; Mije, 179–180; Oaxaca, 240–241; Soke, 179; Totonac, 180, 238; Zoque-Popoluca, 238, 240
—and cycad local names: fat maize, 230, 233t; forest maize, 230, 233t; friend of maize, 239t; hill maize, 235, 236t, 237t, 239t; Huamelula story about cycads as *maíz viejo*, 241; King Maize, 179–180; life of maize, *konlif* as, 227t; life of maize, in Teenek cycad terminology, 227t, 228t; little maize, 242; *maíz, amigo de*, 239t; *maíz, la carne del*, 227t; *maíz, madre del*, 227t; *maíz, tío del*, 237t; *maíz cimarrón*, 236t; *maíz del monte*, 230, 233t; *maíz de los chaneques*, 236t; maize ear father, 235, 236t; maize ear mother, 235, 236t; maize lord, 176; maize of the ancestors (cycad term), 239; maize of the poor, 232; maize's mother, *konlif* as, 226; maize's uncle, cycads as, 175; maize's uncle, *konlif* as, 226; maize's uncle, Nahua term, 237t; *maíz gordo*, 230, 233t; *maíz viejo*, 239t; *mazorca, compañero de la*, 227t; *mazorca, tío de la*, 228t; *mazorcas, el rey de las*, 227t; *mazorca de maíz de monte*, 213; *mazorcas, el rey y carne de*, 227t; *mazorcas, mama de las*, 227t; old maize, 239t, 241; red maize, 241–242; wild maize, 239t
—cycads as (sacred) maize ancestors: in post-conquest Mexico, 214; spread of idea from Mexico to Honduras, 88
—as embodying deities: Chicōmexōchitl/Chicomesintli as maize/cycad deity, 234; Cipactli as maize/cycad deity, 198; God E as maize deity, 210n7; maize/fertility god-cultural hero, 189; Teosintle as maize/cycad deity, 221
—in food: cooking similarity of cycads and maize in 1560s Hernández text, 213; cycad food as better than maize, 232; cycads as bridge/mediator between pre- and post-maize foodways, 169, 170f; cycads as maize eaten by the ancestors, 235; cycads as "maize of the poor," 232; cycads as more important food than maize, 98; cycads eaten together with maize, 242; cycads mixed with maize, 240; cycads used when maize scarce, 99; maize and *Zamia* as most ubiquitous food plants, 122; maize combined with cycads in tortillas, 128; mixture of cycads and maize as Zapotec food, 240; nixtamalization of cycads

and maize, 150, 209n2; substitution of cycads for maize, 129; Teenek ancestors ate cycads as maize, 176; *Zamia* eaten even when maize available, 122
—and *konlif*: as maize companion, 226; as protecting/strengthening maize, 129; terms related to maize, 226–229; vegetative parts homologous to maize, 226–228
—names for: in Chontal de Oaxaca, 239t; in Mazatec (Oaxaca), 239t; in Nahua, 236t–237t; in Totonac, 239t; in Xi'iuy, 230–233; in Zoque-Popoluca, 239t
—as relatives, 176; as ancestors, 129, 175, 176, 178–179, 206, 207f, 241; as deity manifestations, 176; as friends, 129, 178–179; as maize from previous creation, 206; as types of, 129, 221, 235, 242
Maize saint. *See* Mok Santu
"Maize tree," 191
Maltese cross, 103; defined, 124n1; in zamia starch grains, 112
MAM, definition of, 121f
Manatí (Puerto Rico), 99f
Mangroves as Mexican cycad habitat, 127
Manihot esculenta Crantz. *See* Manioc
Manioc, 24, 126, 171; in Caribbean archaeological sites, 114–115; comparison of starch grains of, to *Zamia*, 107; cultivation of, beyond household garden, 210n9; cycads substituted for, when scarce, 99; as dominant in Early Formative, 210n9; in Honduran names for cycads, 219; as less important than maize as food, 98; starch grain morphometry of, 106t; starch grains of, in Puerto Rican archaeological sites, 114; toxins removed from, 99; *Zamia* eaten even when available, 122
Manna, 232
Manu, 227t
Manzanilla (Trinidad and Tobago), 110t, 113, 115
Margaritas, Las (Chiapas), 241
María Cleofas (Nayarit), 254
Maruca (Puerto Rico), 110t, 113, 114
Marunguey, 97, 99; food preparation process of, 121f; starch grains of, in underground stem, 102f
Matuda, Eizi, 48–49
Maya (people): lack of SMA among, in southern Mexico, 241
Maya highlands, raised fields in, 138
Maya Maize God, 187; as cacao tree, 190–191, 191f; emergence of, from bottle gourd, 192f; imagery of, as nearly identical to Arroyo Pesquero celt imagery, 198

Mazatec de Oaxaca: more research needed among, 243; near Nahua/Zoque-Popoluca communities, 240; SMA among, 239t, 240
Mazorca (corncob), 180; emerging from Olmec Maize God head, 203; resemblance to cycad strobilus/cone, 203, 206
Mecayapan (southern Veracruz), 235
Medullosales, 8
Megagametophytes, 83
Megasporangium, 31f, 32; anatomy of, 32; fertilization of, 32
Megaspore mother cell, 31f
Megasporophylls, 8; anatomy of, 30–32; for medicine, 10
Mejía, Isidoro, 216
Merola, Aldo, 49
Mesa de Guaje phase (Ocampo), 145t
Mesoamerica, cycad species richness of. *See* Mexico, cycad species richness of/diversity in
Mesoamerica clade, 77. *See also* Mexican clade
Mesozoic Era, 33, 61
Mestizado, 223
Metabolomes of cycads, 164
Metals, mobilization of, by siderophores, 155
Methylazoxymethanol. *See* MAM
Methylazoxymethanol azoxyglycosides. *See* Cycasin; Macrozamin
Mexica, 214; empire, 217; in Honduras, 217–218
Mexican clade, 37
Mexico, Basin of, raised fields in, 138
Mexico, cycad species richness/diversity in, 35, 41–42, 42f, 63, 126
Mexico, history of cycad research in, 43–50
Mexico, Valley of, cycads not native to, 214
Michaelmas, 237
Michoacán, 88
Microcycas, 20, 27, 35, 36f; arborescent stems of, 30, 41; bark of, 30, 41
Microcycas calocoma, 41; genetic diversity and structure of, 72t
Microlepidopteran moths as cycad pollinators, 32
Microsatellites. *See* SSRs
Microscopy, optic/scanning electron, as technique in experimental archaeology, 153. *See also* SEM-EDS
Microsporangia, 31f; for medicine, 11f
Microspore mother cell, 31f
Microsporophylls: anatomy of, 30–32; for medicine, 11f

Migration, human: patterns of, compared to cycad genetic patterns, 94; related to spread of cycads/maize, 94
Miim, 226, 227t. *See also* Grandmother in maize/cycad stories
Mije, 167, 210n5; compared to Xi'iuy narrative, 185; Corn Master narrative, 184; no SMA among, 240; words for corn/Corn Master, 179–180. *See also* Kondoy
Milpa agriculture, 170, 223, 243; Mazatec (Oaxaca), cycads and, 240; Nahua, and cycads, 234, 235; Teenek, 226, 229
Mim, 227t
Mim konlif, 228t
Minatitlán (S Veracruz), 235
Mineralogical analysis: methods used in archaeology, overview, 154
Miocene epoch, 9, 33; *Ceratozamia* and *Zamia,* 42; and *Dioon,* 39, 83
Miquel, Friedrich Anton Wilhelm, 44
Miqueliana clade, 37
Misión, La (Hidalgo), 52
Missionaries, Spanish, and attitudes toward Xi'iuy, 230
Missions, Roman Catholic: in the Sierra Gorda, 230; and Xi'iuy, 230
Mixe. *See* Mije
Mixe-Zoquean language family, 243
Mixtec (people), no SMA among, 240
Modjadji. *See* Mudjadji
Mokaya sites, 210n8
Mok Santu, 209n1
Molango (Hidalgo), 46
Monkeys, 174
Mono Blanco, Cerro, eruption of, 137
Monocotyledons, 193
Monoculture (agriculture) in the Caribbean, 122
Monster, aquatic, 181
Monte Oscuro (Veracruz), 55; cycad nursery in, 56f, 57, 58–59; value of sales in, 59
Montgomery Botanical Center, 61
Monument 63, La Venta, 183f
Mopán Maya, 242
Moretti, Aldo, 49, 254
Mortars with ancient zamia evidence, 116
Motile sperm cells. *See* Ciliate antherozoids
Mouse, Mexican deer, as *Dioon* seed disperser, 83
Mouth in maize god iconography, 192
Mudjadji: as cycad (see *Encephalartos transvenosus*); as Rain Queen, 15, 18f

Mukoma Mudjadji IV, 4th Rain Queen, 15, 18f
Murillo, Prof. Luis, 48
Musa. See Plantains
Music as encompassed by TMG/GCMH, 188, 190t
Muxi', 177
Mycotoxins reduced in nixtamalization, 209n2
Myōkoku-ji temple (Osaka), 14–15, 16f

Nahoa de Honduras, 88–89; as distinct from Pipil speakers, 217; in *tiusinte* region, 217, 242
Nahua (language): cycad terms/concepts in, shared with Zoque-Popoluca, 240; Hidalgo towns speaking, 175; as not spoken in Honduras, 242; taxonomies, 178; terms for cycads, 219. *See also* Nahuatl (language)
Nahua (people): Cipactli in religion of, 180; connection of, to Xi'uiy narratives, 185; Corn Master narrative of, 167; more research needed among, 243; Nahoa de Honduras as, 217; SMA beliefs among, 233–238; Teosinte maize/cycad deity of, 221; traders in NE Honduras, 217–218; transportation of cycad terms from NE to W Mexico by, 220
Nahuatl (language), 52, 224f; Classical, 212; diffusion of SMA by speakers of, 243; diversification of, correlated to cycads' genetic divergence, 93; etymology of *teosinte* in, 212; in Honduras, 90, 217; speakers of, overlapping with cycad ranges, 214. *See also* Nahua (language)
Nal (Classic Maya word), 172
Nambiyugua, Cerro (Chiapas), 52, 186
Namele, 13–14
NamEo', 233t
Nameu, 233t
Nantōzatsuwa (South Island Chronicle), cycads in, 12
National Botanic Gardens, Glasnevin, 45, 46
National Cycad Collection at Xalapa, 50
Naui Atl (Four Water), 174f
Naui Ehecatl (Four Wind), 174f
Naui Ocelotl (Four Jaguar), 174f
Naui Ollin (Four Movement), 174f
Nayarit (Mexico), 254
Networks, reticulated interaction, 117
Neuralgia and cycad medicine, 128
Neurological damage, long-term: caused by cycads, 51, 99
Neurological disorders caused by cycads, 127
Neurotoxic chemicals in cycads, 127
New York Botanical Garden, 61, 164

Next-generation sequencing. *See* NGS
N fixation. *See* BNF
N-fixing plants. *See* BNF
NGS, 66
Nguyen, Hiep T., 246
NH₃. *See* Ammonia
Nicaragua: as part of *teosinte*-as-grass-region, 218, 221; *teosinte* in toponyms in, 214
Nilssoniales, 33
9 Wind, 184
Nitrogen (N₂). *See* BNF
Nitrogen, isotopic, measurement as cycad-detection method, 155
Nitrogen fixation. *See* BNF
Nitrogen use efficiency (NUE), 130
Nixtamalization, 51, 128; with ash, 90; of cycads, prior to maize, 209n2; of cycad seeds in Honduras and Mexico, 90, 91f, 150; with lime, 90, 128; as potentially detectable by DART-MS, 161; seed-eggs produced via, 206
Nobunaga, Oda (Japanese feudal lord), 14–15, 17f
Nodes (population genetics): definition of, 68; in *Dioon* population graphs, 86f
Nogales Cave (Sierra de Tamaulipas), 144
Nogales phase (Sierra de Tamaulipas), 145t
Norstog, Knut, 33, 245, 250–251
North *D. tomasellii* subgraph of *Dioon*, 86f, 87, 88
Northeast Mesoamerican Periphery, 223
Nostoc, 130
Nucellus, 31f
Nuevo León: *chamal* term used in, 238; Nahua names for cycads in, 236t–237t
Nurseries, community, for cycad conservation in Mexico, 55–60
"Nut on a stick" cycad object, 146; identified as spindle, 165n2
*ña'al, 194; as Chol "god of abundant plants and animals," 172
Ñuhu. *See* Otomí (Ñuhu)

Oak ash used to detoxify cycad seeds, 91f
Oaxaca, Coastal, 210n9
Oaxaca and Northeast subgraph of *Dioon*, 86f, 87
Ocampo region (Tamaulipas), 144, 224f; ancient *Dioon edule* seed fragments recovered in caves of, 146
Ocean: in Teenek mythology, 176, 177t
Ohox. *See* Breadnut tree
Ojo de Agua cave (Ocampo), 144

Ojoshal Scepter (Tabasco), 197f, 205f
Ojox. *See* Breadnut tree
O'k, 228t
Olanchito (Honduras), cycad seeds sold in market of, 90
Olancho (Honduras), 88, 90
Old Thunder God, 190t
Old-time food (cycad term), 239t
Old-timers' food (cycad term), 239t, 241
Olmec deities as representing maize, 192
Olmec God II. *See* Olmec Maize God
Olmec iconography, 196–200; and breadnut tree, 196; cycad representations in, 196
Olmec Maize God, 169, 193f; as a cycad, 203; growth habitat of, compared to cycads, 203; as reified, 192–193; as represented on Arroyo Pesquero Celt, 196
Olotito, 228t
Ōmecíhuatl, 185
Ōmetecuhtli, 185
Ōmeteōtl, 185
OMG. *See* Olmec Maize God
Origin-of-corn/maize mythologies, 188; comparisons across Mesoamerica, 190
Origin-of-corn narratives. *See* Corn Master: narratives of
Orogeny, effects on *Dioon* of, 39
Oshima tsumugi textile tradition, 12
O'tlab, 228t
O'tol, 228t
Oto-Mangue/an, 243. *See also* Chiapanec (culture)
Otomí, 224f
Otomí (Ñuhu), 238
Oto-Pamean (language family), 243
Ovogenesis, 187; in Corn Master narratives, 200; cycad seeds and, 204
Ovule. *See* Megasporangium

Paampi, 179
Pacaya (palm), 179
Pacific Coast (Mexico), SMA as absent from, 241
Pacific Coast Lowlands (Oaxaca), 241
Padre del elote, 235, 236t
Palenque, 182; Temple of Inscriptions at, 201f
Paleobotanical research on cycads, 165n1
Paleoecology, 151; and studies of cycads, 118, 120
Paleoethnobotany: as research method for cycads, 97, 118, 120; of plant starches, 101–102
Paleogene, 33
Paleogenomics, 158, 160f

Palma de chicalite, 53
Palma de Nuestro Señor, 228t
Palmillas phase (Ocampo), 145t
Palms: *Acrocomia*, on Tancoyol church facade, 230; *Brahea dulcis*, fronds of, used for Palm Sunday rituals in Mexico, 55; as consumed in Archaic Mesoamerica, 149, 172; coyol, 179; cycad misinterpreted as, on Basilica di San Marco façade, 256; cycads as, among Zapotec, 240; pacaya, 179; as pollinated by weevils, 83; reintroduction to wild habitat of, in Mexico, 55; *Sabal* spp., as dominant in Early Formative, 210n9; *Sabal* spp., in La Perra horizon, Tamaulipas, 171; in San Bartolo murals, 208; in Teenek names for cycads, 228
Palm Sunday, cycads used on, in the Philippines, 17
Palo maíz viejo, 239t, 241
Pame (Northern). See Xi'iuy
Pamería, 230, 231; district, 51
Panama, 77; ancient *Zamia* in, 136; early maize in, 132; grandmother/young boy antagonism motif in, 210n4. *See also* Isthmus of Panama
Papaloapan/Santo Domingo River basins as *Dioon* hot spot, 41, 42
Papantla (Veracruz), 238, 239t
Paper cutout figures as maize spirits, 209n3
Paradise as aquatic, 190t
Pastry, puff, 123
Pasture expansion and livestock as threat to cycads, 24, 25
Peabody Museum, 164
Pech, 217
Pek mok, 239t
Peñuelas (Puerto Rico), 99f
Peptides. *See* Biomarkers
Permian period, 8, 32, 33
Permian-Triassic extinction event, 8
Peromyscus mexicanus. *See* Mouse, Mexican deer, as Dioon seed disperser
Perra, La (Sierra de Tamaulipas cave and phase), 145t; archaeobotany of, described, 144–147, 171; in context of Greater Caribbean ancient zamia sites, 117; diets of hunter-gatherers in, 147t; *Dioon edule* remains found in caves of, 144, 147t
Persea americana. *See* Avocado
Pestles: with ancient zamia evidence, 116; as artifacts to look for ancient zamia evidence on, 121f
Peyote (cycad), 128
Phallus, 186; in Teenek cycad terminology, 228

Phaseolus. *See* Beans (*Phaseolus* spp.)
Phytoliths, 156; of cycads, experimental archaeology needed to detect, 153; of cycads with low silica abundance, 118; of maize, 132; systematic studies needed of, 119; of *Zamia* in archaeological context, 118
Phytomorphic forms/deities: in Classic Mesoamerican iconography, 210n7; in Formative iconography, 196, 203; Olmec entity as, 194
Pigs, 172
Pilelotsi, 235, 236t
Pilgrimages, cycads used in, 52, 128, 186–187
Piljcintektli, 234
Pilololcintli, 236t
Pinar del Rio province (Cuba), 41
Pinecone as like cycad, 213
Pines in San Bartolo murals, 208
Piña. *See* Pinecone as like cycad
Pipil, 214, 217
Piscine. *See* Fish
Pithecellobium confins, 171
Plantaciones (cycad), 218
Plantains: cycads substitute for when scarce, 99; roots of, as Zapotec famine foods, 240
Plant Genomics Consortium (New York), 164, 253
Playa Grande (Dominican Republic), 110t, 113, 115
Pleistocene epoch: and *Ceratozamia*, 78–79; and *Dioon*, 39; refugia of, in relation to *Dioon*, 84
Pleistocene–Holocene transition, 9, 78, 163, 171
Plum Piece (Saba), 110t, 114
Poisoning from cycads, 99, 229
Po'jodh, 228t
Po'jodh tzamaal inik, 228t
Pojostlil tsamal inik, 228t
Pokomchi, origin-of-maize story of, 190
Pollen, cycad: as carcinogenic, 241; as semen, 186; in Teenek terminology, 228t; transfer, 74
Pollen dispersal, 79; of *Dioon* by insects, as known to Chontal de Oaxaca, 241; by *Dioon* pollinators, 83; long-distance, as contributor to low genetic differentiation in *Cycad debaoensis*, 74
Polyculture (agriculture) in the Caribbean, 122
Polymorphisms, definition, in population genetics, 66. *See also* SNPs
Ponce (Puerto Rico), 99f
Ponytail palm. *See Beaucarnea recurvata*
Popoluca: Corn Master narrative in, 184; Gulf Coast Sierra (people), 194. *See also* Homshuk
Popol Vuh, 167, 180, 188, 189; Hero Twins and

maize in, 181, 185; as not akin to TMG/GCMH narratives, 190

Population bottlenecks: related to cycad diversity, 70, 74, 75; in *Dioon angustifolium*, 84; in *D. caputoi*, 84; in Tomasellii clade *Dioon*, 87–88

Population genetics of *Ceratozamia*, 78–81; current divergence process of, 79; history of studies of, 79; as related to anthropogenic activity, 79, 81

Population genetics of cycads, 70–76, 158–159; as connecting humans and cycads worldwide, 94–95; as dependent on markers chosen, 75–76, 95; genetic differentiation between populations in, 71t–74t; history of research on, in Mesoamerica and the Caribbean, 76; limitations of early studies of, 70; in Mesoamerica and the Caribbean, 76–92; overview of, 63; prematurity of generalizing patterns in, 75; as related to human DNA, 94; role of geographic factors in, 70–71

Population genetics of *Dioon*, 75, 81–92; pairwise Fst, 89f; population graphs applied to, 85–90; as related to human-induced dispersal, 88, 90, 93; as related to linguistic diversification, 93; scale and, 67; subgraphs of, 87–88

Population genetics of *Zamia*, 75; in Mesoamerica and the Caribbean, 76–78; as potentially explaining movement from Jamaica to the Cayman Islands, 78; as potentially related to human activity in the Caribbean, 77–78; scale and, 67

Population genetics: aims of, 62; coalescent theory-based models and, 69; and "complete population structure," 64; definition of, 64; emergence of population structures in, 64; description and evaluation of methods of, for estimating/modeling structure, 66–70; effective population size related to, 69; incorporation of geographic data into, 65; integration of, with archaeology, 63; intersection of, with social sciences, 62, 63; introduction to, 64–70; markers of, 65–66; for modelling population structures, 65; patterns, models of, 65; population graphs in, 68–69; role of biological traits/environmental characteristics in, 64–65; role of genetic drift in, 64; role of habitat in, 64–65; role of mutation in, 64; role of reproductive strategies in, 64–65; role of seed and pollen dispersal mechanisms in, 64–65; as useful for conservation, 65; usefulness of landscape ecology tools for, 68; as useful for correlating human cultural record and plant evolution, 65; uses for, 65

Population graphs: construction of, 68; defined, 68; for *Dioon*, 85–90; usefulness of, 68–69

Portable X-ray fluorescence. *See* pXRF spectrometry

Postclassic (Mesoamerica), 134, 138; Cipactli mythology in, 181–182

Potato, sweet: as less important than *Zamia* as food, 98; starches of, in archaeological context, 111; starch grains of, in ancient dental calculus, 114; *Zamia* eaten even when available, 122

Pots, cooking, for zamia dough, 115

Poua, 239t

Poyotl, 237

Pozol, 52

Prestige items, ancient Caribbean zamia as, 117, 120

Prince Charles, 14, 14f

Principal Young Lord, 188

Proteins. *See* Biomarkers

Proteomes (cycads), 164

Proto-Classic, 208

Proto-Mije-Sokean, 179

Pseudobombax, 51

Puebla (Mexico), 82f; *chamal* in, 236t, 238

Puebloan (culture), 189

Puerto Ferro (Puerto Rico), 110t, 112–114

Puerto Rico: African slavery in, 123; archaeological sites with zamia starches in, 110t, 111t, 112–113; archaeological zamia outside current range in, 118; cycads as food in, 97; cycads of, 97; history/loss of *Zamia* culinary tradition in, 123–124; plant evidence in late ceramic age sites of, 115; vicariance of cycad populations suggesting multiple introductions to, 77; *Zamia* in, 108; zamia toponyms in, 118

Punta Candelero (Puerto Rico), 110t, 113, 115

Punta Macao (Dominican Republic), 110t

Purpurea clade, 40

Purpusii clade, 39, 83; population graph of, 86

Purrón (Tehuacán Valley rockshelter) containing ancient plant remains, 148, 148f

pXRF spectrometry: explanation of, 157f; as method for detecting cycads, 154, 155

Pyramid Builders, 177t

Quaternary Period, 74

Quercus, 51

Querétaro, 83, 88, 90, 93; *chamal* in, 236t; Nahua terms for cycads in, 236t

Quetzalcóatl, 174f; as cultural hero, 177

Quiotamal, 237t
Q'uith, 227t

Raccoon: humans transformed into, in Teenek mythology, 210n6; in Teenek mythology, 176, 177
Rainforests: as *Ceratozamia* habitat, 78, 127; as *Dioon* habitat, 38; as *Zamia* habitat, 40, 127
Rain-maize-ear, 239t
Rajania. See Yams, domestic, secondary role of, in pan-Caribbean foodways; Yams, wild
Raptorial elements of Thipaak stories, 187
Rattles, 190t
rbcL, 159
Real del Monte (Hidalgo), 45
Real del Monte Mining Company, 45
Reana luxurians. See Zea luxurians
Red (color), significance of, in Maya palms and cycads, 241–242
Rees, John, 50
Regeneration, 185–187
Rheumatism and cycad medicine, 128
Rhizobia, 130
Rhizosphere, 156
Rice, 62–63; origins and diversification of, 63
Ridge-and-furrow systems, 137–138
Riego, El (Tehuacán Valley rockshelter), ancient plant remains in, 148, 148f
Río Balsas Basin, 39
Rituals involving early maize, 195
Rock shelters with cycad remains, 90. *See also* Cave sites with cycad remains
Romero's Cave (Ocampo), 144
Roots in Tenochtitlan 5 ages sequence, 174
Royal Garden at Hampton Court Palace, 43. *See also* England
Royal Horticultural Society, 43
Ryukyu Islands, genetic diversity of *Cycas revoluta* in, 74

Saba, archaeological site with zamia starches in, 110t, 112, 114
Sabal. See Palms
Sabato, Sergio: field work in Colombia, 245; field work in Mexico, 49; work on New World cycad taxonomy with D. Stevenson, 49
Sacred ear as name for maize/cycads, 173
Sacred-maize-ancestor concept. *See* SMA
Sacrifice, human, 178
Sago, 179
Saint John (Trinidad and Tobago), 114, 115, 117; as earliest Caribbean site, 112; as important early zamia archaeological site, 110t, 111–112
Saint Lucia, archaeological site with zamia starches in, 111, 115
Saint Michael. *See* San Miguel as Teosintle maize-cycad god
Saint Vincent: archaeological site with zamia starches in, 110t, 113; zamia grains in ancient dental calculus at, 115
Saint Vincent and the Grenadines. *See* Saint Vincent
Saladoid, late, 113
Saliva, human, applied to plant starches, 107
Salta, La, phase (Sierra de Tamaulipas), 145t
San Andrés, 210n9
San Antonio phase (Ocampo), 145t; *Dioon edule* remains in caves of, 146
San Bartolomé Ayautla (Oaxaca), 240
San Bartolo murals, 208; Sub-1A North Wall mural of, 192f, 192
Sand as root of *konlib* and allied terms, 179, 227t
San Juan Ixcatlán, 179–180
San Lorenzo phase (Ocampo), 145t
San Lorenzo, 195; early nixtamalization in, 209n2; maize in diet of, 195
San Lucas Yautepec (Oaxaca), 241
San Luis Potosí, 82f; *chamal* term among mestizos in, 238; Nahua names for cycads in, 236t–237t; relatedness of cycads in, to Honduras, 93
San Marco, Basilica di (Venice), 255–256, 256f
San Marcos (Tehuacán Valley rockshelter), ancient plant remains in, 148, 148f
San Miguel as Teosintle maize-cycad god, 237
San Pedro Huamelula (Oaxaca), 241
Santa Catarina (Oaxaca), 240
Santa María Acapulco (San Luis Potosí), 230, 231, 232
Santa Marta cave (Chiapas), 148
Santiago Ecatlán (Puebla), 239t
Santiago Textitlán (Oaxaca), 240
Sarcotesta: of *Dioon,* 83; eaten, by Chontal de Oaxaca; as edible, 83; as not requiring detoxification, 233; removal of, during tamale preparation, 91; in Teenek cycad terminology, 228t
Savannas: as Caribbean *Zamia* habitat, 41; as Mesoamerican *Zamia* habitat, 127; as Mexican cycad habitat, 127
Scale (spatial): in population genetics, 67–68
Scale (time): compared for usefulness in cycad

genetic studies, 95; problem of, in genetic data, 159
Scales, female cycad cone, in Teenek cycad-maize terminology, 227t, 228
Scanning electron microscopy-energy dispersive X-ray spectroscopy. *See* SEM-EDS
Scepter. *See* Staff
Schutzman, Bart, 50
Sciurus. See Squirrels as *Dioon* seed dispersers
Sclerotesta as hatching egg, 202
SCNGs: in Caribbean *Zamia*, 77; as used in cycad genetics studies, 76
Scrapers, 115, 116; as artifacts to look for ancient zamia evidence on, 121f; lithic, coral, and shell, 121f
Scrub as habitat for *Microcycas calocoma*, 41
Scrublands/environments, xeric, as habitat for *Dioon*, 42, 83
Secondary contact affecting evolution of cycads, 42
Seed dispersal in cycads, 70, 74; in *Ceratozamia*, 79; via gravity, 79, 80f
Seed-egg imagery, 206
Seedling-guardian, 239t
Segovia, Río, 221
SEM-EDS: as applied to cycads, 156; explanation of, 157f
Semen, 186; cycad pollen as, in Teenek thought, 228; in Xi'iuy thought, 232
Sepultura Biosphere Reserve, La, community cycad nurseries in buffer zone of, 57
Serpentine: celt of, 199f; as *Zamia* habitat in Puerto Rico, 114
Serra, Fray Junípero, 230
Setaria: (foxtail millets) as Central America *teosinte*, 214, 215f; as possibly occurring in *tiusinte* region, 218; "teosinte" term applied to, 172
Seven-cob. *See Chicome-Sintli*
Seven-ear, 235, 236t
Seven-flower, 235, 236t. *See also* Chicōmexōchitl
Shamanism related to *Zamia fischeri*, 229
Shark, 170, 181, 182, 207; and Arroyo Pesquero celt imagery, 198; Cipactli as, 198. *See also* Fish
Shima uta folk songs, 12
Siderophores: culturing of, in ancient soils, 156; overview of, 155–156; in relation to detection of ancient cycads, 156
Sierra de los Tuxtlas, 137
Sierra de Tamaulipas, 53, 143–149, 224f

Sierra Gorda, 83, 224f, 231; Teenek relationship to cycads in, 229; Xi'iuy in, 230
Sierra Madre de Chiapas, 39
Sierra Madre del Sur: as barrier to *Dioon* dispersal, 88; inhabited by Chontal de Oaxaca, 241
Sierra Madre Occidental, 39; SMA as absent from, 223, 241
Sierra Madre Oriental, 39, 143–149, 243; maize/cycad deities in, 187; *teocentli* in, 213; *teosinte* in, 221
Sierra Norte (Puebla), 179, 224f, 239t; discussion of SMA in, 238; *tepecentli* in, 213–214
Sierra Otomí-Tepehua (Hidalgo), 238
Sierra Popoluca (people). *See* Zoque-Popoluca
Sierra Teenek (place), 229
Silica in Caribbean zamia phytoliths, 118–119
Simple Sequence Repeats. *See* SSRs
Sindiopi, 235
Single-copy nuclear genes. *See* SCNGs
Single nucleotide polymorphisms. *See* SNPs
Sintiopiltsin: as Corn Master narrative, 167; and ovogenesis, 200; translation of, 200
Sintli, 212. *See also Centli/Cintli*
Sipacna: as K'iche' telluric deity, 180, 185; Zipacna, Hero Twins' defeat of, 181
Skull imagery, 180, 181, 189f, 206
Skull-waterlily complex. *See* Waterlily/lotus-skull motif
Slab, milling: with ancient zamia, 116; for zamia dough, 115
Slash-and-burn farming: and early maize, 132; in Nahua milpa, 235
Slavery, African: in Dominican Republic, 123; in Puerto Rico, 123
SMA, 166, 212, 223–243, 224f, 225f; absence of, among Oaxaca ethnic groups, 240; absence of, among Zapotec, 240; as aligned with Fertility Deity Complex, 169; areas absent from, 223; in Chontal de Oaxaca culture, 239t, 241; diffusion of, to Oaxaca, 240; as imported from Mexico to NE Honduras, 217; impoverishment of, 243; interchange of, among Xi'ui, Nahua, and Teenek, 238–239; lack of evidence for, in *teosinte* grasses, 222; in Mazatec (Oaxaca) culture, 239t, 240; in Nahua culture, 233–238; as not found in Belize/Colombia despite cycad/maize terms, 242; as originating in NE Mexico, 221, 223; origin of, tied to Nahua diffusion, 243; origin of, tied to origin/diversification of maize, 243; prevalent

SMA—*continued*
    among first-language Indigenous speakers, 223; research needed in, 223, 225; in Teenek culture, 225–230; *teosinte* as term for, in NE Mexico, 219; *teosinte*-related terms for, in the Huasteca, 219; in *tiusinte* region (Honduras), 242; in Xi'uiy culture, 230–233; Xi'iuy and Honduran parallels of, 242; in Zoque-Popoluca culture, 238–240

Snake in Kondoy narrative, 184

SNPs: advantages of, over SSRs, 66; in studies of *Dioon*, 81; as used in studies of Mesoamerican and Caribbean cycads, 76; uses for, 66

Soconusco (Chiapas), 196; Cenozoic floristic refuge, 42; region, 48

Sodomy, 176; in Flat Earth Teenek world age, 177t

Soil: edaphological studies/soil chemistry analyses of, as useful for detecting ancient cycads, 154, 163; effects of cycads on, in research designs, 155; quality of, as related to *Dioon* distribution, 38; quality of, related to where cycads thrive, 154; and rhizosphere, 156; and siderophores, 155–16

Soke, 167; connection of, to Xi'uiy narratives, 185; *kom* in, 179. *See also* Zoque-Popoluca

Sokean, Soteapan Gulf, 179

Sonora, 88; SMA as absent from, 241

Soteapan (Veracruz), 238

Soteapanec. *See* Zoque-Popoluca

Sotetsu, 11, 12. See also *Cycas revoluta*

Sotetsu jigoku, 25

Soul of maize (ear): *konlif* as, 227t; Nahua cycads as, 236t

South Africa, cycad species richness in, 35

South America: and early maize in N, 132–133; problem of archaeological zamia in NE, where no modern distribution, 116–117

South American clade, 77

South *D. tomasellii* subgraph of *Dioon*, 86f, 87, 88

Southeastern Mesoamerican Periphery, 223

Speciation: in *Dioon*, 39; in Mesoamerican cycads, 42–43

Species richness in Mexican cycads, 41–43

Spectrometry, mass, as method for detecting cycads, 156, 161. *See also* DART-MS

Spermatazoids, flagellate, 32

Spindle, 165

Spine: red, 241; spiny leaf (cycad term), 240

Spinulosum clade, 38, 39, 83; population graph of, 86

Spiny palm tree (cycad term), 240

Spirit of maize, 176; in Ch'orti' Maya, 209n3; cycads as, among Nahua, 235

Spirit of maize, *konlif* as, 227t

Spirit of the forest, *konlif* as, 227t

Sporophylls: of *Ceratozamia*, 203; resemblance of, to scales, 206

Sprouting imagery, 181, 188, 189f; as connected to staff/scepter, 204; discussion of, in Olmec imagery, 193–194; in Olmec art, 192, 196; as related to Olmec Maize God and cycads, 203, 206

Squash, 141; as basic Mesoamerican food, 125, 133; bottle gourd, in Tamaulipas caves, 145t; bottle gourds in San Bartolo murals, 208; butternut, in Tamaulipas caves, 145t; cushaw, in Tamaulipas caves, 145t, 147; early, in Tamaulipas caves, 144; early, in Tehuacán Valley caves/rockshelters, 148; maize god's emergence/sprouting from bottle gourd, 192, 200; pepo, in Tamaulipas caves, 145t

Squirrels as *Dioon* seed dispersers, 83

SSRs: in Caribbean *Zamia*, 77; definition of, 66; as markers in cycads, 71t–74t; in *Dioon* studies, 81, 86, 86f, 95; as tools for sequencing SSRs, 66; use of, in studies of *Ceratozamia*, 79, 80f, 81; use of, in studies of *Cycas*, 74; use of, in studies of Mesoamerican and Caribbean cycads, 76

Staff: Tenaspi Egg as attached to, 204

Stangeria eriopus, 10–11

Starch (cycad): in experimental archaeology, 153–154; in reference databases, 153

Starch (plants): definition and overview of, 101–103; grain description, 101; longevity of, in archaeological contexts, 107; morphometry of, 103; survival of, in archaeological contexts, 102; taxonomy of, 102–103; transitory versus storage, 101

Starch (*Zamia*): ancient food preparation with, 112; ancient, morphometry of, 113; in archaeological contexts, 78, 100, 136; archaeological sites where found, 110t; as evidence for ancient human use, 107–116; for food, 51; grain characteristics of, distinguished from other plants, 106–107; methods for detecting, 152; morphometry of grains of, 103–107; for new foodways in the Dominican Republic, 123; in Panamanian archaeological context, 136; reference collection of, for archaeology, 103, 112; on tools, 109, 116; unidentifiable to species, in archaeological context, 112, 114; used in Florida for commercial purposes, 51; in Yucatán archaeological context, 78, 136, 152

Starry-deer-crocodile, 182
Stevenson, Dennis: cycad work of, 245, 246f, 247, 250–256; work on New World cycad taxonomy with Sabato, 49
Stockpiling of cycad seeds and cones, 90
Stomata of *Dioon*, 38, 83
Stone: green, 197f; magic green, 177; milling, lithic/coral, as artifacts to look for ancient zamia evidence, 121f; semiprecious, 116; used for ancient cycad processing Australia, 152
Storms, tropical. *See* Hurricanes
Strobilus, female (*Ceratozamia*), 203; as compared to Olmec sculpture, 203–205
Subgraphs, of *Dioon* population graphs, 87
Subpopulations, in population genetics, 64, 67
Suchiapa (Chiapas), 52, 186
Sugar in *Zamia* food, 121f
Sustainable utilization as conservation strategy, 54–55
Swamps, tidal, as Mesoamerican *Zamia* habitat, 127
Swidden, 235
Symbolic uses of cycads: as markers of chiefly power, 13–14, 15, 18f; in Mesoamerica, overview of, 52–53; in sacrifices, 14; in South Africa, 15, 18f; as tabu, 13; in Vanuatu, 12–14
Syncretism: of Roman Catholic and Indigenous religious practice involving cycads, 52, 64; of Teosintle/San Miguel, in San Juan Huazalingo, 237

Tabasco (Mexican state), 45, 48; SMA concept as absent from, 241
Taiwan, genetic diversity of *Cycas taitungensis* in, 74
Tajín, El, 210n6
Tamakastiin, 235
Tamales: of *chamal*, 238; as cycad food, 51, 90, 128, 150; of *Dioon edule*, among Teenek, 230; of maize, 150; preparation of, 91f; sent to Xi'iuy relatives in US, 231; *tiotamal* as "Sacred tamale," 219, 237t; among Xi'iuy, 231, 233t
Tamaulipas (Mexico): cave sites in, 144–149, 164, 171; *chamal* term among mestizos in, 238; loss of cycad knowledge in, 243; Nahua names for cycads in, 236t–237t
Tancoyol de Serra (Querétaro), 230
Tancuime (Aquismón, San Luis Potosí), 178
Tanki Flip (Aruba), 111t, 115
Tatatlcintli, 236t
Taube, Karl, 188, 193, 194
Tawahka, 217

Taxonomies, Indigenous Mesoamerican, 166
Tecorral (Tehuacán Valley rockshelter), ancient plant remains in, 148, 148f
Teenek, 2, 224f, 225, 243; Christian God in, 178–179; compared to Xi'iuy narrative, 185; cycad/maize terms of, interrelated with Mije/Soke/Totonac, 179–180; epistemologies of, compared to Western, 168; in experimental archaeology, 153; and *konlif/konlib* cycad, 129, 176; maize and cycads in nomenclature of, 178–180; mythology of, 175–180; negative views of cycads as food among, 51; SMA among, 225–230; Thipaak Corn Master narrative of, 167, 176–178; world creation, correspondence to Tenochtitlan 5 ages sequence of, 176
Teenek, Sierra (people), cycad knowledge loss among, 226
Teenek Maya. *See* Teenek
Teeth, isotropic nitrogen in, 155. *See also* Dental calculus (human)
Tehuacán-Cuicatlán: Desert, as *Dioon* hot spot, 42; Valley, 84
Tehuacán Valley, 85; cycads in rockshelters of, 148–149, 164, 171–172, 224f; precolumbian human effects on cycad populations in, 85
Temple XIX (Palenque), Southern Platform, 182, 183f
Tenaspi Egg, 200, 202, 202f; as portraying cycad, 202
Tenochtitlan, 214; sequence of the five ages, 174f
Teocentli (cycad), 213
Teocentli (grasses): *cencocopi* as, 174f; *cincocop* as, 174f
Teocinhuatl, 236t
Teocinishuatl, 236t
Teocinte, El, as Jalisco watering place/hacienda, 220
Teocinte: in colonial Honduran land titles, 217; as *Dioon mejiae* in NE Honduras, 217
Teocintle (cycad), 175, 207f, 234, 236t
Teo- prefix and variants, 212
Teosinte (maize): as related to cycads, 51–52; wild, as progenitor of maize in Balsas River valley, 132
Teosinte (Nahua term for cycads), 175; for *Ceratozamia fuscoviridis*, 172; as derived from similarity of appearance to maize, 212; in Honduras, 216–217; as name for *Zamia*, in northern Guatemala, 219; in postconquest Mexico, 213
Teosinte (term/concept referring to either maize or cycads), 93, 176, 212; etymology of, 212

Teosinte (term for cereal grasses in Central America), 214–216; as human food/livestock forage, 214; no SMA concept known for, 216; in toponyms, 214
Teosinte (term for Jalisco *Tripsacum*), 220
Teosinte (term for *Tripsacum* introduced to Coscomatepec, Veracruz), 220
Teosinte (term for various grasses or cycads in W. Mexico), 220–221; toponyms containing, 220
Teosinte (*Zea luxurians*), 215f; as trade name, 216
Teosinte (*Zea mays parviglumis*), 212; misconception that it is autochthonous Mexican term, 216; name of, re-applied from Guatemala to Mexico, 216
Teosinte/tiusinte (Nahua terms for cycads as maize ancestors), 178
Teosintes, Los (Jalisco), 220
Teosintle (cycad) as Nahua term for *Ceratozamia fuscoviridis*, 235, 236t
Teosintle (maize/cycad deity), 221, 229, 237; as San Miguel, 237. *See also* Chicōmexōchitl
Teotl, 212, 237
-téotl, 52
Teoxintli, 220
Tepe[t]zintle, 237t
Tepecentli, 237t; etymology of, 213–24; in Huayacocotla, 1560s, 213
Tepecintle, 237t
Tepecintli, 237t
Tepehua, 224f; apparent SMA among, in Sierra Norte, 238; no SMA among, in Hidalgo
Tepetlcintle, 237t
Tepetzintle (Totonac/Nahua term), 239t
Tepetzintli (Totonac/Nahua term), 238, 239t
Tepexi, 236t
Terraces. *See* Fields, terraced
Téutle, 52
Texquitote (San Luis Potosí), 235
Tezcatlipoca, 174f, 184; battling a fish/crocodilian, 182
Theobroma bicolor. *See* Cacao
Theobroma cacao. *See* Cacao
Thimaloon poko, 228t
Thipaak, 2; as child-maize god, 176; and Cipactli, 180–184; comparison of, to other maize/cycad deity elements, 187; connection of, to Xi'iuy cycad-origin stories, 233; as Corn Master narrative, 167; as cultural hero, 176–177; definition of, 176; as equated with Cipactli creatures, 206; as fertility hero, 188; identity of, as shared with *konlib,* 179; as introducer of maize, 178; *konlif* as friend/brother/helper of, 226; as maize and cycad, 176, 207f; as maize/cycad/earth/fertility/calendar deity, 180; overview of Teenek stories about, 226; as related to alligator gar *zipac,* 181; as related to Popoluca and Mije narratives, 184; in Teenek SMA, 225–229; as term for cycads, 178. *See also* Cipactli
Thipaak (Teenek term for cycad), 228t
Threats to cycads, 53–54, 55, 127
Three Sisters, 125, 133
Thrips: as cycad pollinators, 32; as *Dioon* pollinators, 83
Throne, 190t
Thunder: as associated with TMG/GCMH, 188, 190t; in Homshuk story, 184; Old Thunder God, 190t; in Teenek mythology, 177, 179. *See also* Maam
Thunderbolt (Popoluca deity), 184; thunderbolt deities, 187
T'ichol, 228t
Tierra, planta dueña de la, 228t, 229–230
Tiñuk, 217
Tío de la mazorca, 226
Tionishuatl, 219–220
Tiosin[is]kuatl, 236t
Tiotamal, 219, 237t
Tiozin[is]huat, 236t
Tissues (cycad), experimental archaeology related to, 153, 156
Tiucsinte, 217
Tiusintales, 218
Tiusinte, 218, 231, 242; in colonial Honduran land titles, 217; origin of term, 217; region, 215, 215f, 242; as term to disambiguate cycads and grass teosinte, 172; Tolupan term for, 217. *See also Dioon mejiae*
Tlahuizcalpantecuhtli battling a fish/crocodilian, 182, 183f
Tlaloc, 174f
Tlalocan, Lord of, GCMH as, 190t
Tlalocan-Tamoanchan, 190
Tlanchinol (Hidalgo), 52
Tlanchinola (Veracruz), cycad nursery in, 57
Tlapacoyo, 209n2
Tlatecli. *See* Tlatecuhtli
Tlatecuhtli, 182
Tlatilco, 209n2
Tlaxcala, 214
TMG, 188–190; uncertainty of identity with maize, 192. *See also* GCMH

TMVB. *See* Trans-Mexican Volcanic Belt
Tobacco, 177
Tobago. *See* Trinidad and Tobago
Tohuaco II (Hidalgo), 175, 235
Tolupan, 217
Tomasellii clade, 38, 39, 83; population graph of, 86
Tomentum (on cycad cones): of *Dioon*, 38; in Teenek cycad-maize terminology, 227t, 228
Tonacācihuātl, 185
Tonacātēcuhtli, 185; Xochipilli as, 190t
Tonatiuh, 174f
Tonsured Maize God. *See* TMG
Tools: for food processing, starches preserved in, 102, 116; preceramic, *Zamia* starches on, in Panama, 136; stone, zamias starches preserved on, in the Yucatán, 136; as unanalyzed for ancient cycad evidence in Mesoamerica, 142; used in ancient cycad processing, detection of, 153–154 for zamia processing, 109f, 111–112, 115
Toponyms: with *teosinte* in Central America, 214; with *tiusinte/teocinte* cycads in Honduras, 217
"Torches," 206
Toro, 228t
Tortillas: of *chamal*, 238; as cycad food, 51, 128, 150; of cycads in Chontal de Oaxaca culture, 241; of cycads among Xi'iuy, 231; of *Dioon edule* among Teenek, 230; of maize, 150; red (Yucatec Maya), 241
Totonac, 224; more research needed among, 243; names for cycads among, 214, 239t; overview of SMA in, 238; SMA parallels of, with Teenek/Nahua, 238
Totonacan (language family), 243
Totonacapan (Veracruz), 238, 239t
Toxicity (cycads): immunity to, in *Dioon*, among birds, butterflies, and rodents, 83; overview, 127; as related to lack of scholarly focus, 24; in relation to livestock, 25; of *Zamia*, in the Caribbean, 97
Trade routes/corridors: and *Dioon* distribution in Honduras, 218; and SMA spread, 223
Trans-Mexican Volcanic Belt, 37, 39, 85
"Tree dioons." *See* Spinulosum clade
Trees, human ancestors turning into: in Classic Maya imagery, 200, 201f; in Formative/Olmec imagery, 196, 197
Tree-seedling-guardian, 239t
Tres Marías, Las, 254
Trinidad and Tobago: archaeological site with zamia starches in, 110t, 114; archaeological *Zamia* in, 108, 122; early maize in, 132; underground stems of zamia eaten in, 115; why no natural distribution of *Zamia* in, 108
Trinity (agriculture), 125
Trinity College Botanical Garden, 45, 46
Tripsacum: as introduced from Guatemala to Mexico, 220; as most ubiquitous "teosintes" in W Honduras, 221; as possibly occurring in *tiusinte* region, 218; "teosinto" term applied to, in Central America, 172; teosinte as, in Central America, 214–216; as *teosinte* in Mexico, 215f, 220
Triqui (people), cycad use by, 240
Triunfo Biosphere Reserve, El, community cycad nurseries in buffer zone of, 57
Tsakam way', 228t
Tsalaam Thipaak, 166, 178
Tsalam Thipaak, 228t
Tsamal, 228t
Tsamalib, 228t
Tsamay, 228t
Ts'een Thipaak, 228t
Tsubal, 228t
Tuberculosis and cycad medicine, 128
Tubers, 115
Tuerckheimii clade, 40
Turkeys, 174f
Turks and Caicos Islands, archaeological evidence for zamia as food in, 115
Turtle in Homshuk story, 184
Tush-kjù, 239t
T'uul, 228t
Tuxtepec (Oaxaca), 240
Tuxtlas (Los): as Cenozoic floristic refuge, 42; *Ceratozamia* spp. endemic to, 202–203; concentration of maize/cycad deities in, 187; as Nahua region, 173
Twin element: of corncobs, 187; of cycad seeds, 204, 206
Tzalam-thipac, 228t
Tzama(a)l, 228t; *dameu* as derived from, 230
Tzamaal, inik, 227t
Tzamaal, ushum, 228t
Tzamaalib, 228t, 229–230, 233t; "chamal" as derived from, 229, 236t; as plant associated with cold climate, 229. *See also Chamal*
Tzamalib, 228t
Tzamay, 228t
Tz'itziin, 176
Tzitzimeme (celestial monsters), 174f
Tzom, 237

Tzompollo, 52
Tzompoyo, 234, 237t; as interchangeable with *teosintle;* meaning of, 237; as related to *chamal,* 237
Tzotzil Maya (language), 191

Ukiyo-e woodblock prints of cycads, 15, 17f
Ulúa, Río, basin, 221
UMA cycad nurseries, 57
Uncle Tamal, 219
Understory as *Ceratozamia* habitat, 37
Unidades de Protección para la Conservación de la Vida Silvestre. *See* UMA cycad nurseries
Universidad Veracruzana, role of, in cycad nurseries for conservation, 57
UPLC-MS as proxy for ancient zamia detection, 121f
Utuado-27 (Puerto Rico), archaeological site with *Zamia,* 113
Uxum, 185. See also *Konlif*

Vagina, female cone and, in Teenek thought, 228–229
Valenzuela's Cave (Ocampo), 144
Vázquez Torres, Mario, 49
Vega Nelo Vargas (Puerto Rico), 111t
Venezuela, 254
Venta, La, 183f
Vertices. *See* Nodes (population genetics)
Vieques (Puerto Rico), archaeological sites with zamia starches in, 110t, 114
Vietnam, 246
Volcanic activity: preserving ancient agricultural landscapes, 137; related to *Dioon* adaptations, 83
Volcanic substrate; cycads growing in, 99f, 114
Vovides, Andrew: establishment of National Cycad Collection, 50; taxonomic work on Mexican cycads, 44, 45, 49

Waterlily, 180, 181, 184. *See also* Imix
Waterlily/lotus-skull motif, 180; waterlily-skull imagery complex, 181, 189f, 198
Weevils as palm pollinators, 83
West African cycads, 74
Western Pacific region, 51
Womb, 185; in Xi'iuy cycad mythology, 232
Worms, 184
Wright, Sewell G., methods in population genetics of, 66–67
Writers, Corn Masters as, 187

Writing, encompassed by TMG/GCMH, 188, 190t

Xalapa (Veracruz), 48, 49, 50; dioons as ornamentals in, 59; Nahua cycad terms in, 236t–237t; *tiotamal* in, 219
Xanthosoma. See Cocoyam
Xibalba, 181
Xi'iuy, 224f, 230–233, 243; attitudes toward cycads among, 231; Corn Master narrative of, 196; cycad/maize knowledge among, compared to Honduras, 231; cycad origin stories among, 232–233 as dependent on wild resources, 230; interaction of, with Nahua and Teenek, 230; mixed culture of, and Nahua in NE Mexico, 218; detoxification of cycads among, as identical to Honduras, 218, 231; women, as cycad knowledge preservers in Honduras, 218
Xochipilli, 190t

Yahtuchó, 239t
Yams, domestic, secondary role of, in pan-Caribbean foodways, 122
Yams, starch grains of, in Puerto Rican archaeological sites, 114
Yams, wild: secondary role of, in pan-Caribbean foodways, 122; starches of, in archaeological context, 111
Ya-tuj-cho, 239t
Ya-tuj-cho-chu, 239t
Ya-tuj-cho-shi-i, 239t
Yax Naah Itzamna, 182
Yazn-goag, 240
Yokai. See Cycads: and snake deities
Yoro (Honduras), 90
Yoshitoshi (Edo period artist), 15
Yoshitsuya, 17f
Young Lord sculpture, 196, 197f, 198, 199f; as Cipactli, 196–197; compared to Arroyo Pesquero celt, 198
Yucatán (Peninsula): ancient zamia in, 117, 136, 152; as avoided by Nahua-speaking travelers, 219; *chamal* in, 236t, 238; Nahua names for cycads in state of Yucatán, 236t–237t; Northern, Classic Maya iconography in, 189f; SMA as absent from, 241. *See also* Escalero al Cielo (Yucatán), cycad starch on tools at
Yucatec (Maya): *chak waj* cycad among, 241; religion, 168
Yucatec Maya (language), 191
Yuto-Aztecan (language family), 243

Zacate Guatemalteco, 220
Zamia, 20, 27, 35, 36f, 63, 76–77, 126; ancient, found outside current Puerto Rican distribution, 118; ancient exchange networks of, 116–117; artifacts potentially containing signatures of, 121f; beetle pollination of, 32; as common name in the Greater Antilles, 97; cones of, compared to maize cob, 151f; as conferring power to maize ears, in Nahua thought, 235; in Cuba, morphometry of storage starch grains of, 103; cultural relationships of, with maize, 225f; distribution of, in Mesoamerica, 215f; first American specimen of, in Europe, 43; as food, stems of, 51, 97–98; as food, tools in processing sequence of, containing starch grains, 109; food preparation of, in ethnohistoric/ethnographic sources, 121f; historical ecology of, in the Caribbean, 97–124; as involved in SMA, 234; Mesoamerican hot spots of, 41–42; moth pollination of, 32; Nahua names for, 236t–237t; overview of Caribbean species of, 40–41, 63; overview of Mesoamerican species of, 39–40; proxies for detecting signatures of, 121f; red-seeded cones of, in Maya cycad terminology, 241; seeds of, in El Gigante rockshelter (Honduras), 135, 163; as SMA, in NE Mexico, 219; as spirit of maize/Maize Lord, 176, 207f; as staple food in ancient/modern Caribbean, 117, 120, 122; stems of, 30; symbolic values of, in ancient Caribbean, 117; Teenek terms for, 180; as *teosinte* in Guatemala, 219; *Thipaak* term as applied to, 178; toponyms containing, 118; Totonac names for, 239t; *tsalaam Thipaak* as, 166
Zamiaceae, 27, 39, 63, 99, 126, 170; in Veracruz, 45
Zamia cf. *portoricensis.* See *Zamia portoricensis*
Zamia chigua, 53
Zamia cremnophila, 40; genetic diversity and structure of, 72t
Zamia decumbens, 40; as corn palm, 242
Zamia erosa: morphology of, 99f; morphometry of storage starch grains of, 103, 104t, 105t; starch of, in reference collections, 103; starch grains in archaeological context, 112, 114; as starch source in WWII Jamaica, 248
Zamia fairchildiana, 84
Zamia fischeri, 40, 229; Nahua names for, 236t–237t; as related to *konlif* in Teenek thought, 229

Zamia furfuracea, 5f, 40, 43f; cones of, compared to maize cob, 151f; extraction of, for landscaping trade, 54; as first Mexican cycad cultivated in Europe, 43; genome sequencing of, 253; Nahua names for, 236t–237t; as ornamentals on coastal promenades, 59; as produced/sold sustainably in Mexico, 56f, 57, 58; seed/pollen dispersal of, 79, 84; as weevil pollinated, 253–254
Zamia grijalvensis, 35, 40
Zamia herrerae, 40; in Central American *teosinte*-as-grass region, 218
Zamia inermis, 35, 46f; genetic diversity and structure of, 73t; *Z. inermis* Vovides Rees, & Vázq. Torres, as new species, 45
Zamia integrifolia, 40
Zamia katzeriana, 40; genetic diversity and structure of, 72t; as a hybrid, 40
Zamia lacandona, 40; genetic diversity and structure of, 72t
Zamia lecointei, 254
Zamia loddigesii, 40; as *chak waj* (Yucatec Maya), 241; as *Chicōmexōchitl/Chicome-Sintli,* 175, 234; as described by Miquel, 44; genetic diversity and structure of, 72t; in Nahua milpa, 235; Nahua names for, 236t–237t; as origin of maize, in Papantla, 238, 239t; as a *Zamia* species in the Yucatán, 46–47, 49; in Zoque-Popoluca SMA, 239t
Zamia lucayana, genetic diversity and structure of, 72t, 77
Zamia meermanii, 40
Zamia obliqua: as *chigua,* 247, 247f; as *maicito,* 242
Zamia onan-reyesii, 40
Zamia oreillyi, 40
Zamia paucijuga, 40, 254; chromosomal fissions reported in, 49; cytotaxonomic studies of, 49
Zamia pollen, in archaeological context, 118; identification of grains of, 118; useful for phytogeographical research, 118; problem of low counts because insect-pollinated, 118
Zamia polymorpha: segregation from *Z. loddigesii* in the Yucatan and Belize, 47; *Z. prasina's* nomenclatural priority over, 47
Zamia portoricensis: morphometry of storage starch grains of, 103, 104t, 105t; starch grains in archaeological context, 112, 113, 114; starch of, in reference collection, 103
Zamia prasina, 35, 47; as published by Bull, 47; chromosomal fissions discovered in (as *Z. loddigesii*), 49

Zamia pumila: as described by de las Casas for food, 98; morphology of, 99f; morphometry of storage starch grains of, 103, 104t, 105t, 106f; starch grains of, in archaeological context, 112–114; starch grain in seeds of, 103; starch grain in underground stem of, 102f; starch of, in reference collection, 103

Zamia pumila complex, 97; genetic diversity and structure of, 73t, 77

Zamia purpurea, 40; genetic diversity and structure of, 72t; *Z. purpurea* Vovides Rees, & Vázq. Torres, as new species, 45

Zamia sandovalii, 40

Zamia skinneri in Panamanian archaeological context, 136

Zamia soconuscensis: community cycad nursery with, 57–58; discovery and naming of, 48; sympatry of, with *C. matudae*, 48

Zamia spartea, 40

Zamia splendens, 40

Zamia standleyi, 40

Zamia stevensonii, 251f

Zamia tuerckheimii, 40

Zamia variegata, 40; genetic diversity and structure of, 72t

Zamia vazquezii, 40; as Totonac SMA, 239t

Zamia wallisii, 245, 251, 253

Zapotec (people), 186; cycad terms/concepts among, 240; in Kondoy myth, 210n5

Zea, 214; not traditionally named *teosinte* in W. Mexico, 220; among Tojol-ab'al, 242; wild, as absent from *tiusinte* region, 218; wild, in Mexico, 216

Zea luxurians, 215f, 216, 220; as referred to by Standley/Williams, 216; as *teoxintli* in W. Mexico, 220

Zea mays ssp. *parviglumis*, 132, 172, 175; as maize's genetic ancestor, 212; as *teosinte* in Mexico, 216

Zea perennis possibly called *teoxintli*, 220

Zempoaltépetl, 210

0–1 Fst scale, 67

Zinc in siderophores, 155

Zipac as related to Thipaak/Cipactli, 181

Zipacna. *See* Sipacna

Zn. *See* Zinc in siderophores

Zoomorphic, Olmec entity as, 194

Zoque-Popoluca, 224f, 243; cycad terms among, 239t; Nahua-language cycad terms/concepts shared with, 239t, 240; SMA in, 238–240. *See also* Popoluca; Soke